Science in the Study of Ancient Egypt

Egyptology has been dominated by the large quantity of written and pictorial material available. This amazing archaeology has opened up a wonderful view of the ancient Egyptian world. The importance of hieroglyphics and texts, and their interpretation, has led to other areas of archaeology playing much less prominence in the study of Egypt. Perhaps most notable in the relatively infrequent application of analytical science to answer Egyptian questions. This problem has been compounded by difficulties in accessing the material itself. In recent years, however, new research by a range of international groups has overturned this historic pattern, and science is now being routinely incorporated into studies of the history and archaeology of Egypt.

Science in the Study of Ancient Egypt demonstrates how to integrate scientific methodologies into Egyptology broadly, and in Egyptian archaeology in particular, in order to maximise the amount of information that might be obtained within a study of ancient Egypt, be it field, museum, or laboratory-based. The authors illustrate the inclusive but varied nature of the scientific archaeology being undertaken, revealing that it all falls under the aegis of Egyptology, and demonstrating its potential for the elucidation of problems within traditional Egyptology.

Sonia Zakrzewski is an Associate Professor in Archaeology at the University of Southampton, where she runs a masters course in bioarchaeology and osteoarchaeology, synthesising both human and faunal studies. She publishes widely in bioarchaeology, physical anthropology and science journals and has edited two books.

Andrew Shortland is Professor of Archaeological Science at Cranfield University. He is Deputy Director of Cranfield Forensic Institute, where he runs a group that specialises in the application of scientific techniques to archaeological and forensic problems.

Joanne Rowland is a Junior Professor in the Ägyptologisches Seminar of the Freie Universität, Berlin. She directs the Egypt Exploration Society Minufiyeh Archaeological Survey and Imbaba Prehistoric Survey, and previously was a researcher in the Egyptian Collection at the Royal Museums of Art and History (Brussels) and at the Research Laboratory for Archaeology and History of Art (RLAHA) at Oxford University.

Routledge Studies in Egyptology

Science in the Study of Ancient Egypt

Sonia Zakrzewski, Andrew Shortland
and Joanne Rowland

Routledge
Taylor & Francis Group

LONDON AND NEW YORK

First published 2016
by Routledge

2 Park Square, Milton Park, Abingdon, Oxfordshire OX14 4RN
52 Vanderbilt Avenue, New York, NY 10017

Routledge is an imprint of the Taylor & Francis Group, an informa business

First issued in paperback 2019

Library of Congress Cataloging-in-Publication Data
Zakrzewski, Sonia R., author.
Science in the study of ancient Egypt / Sonia Zakrzewski, Andrew
Shortland and Joanne Rowland.
    pages cm.—(Routledge studies in Egyptology ; 3)
    Includes index.
    1. Egyptology—Technological innovations. 2. Egypt—History—
To 332 B.C. 3. Egypt—Antiquities. 4. Archaeology—Methodology.
I. Shortland, Andrew J., author. II. Rowland, J. (Joanne), author.
III. Title. IV. Series: Routledge studies in Egyptology ; 3.
DT60.Z35 2015
932.010721–dc23   2015016084

ISBN: 978-0-415-88574-4 (hbk)
ISBN: 978-0-367-86563-4 (pbk)

Typeset in Sabon
by HWA Text and Data Management, London

Contents

Figures

Boxes

Acknowledgments

This volume has been many years in its gestation. We choose the term gestation with intention – as it was indeed delayed by actual pregnancy.

The ideas in this book started out as discussions that developed during the very first Current Research in Egyptology symposium, organised by Angela McDonald and Christina Riggs in 2000. From there, the ideas (especially in terms of the intersection and interactions between science and archaeological theory in Egyptian archaeology) were discussed many years ago with Rachael Dann (University of Copenhagen), David Wengrow (University College London) and the late Dominic Montserrat. Although we thought it important to write a book that synthesised archaeology in Egypt with the recent advances in archaeological science, the actual idea kept being put on a back-burner until we could ignore it no longer. We feel that it is imperative that the potential of archaeology in Egypt is maximised and hope that this book might help in some small way.

We owe a debt of thanks to a great many different people. We would like to start by acknowledging the help and support of the Egypt Exploration Society in London, the Institute of Archaeology (UCL), the Ägyptologisches Seminar (Freie Universität Berlin), the Sackler Library and Oriental Institute (University of Oxford) and the University of Southampton. Over the years, advice has been sought and very gratefully received from a whole series of people during the thinking about and the final writing of this book. There are far too many people to mention, but we would like to specifically include (in alphabetical order): John Baines (University of Oxford), Pearce Paul Creasman (University of Arizona), Aidan Dodson (University of Bristol), the late Ahmed Gamal el-Din Fahmy, Angus Graham (Uppsala University), Wolfram Grajetzki, Liz Jones (UCL), Claire Malleson, Mary Anne Murray, the late David Peacock, Jan Picton (Petrie Museum), Kristian Strutt (University of Southampton), Ian Shaw (University of Liverpool), G. J. Tassie, and John Taylor (British Museum). There are many others who have aided and supported us through the drafting and completion of this book. It would not have been possible without you and we hope that you know who you are! We would also like to thank all our colleagues at the Ministry of Antiquities and all inspectors with whom we have worked.

We also gratefully acknowledge the support of all the staff at Routledge and Taylor & Francis, including Lola Harre and Stacy Noto, who have helped, supported and cajoled over the years. You have all been very understanding when family issues have intervened. We thank everyone, and we apologise for any errors or omissions.

Lastly we would like to take this opportunity to dedicate this book to people with whom we have worked who have sadly passed away. We have admired your academic works, are grateful for the support you have given us and will miss you. We therefore dedicate this book to Ahmed Gamal el-Din Fahmy (who has contributed a case-study box), Michel Wuttman Noël Gale and, last but by no means least, the wonderful David Peacock whose enthusiasm for archaeological science in Egypt knew no bounds.

1 Introduction

Biographies and Lifecycles

Ancient Egypt, with its impressive and unique series of monuments, representations, and physical remains (including mummies and skeletons), has fascinated the world and been a major source of archaeological inspiration. But how is the archaeology interpreted? How were today's archaeological remains created originally? Any why were certain methods used in preference to others by the ancient Egyptians to make those things that have become today's archaeological remains? How is Egyptian archaeology understood? These and many other questions continue to inspire as well as perplex and this book aims to help provide possible ways in which to answer some of those questions. The following chapters present methods that can be applied in Egyptian archaeology to develop and deepen our understanding of the past individuals, groups and communities who inhabited Egypt and travelled through its deserts from the earliest times onwards. It focuses on the application of scientific methods to questions that are specifically pertinent to Egyptian contexts. This book is not a methodological guide for Egyptian archaeology, nor is it a synthesis of Egyptian archaeology, but rather it is a series of signposts to facilitate the integration of appropriate scientific techniques. This book will hopefully fire the archaeological imagination and allow those engaged within Egyptology, as professionals and students, to explore the varied scientific approaches being used today. It is also hoped that this book will inspire new research questions and bring a deeper questioning of those archaeological and Egyptological concepts we *know* right now.

1.1 What is Egyptian Archaeology?

This apparently easy question underlies a deep paradox – the great difficulty in *actually* precisely defining the discipline. In principle, we, as archaeologists or Egyptologists, all feel that we intrinsically *know* what is meant by the term, but is it easier to state what archaeology is not? Archaeology is not 'the past'. In this sense, the past did not happen, but rather is what we think we understand. Hayden (1993: 10) has argued that archaeology renews a sense of wonder at the complexity and uniqueness of contemporary and past societies. Certainly archaeology brings together multiple voices to write multiple histories with multiple interpretations. "The questions that archaeologists ask are inherently subjective. Here lies the paradox of archaeology" (Andrews and Doonan 2003: 13).

Archaeology is the study of the material remains of past cultures, and has been argued not to follow scientific laws in exactly the same way as the 'harder' sciences of physics and chemistry (Bard 2008). Archaeology does not (usually) have the ability to develop hypotheses that can be tested by repeat experimentation. The archaeological resource is finite, and archaeological sites can only be fully excavated once. We argue, however, that this actually means that we, as archaeologists, need to adopt a more scientific approach, and maximise the use of that finite and limited resource. But archaeology is, like all sciences, a body of arguments advanced to make sense of the world. There are often multiple acceptable ways of interpreting the archaeology. Scientific analyses of Egyptian archaeology can aid in the selection of the most likely or suitable pathways for interpretation when placed in broader Egyptological frameworks.

"Archaeology is a total study" (McIntosh 1999: 2). It involves analysing everything that remains from the past, integrating it with the written record, and then using these data to reconstruct past life as fully as possible. As Kristiansen (2009) argues, archaeology has a dual practice, so that logical research leads to conservation in the present. This volume attempts to show how scientific methods and approaches can be integrated into broader archaeological work in order to maximise the Egyptian archaeological resource, and maintain or indeed develop it for future generations.

In Egypt from the third millennium BC onwards, archaeology has the good fortune of being situated within a written record of sorts. Archaeology is often said to be the handmaiden of history, but the relationship between the two disciplines is not a simple relationship of master and servant but is rather more complex. This is even more the case in Egypt. History relies on written records: these, however, are usually incomplete and may be biased or inaccurate. Archaeology, by contrast, focuses on material that is often not mentioned in written records – including, with some exceptions, many mundane details of everyday life. In Egypt, the written record is diverse but fragmented. Although only a small portion of the Egyptian population is thought to have been literate, the study of ancient Egypt comprises the study of all those who lived within and travelled within its borders.

The sources that archaeologists work with in Egypt are predominantly tangible items, be they artefacts or ecofacts. Modern archaeological research in Egypt increasingly involves the integration of a variety of methods, of which the scientific methods described in this book play an important part. The textual evidence from Egypt provides a framework onto which archaeological interpretations can be mapped, although the archaeological evidence can contradict the historical texts, illustrating the complexity of textual interpretation. It is hoped that this volume will demonstrate that scientific analyses can strengthen these connections and enable further nuance to be brought to understandings of ancient Egypt.

1.2 Archaeology in Egypt: an Egyptian Archaeology

Howard Carter's words, when glimpsing the treasure of Tutankhamen by candlelight through a small hole in the wall sealing the tomb, are well known. But archaeology in Egypt is about much more than just 'wonderful things'!

The focus of the early years of Egyptian archaeology comprised the unsystematic collection, or looting, of Egyptian antiquities. Indeed, in the early nineteenth century, the British consul Henry Salt and his agent Giovanni Belzoni collected and transported items such as the colossal bust of Ramesses II from Thebes to the British Museum (Moser 2006). Although this early focus was on acquisition of antiquities for European and other markets, Egyptian archaeology gradually adopted new methods and new approaches. Flinders Petrie was one of the first archaeologists working in Egypt to employ modern scientific methods within his excavations. His scientific methods included the documenting and drawing of his excavations and the seriation of ceramics for relative dating purposes (Petrie 1901; see Section 2.1: Time: Dating Methods). In turn Gertrude Caton-Thompson was the first to apply seriation dating to vertical stratigraphy within Egypt during her work at Hemamieh (Brunton and Caton-Thompson 1928) – Petrie having developed methods of stratigraphical recording earlier at Tell el-Hesy (Petrie 1891). Before Petrie's work, items such as potsherds were routinely discarded, whereas Petrie recognised that ceramic diversity could be used in association with stratigraphy for seriation and hence relative dating (see Section 2.1.1: Relative Chronology and Relative Dating). Amongst his other scientific approaches or methodologies were the maintenance of copious field notes and the precise detailing of different strata in his drawing of archaeological sections. Even very traditional Egyptology, which many may have considered to focus exclusively on the language of ancient Egypt, and hence might not engage directly with scientific archaeology, has laid great store in the ability to seriate, age, and date the past. Material culture cannot be divorced from its own history of consumption, presentation and indeed representation (Moser 2006), and Egyptology *is* the study of ancient Egypt and, as such, is equally focused on the study of archaeological contexts as well as the artefacts and ecofacts that derive from these.

Population estimates vary considerably, but past Egyptian society comprised probably over a million people by the middle of the third millennium BC (e.g. Butzer 1976), and each person was a distinct individual who made his/her own choices, albeit within the framework of the Egyptian norm for society at the time. Each person lived within groups and communities, and acted out his/her own life within the wider society, and was in turn variably affected by it. Within the social system, these individuals and groups were affecting their local environment, as well as being affected by it, and producing, using, reusing and discarding objects within their daily lives. These very individual processes form those artefacts and ecofacts that are used to develop an understanding of the ancient Egyptian past. This diverse group of people each had their own personal life history and made use of their own, or shared objects, either on a daily or less frequent basis. Each of these objects has its own story to tell, leading to the multiplicity of 'pasts' or voices. How to interpret and translate these stories from texts is the work of the Egyptologist; how to interpret from the objects is also the role of the Egyptologist – albeit usually considered to be an Egyptian archaeologist. This latter aspect might require the interlinking of results obtained from various and different archaeological specialties. This dichotomy between philology and archaeology within Egyptology is not productive and all can benefit from the application of scientific methods.

1.3   Archaeological Science and the Study of Egypt: Egyptological Science

Academics studying the history of archaeology are very aware of conflicting views as to whether archaeology should be viewed as a science or as an art, and the potential conflict therein. Here, the intention is to address the arts and the sciences, as popularised by C.P. Snow in 1959, not as distinct entities, but together within an Egyptian framework. Clearly it is a 'given' that archaeology is a humanity in that it comprises the study of both humans and human societies, but it necessarily also draws upon a range of other disciplines. The research questions that are of importance within Egyptian contexts are the questions that are important to anyone working in the broader Egyptian field. In that sense, we argue here that science *can* and *should* be used within both Egyptian archaeology and Egyptology to add nuance and understanding.

Archaeological science is commonly thought of in terms of artefact analysis, ancient technology, dating materials, trade, exchange and provenience work. But archaeological science is much more than this. It can be defined as the systematic study of the nature and behaviour of the material physical universe based upon experiment, measurement, and analytical testing. Archaeological interpretation involves building and reconstructing representations and ideas as to how people lived and behaved. It is rooted in the study of the material objects and remains of these people and requires both theoretical and scientific approaches to further our understanding. The focus of many archaeological science books has been on the various materials encountered by archaeologists and/or on the different techniques that can be employed. Here, the main interest is in archaeological science within an Egyptian framework, and how scientific methods can be used to develop and improve the questions that archaeologists can ask, and to come up with ways to answer those questions. Despite all its tangible material culture, ancient Egypt remains a world relying on archaeological interpretation and imagination. It is how we develop that archaeological interpretation and imagination that is of vital importance, and this is where this book fits in.

In the past, it has been noticed that few archaeologists attend archaeological science conferences and that few scientists attend archaeology conferences (Andrews and Doonan 2003). This is even truer when considering Egyptian archaeology and Egyptology. We hope that this book will bring a link between the disciplines and demonstrate how archaeological finds can be used to aid Egyptian archaeology and Egyptology. Furthermore, science was of considerable importance to the Egyptians themselves. For example, Egyptian astronomical and mathematical knowledge has been demonstrated through studies of the structures of the Old Kingdom pyramids (Spence 2000). The role of the Egyptological scientist is thus to mediate between the ancient Egyptians and their material culture, and to aid in its translation within Egyptian archaeology.

1.4 Studying Objects, or 'Material Culture'

We understand and identify the objects that we and others use every day (Caple 2006). The question is how to do this using the material past. As Caple (2006: xv) explains, "My car is a functional object to get me from A to B, but it also identifies the century in which I live and the material culture/society to which I belong. Attributes such as the size, model and colour of my car, plus its accessories and contents, demonstrate my personal taste and wealth, or lack of it! Objects are a reflection of the society and the individuals who make, own and use them; a physical representation of our desires within the limitations imposed by technology, economic circumstances and social acceptability. Objects also indicate the use we make of them. Every scratch and bump on my car tells a story of how good or, more accurately, how bad a driver I am. People in present-day society are familiar with cars, the different makes and models, their prices and reputations, even collision damage, but when they see my car they can 'read' it. They quickly gain an idea about my wealth, status, aspirations and driving ability." Caple (2006) here demonstrates the deeper meaning and understanding that can be drawn from such an object, moving beyond the functional, typological and descriptive, and instead delving into issues such as individual prosperity, and personal taste. Furthermore, he examines what the physical state of the car says about his abilities, and how people can 'read' his car and the interpretations that they make about him not only as a driver, but also as a member of society (Caple 2006: xv). This example demonstrates the importance of the life-cycle of an object. But it is only recently that the 'histories' (or '*chaîne opératoires*') of objects are being seriously considered from manufacture to our interpretation (Moser 2006: 6). "Objects are ... [now being] recognised as having their own complex 'histories,' their original creation function constituting only one chapter in a trajectory of existence that include subsequent reuse, discovery, presentation, and interpretation" (Moser 2006: 6). Material culture has the potential to provide information as to its manufacture, its provenance (both of the raw material and the object itself), its creator, its location in time and space, its function, its use and reuse, its symbolic or other meanings, the damage it sustained during its lifecycle, its alteration and any subsequent reuse, its discard, disposal and/or burial. As Caple (2006: xv) writes, "Objects are reluctant witnesses to the past; they have to be questioned carefully and closely if they are to provide accurate information". The life course and *chaîne opératoire* approach to analysis is followed in this volume, within the broader archaeological context.

1.5 Archaeology in Egypt and How to 'Read' This Book

A few concepts and ideas permeate all sections of the book. These include an appreciation of scale and patterning. In order to understand what the result of a scientific test actually means involves knowing the amount of variation possible and comparative data is normally required. For example, when looking at isotope values of bone collagen in order to understand diet (Section 3.4.2: Isotopic Markers), it is important to know what the normal values (or baseline values) are within Egypt for herbivores, carnivores and fish. This is vital for isotope values of both humans and animals as either might have an unusual diet or component thereof. A further issue is an appreciation that taphonomy and diagenetic change may affect results. Some of these changes are predictable and compensation or scaling of results can be undertaken so as to render them meaningful. But for other analyses, diagenetic processes are not fully understood and so these results should be viewed with caution. For example, iron as measured in human bone may be real or may be the result of contaminating clay inclusions as soil concentrations of iron are several orders of magnitude greater than the iron concentration in bone (Ezzo 1994). Thus scientific techniques require awareness of the variance, patterning and distributions normally expected, in order to understand usual or unexpected results. It is far too easy to simply allow a black box scenario to exist in which a scientist undertakes analyses but the results cannot be understood or interpreted as the context is missing. It is always important that results are placed in the context of their precision and accuracy. A result may be accurate but not precise, or precise but not accurate. These are very different and suit different contexts and situations, but the terms are often used interchangeably in colloquial usage.

It is also important to be aware of the mechanisms by which archaeological missions working in Egypt are assessed and governed, and which are subject to change over time. This book has been written taking into account the current rules and regulations, although readers should ensure that they check for alterations in these at any given time. Some of the details within this text may, therefore, change in the coming years, but the key issues and their relevance will remain. At present where facilities exist, scientific analyses must be conducted within Egypt. These will also change and improve over time, and currently available facilities should be checked at the time of work. Permissions for all analyses to be carried out in Egypt, but away from the main excavation site, have to be stipulated in advance for consideration within the main application document. At present, archaeological samples cannot be exported from Egypt. To some extent, this limits archaeological research in Egypt as it inhibits the application of some novel scientific techniques; new and developing methods can thus only be applied to items curated in overseas museums or other collections. Unsurprisingly, certain items of large-scale or fixed equipment have varying degrees of availability, accessibility, suitability, precision and accuracy within Egypt. Such facilities are not necessarily at the standard of the newest and most cutting-edge facilities in other parts of the world. This is not a criticism as these facilities are extremely expensive to construct, develop and maintain. We hope that, in the future, the export of a limited number of samples might be enabled in order that such facilities and their potential might be employed for the betterment of Egyptian archaeology and its conservation.

During the writing of this volume, there have been changes in the name of the governing organisation within Egypt that oversees all archaeological work/work relating to Egyptian objects inside Egypt. In July 2014, this body was renamed the Ministry of Antiquities and Heritage, and more recently named simply the Ministry of Antiquities (hereafter MA). Previous names, which might be encountered in other texts, include the Supreme Council of Antiquities (SCA), the Ministry of State for Antiquities (MSA) and the Ministry of Antiquities. For simplicity, throughout this book, we have continued to refer to this vital ministry as the MA. The ministry needs support and more active engagement by academics in order to further promote Egyptian culture and conservation worldwide.

The volume itself will hopefully be seen as a reasonably innovative and integrated approach to the use of scientific techniques and methodologies within the study of ancient Egypt, presenting, in an accessible manner, both the scientific methods of analysis available and their potential applications to Egyptian objects and contexts. To do this, the text is written as if engaging in a conversation, with the presentation of key questions and remarks as new topics and methods are discussed. By our own admission, this approach has been more successful in certain sections of this book than in others, and hopefully this variability will be forgiven. In adopting this approach, it is hoped that the explicit integration of scientific methods with case studies from within Egyptian history are clarified in order to exemplify best practice within archaeological science. Furthermore, it is hoped that the reader will reply to the text so that it becomes the dialogue or conversation that is both desired and needed between those working within Egypt.

The book is organised around three main themes each forming a long chapter. These themes are:

1 time and space,
2 people and,
3 objects.

An unusual approach has been taken through the attempt to develop these three themes within a framework of biography or the lifecycle. In this sense we have tried to delineate objects found within Egyptian contexts in terms of understanding aspects of their lifecycle or have explored the biographies of archaeological sites. In some places this approach has been more successful than in others, but hopefully it still may aid and structure thought processes within Egyptological science.

The second chapter of this book represents the first theme, and consists of the biography of time, space and place. Within this first theme, focus is placed upon time and dating, finding sites, survey approaches, the environment and the organisation of burial grounds. Key questions are: 'How old is this site?' 'How long was it occupied?' 'Without excavation, how is it possible to identify the towns, villages and houses in which people lived and the cemeteries in which they were buried?' 'What was the local environment like at the time of occupation?' 'What plants and animals lived in the area?' The last portion of this theme forms a link to the second theme (people) by considering methods specifically applicable to studies of cemeteries. With such a predominance of evidence coming from cemetery sites in Egypt, mortuary analysis has been vital since the days of Flinders Petrie, and the development of modern scientific techniques enables a greater diversity and breadth of questions to be answered.

The second theme of the book deals with the biography of people and the analysis of human remains, both mummified and skeletonised. As cemetery sites are much better represented in the Egyptian archaeological record than settlement sites, it is vital that the maximum potential be obtained from these contexts. When archaeologists uncover human skeletons or mummies, the first questions asked usually are 'how did she or he die?' and 'what was her or his life like?' These can be broken down into several components which are critiqued in a lifecourse dimension. The first portion considers scientific approaches to the classic aspects of ancient Egyptian studies of death and burial. Aspects of this section link back to the last portion of the preceding theme (spatial analysis of cemeteries). 'What did the person do during their life?' This forms the basis for the next section evaluating activity and occupation. This type of bioarchaeological analysis can lead to the identification and understanding of individuals with bodily modification, physical impairment and/or disability. It also links to studies of health and disease. 'What did people actually eat and drink?' Information as to actual diet provides a great source of potential knowledge of subsistence practices. Data obtained from skeletal or mummified material can be linked with more classical Egyptian archaeological or Egyptological knowledge, such as that derived from texts or artistic representations. 'Did this person live here all their life?' This part of the chapter shows how scientific methods can be used to study migration patterns and hence the lifetime mobility of individuals. This section also explains the potential of other (potentially non-destructive) analytical methods for assessing movement and interacts with concepts of population affinity and ethnicity. The last portion of the theme addresses the biography of individuals in terms of how science can aid in the understanding of the development and attribution of social rank and ethnicity.

The final theme comprises the biography and analysis of objects, and works through aspects of an object and its usage from its manufacture to its use, discard and museum display. It includes sections considering raw materials, production technologies including pyrotechnologies, deposition and taphonomy, and ends with an evaluation of reconstruction and display. Raw materials are considered both in terms of their extraction and recovery, with focus placed upon mining and other extraction methods for different materials and fabrics, but also the location of manufacture. 'Where does this come from?' is one of the questions most frequently asked of the archaeological materials scientist specialising in objects. In an ancient society, the vast majority of objects found tend to be locally produced. Rare imports from elsewhere can have value and prestige simply because of their rarity. Foreign objects provide evidence of external contact and provide important 'pins' for dating. This part of the objects theme ties together with the first theme focusing upon the biography of time and space. The section considering aspects of deposition and taphonomy critiques the effects of the Egyptian context, including the methods of disposal, upon the lifespan of an object. These effects modify the potential of other analyses and so are key to maximum exploitation and minimum misinterpretation (or over-interpretation) of the available scientific data. The last section of the chapter, comprising reconstruction and display, concentrates upon what archaeologists and museum specialists make of objects after excavation and their use in explaining ancient Egypt to the wider public. This last portion of the third theme links back to both the previous themes as the objects being conserved may include human remains, and reconstruction, reassembly and display may involve the use of three-dimensional reconstruction, as considered in the context of whole sites in the first theme.

The concluding chapter uses the case study of the site of Akhetaten, modern Tell el-Amarna, as an example of how scientific archaeology may be linked with more traditional forms of archaeology in order to maximise the potential of archaeological excavation within Egypt. Amarna has been selected as an example of best practice, where a wide range of scientific techniques have been and continue to be applied. There are, of course, other sites and archaeological teams working in Egypt who are making equally sound use of some of the techniques that are described in this book. It is hoped that the exemplar of Amarna demonstrates the potential of an Egyptological science.

It is important, when reading this volume, to consider the scientific facilities available in Egypt at the time of writing. For example, radiocarbon dating is available at the Institut Français d'Archéologie Orientale (IFAO) in Cairo. Such facilities may change and/or be upgraded and improved. Other analyses are currently only available abroad, and as such, their use is currently restricted to material that has already been exported or for which special permissions have been obtained. We hope that this book demonstrates the potential of such analyses for Egyptian archaeology and hope that more scientific studies might be possible in the near future.

2 The Biography of Time and Space

The location, environment, size and dating of archaeological sites underpin all of our investigations into the prehistory and history of Egypt. This information, however, comes as the result of many different types of analyses – analyses that will be discussed in this chapter. These methods vary from those that require specialised training, with specific equipment, to those that can be carried out in the field by students and professionals alike – as long as all adhere to systematic methods of recording. Some techniques discussed require quite substantial funds to be raised, whereas others can be carried out very economically. It is important, particularly during training, that everyone is aware of what can be achieved cheaply and easily in the field with the minimum of equipment, as well as the full range of possibilities that might be applicable and appropriate. Local archaeological missions do not always have access to extensive equipment, so alternative methods should be sought where possible, with local sources located for purchasing required supplies for these various analyses (Orton & Hughes 2013: 51). In addition, due to the lack of availability of certain non-portable and fixed equipment in Egypt, there are some analytical methods that cannot currently be applied to freshly excavated material.

Certain methods can be carried out in the field, museum or laboratory, in Egypt or overseas, whereas other methods have more restricted applications. Whatever methods are used, it is important that they are employed to achieve specific aims and objectives. In addition, meticulous care must be taken in all recording, whether separating finds from different archaeological contexts in the field, or labelling small samples for radiocarbon dating in a museum storeroom. Good methodology and practice have to be in place so that objects can be associated back to their original contexts, and therefore achieve meaningful results.

Although the most common type of archaeological site found in Egypt is the cemetery, there are a diverse range of different forms of Egyptian archaeological site. Furthermore, these archaeological sites need to be placed into the broader archaeological context or landscape. There are various means by which the archaeologist can set out to examine wide regions and attempt to locate sites of all types in varied environments. The modern environment in Egypt often affects our ability to locate and investigate such ancient sites, and it is important to investigate sites within the context of the ancient environment as this is a crucial factor in its initial and original location. Furthermore, local environmental conditions greatly affect a site's longevity and its specific functions, and both past and present environmental conditions affect its preservation. It is through the study of the plant and animal remains in addition to sediments that archaeologists can start to interpret the palaeoenvironment and its temporal changes. Such features include the proximity of the site to water sources, the changing local environment over time, and local flora and fauna. As archaeologists, we need to find out the chronological range of the sites in order to understand how they fit into and relate to the previously existing sources of archaeological and textual evidence. Sites may also change in size through time, thus necessitating the examination of potential social, political and/or environmental causes in addition to the timing of such changes. A site or group of objects is commonly dated to deriving from within the reign of a certain queen or king, or within what Egyptologists might refer to as a certain 'kingdom' or a 'dynasty', but we can also assign absolute calendrical dates to objects and sites, and then use the objects to help answer the questions: what happened before what, when and even why.

2.1   Time: Dating Methods (Relative and Chronometric Dating)

How old is it? How long was it occupied? These are very common questions that the layman will ask the archaeologist. The archaeologist might ask slightly finer tuned questions such as, which site/context was earlier? Or, how can we relate these contexts to order events in time? To answer these sorts of questions, there are a number of methods that may be used, all of which vary in precision in terms of exactly what information they can provide. Flexibility, a good knowledge of available scientific techniques, and an open mind are required, since the archaeological context, as well as the research questions themselves, will determine which methods we can and cannot use, as well as which types of data outputs will be meaningful.

Within archaeology, there are a variety of methods currently used which help the archaeologist and Egyptologist to identify the point in time to which an object, an archaeological feature, or a context dates. Such methods can be divided into two basic forms: those which give relative dates, concentrating on the relative chronological position of objects within contexts, and those which can provide chronometric dates (sometimes described as absolute), and provide an absolute point in time. Both types of dating play a crucial role in understanding the chronological sequence of events and the question of how it is possible to reconstruct a series of events within the archaeological record can often result in very complex processes. Relative dating will be examined first, followed by chronometric dating. Relative dating methods allow us to answer questions such as 'which archaeological context (or site) is earlier?' 'how do the contexts relate to each other?' Relative chronology is examined here as a means by which to assign dates to archaeological contexts encountered in the field. It is the seriation of objects, described below, that allows archaeologists to place objects, and in turn contexts, in chronological position relative to each other. When a relative date is assigned, it places the feature or context in relation to other features or contexts. This may be at the site under investigation or in relation to other archaeological evidence over the broader landscape, either within Egypt or beyond its borders. The context, or object, essentially fits within a series of events. The most widely used relative chronologies are based on ceramic sequences (seriation), described below, which were developed through the study of ceramic typologies. A similar approach is used on a wider range of objects, both objects with great longevity and objects with shorter chronological spans (Orton & Hughes 2013: 226–232).

The second main method of dating is through achieving absolute/chronometric dates to provide an actual point of time in the past. These are dates shown as dates BP ('before present'), which are often (but not always) calibrated into dates 'cal BC' ('before Christ') or BCE ('before the common era'). For prehistoric periods, particularly the Palaeolithic, dates are often given as 'kyr BP' (kyr referring to a unit of a thousand years); thus the start of the Palaeolithic in Egypt can be written as 700 kyr BP or 700,000 BP. Chronometric dates are real points in time, given in ranges of years. One of the most common of these is the radiocarbon (^{14}C) method (Arnold & Libby 1949; Manning 2006). The main benefit of absolute dates is that they enable us to look at the real-time, as well as relative, sequence of events. Ceramics on archaeological sites, for example, help to tell the archaeologist what came (or happened) before what. This is a necessary part of detangling complex stratigraphic sequences. Sequences of events are not simply and neatly overlain, but consist of pits cutting into earlier layers, with the fill of these pits being re-deposited in a new context, and the simultaneous use of fresh material alongside older, re-used materials. Within such archaeological contexts, there is not always a suitable object, such as a ceramic vessel or a diagnostic sherd, which relates to the context that the archaeologist hopes to date. There may, however, be fragments of organic (or inorganic) material which could be sampled for application of an absolute dating method. Depending on the quality of sample and the integrity of the relationship between the sample and the context for which a date is required, these methods can be extremely accurate. Relative and chronometric dating can complement each other and it may be possible to assign absolute date ranges to certain ceramic types, as opposed to, or in addition to referring to ceramic types in relation to dynasties or particular Pharaonic reigns. Context is all important, and investigations may show that the same ceramic types represent differing absolute date ranges subject to their horizontal distribution. One of the great benefits of chronometric dating is its ability to act as a lynchpin between events in different geographical regions (e.g. Marcus 2013; Quiles et al. 2013; Zohary et al. 2012: 17).

2.1.1   Relative Chronology and Relative Dating

How do we arrive at relative dates? And how can they help us to tell which site is earlier than another, or which areas of a single site were used when? How can we tell whether a certain archaeological feature was constructed more recently than another one, and which burials are earlier than others? At some times it can be much more obvious than at others; this could be, for example, when a feature clearly cuts through an earlier feature or where there are datable objects in features within a site. But can we also compare *between* sites? This section will introduce seriation and typologies, and their practical implementation in the field as we attempt to reconstruct the timing of events.

The investigation of chronology within modern archaeology is indebted to the work of Sir William Matthew Flinders Petrie (Petrie 1901; 1920; 1921). Although the 'typological phase' began already in the 1880s, it was Petrie who began looking at pottery within archaeological stratigraphy (Orton & Hughes 2013: 8). Petrie, working on the typology of ceramic vessels from the cemeteries of Abadiyeh, Hu, Naqada and Ballas, developed ceramic seriation (Petrie 1899; 1901). This system enabled him to examine which ceramic types co-occurred with which other types in graves, and in which combinations, and which types never co-occurred (see Hendrickx 2006: 60–64 for a comprehensive review of seriation and Predynastic–Early Dynastic chronology). There were two key factors on which chronological ordering was based: 1) Petrie's observations concerning the evolution of the Wavy-handled class of ceramic (W-ware), and 2) the observation that a) White Cross-lined vessels (C-ware), and b) Decorated (D-ware) and Wavy-handled vessels, hardly ever co-occurred. At this time it was simply 'accepted' that the shapes of vessels evolved from more globular to cylindrical shapes and that the handles became less functional and more decorative (Hendrickx 2006: 61), and it was noted that the early and late forms of Wavy-handled types did not co-occur in a single grave (Petrie 1921: pl. XXVIII–XXX; Hendrickx 2006: 61–62). The ultimate result of this analysis was the creation of a system that placed ceramic types in a chronological sequence, with each type being assigned to a Sequence Date (SD) or range of Sequence Dates. These Sequence Dates originally started at SD30 because Petrie considered it inevitable that earlier archaeological evidence would be found; Sequence Dates ended at SD87 (Petrie 1953: 27). The pre-SD30 dates were later assigned to material attributed to the Badarian culture, but there were problems with the use of the Sequence Dating system on this material and it was never fully implemented (Brunton & Caton-Thompson 1928:

26; Hendrickx 2006: 62). Petrie's system can be criticised for the small number of cemeteries included in the initial analysis, for the fact that his classificatory system crosscuts fabric type, finishing techniques (including decoration) and forms, and because he does not seem to have taken the horizontal distribution of the graves into account (Petrie 1901: 31–32; Hendrickx 2006: 63). Nonetheless, the system of seriation has remained a well-used tool by archaeologists and scientists across the globe, and, within studies on Predynastic and Early Dynastic Egypt it is still referred to within the more recent chronological revisions, including the 'Stufen' chronology of Werner Kaiser (1956; 1957) and revisions by Stan Hendrickx (2006).

Typologies exist for other object types as well as for ceramics, although the latter remains the most frequently used method applied to dating archaeological objects and associated contexts during fieldwork and for object-based research in museums worldwide (Orton & Hughes 2013: 219). Some ceramic forms have very brief lifespans, providing quite a restricted time range for a given context, whilst others, such as Egyptian amphorae of brown alluvial clay, have a much longer period of use, leading to less precision. One single sub-type of amphora, AE3, for example, is used throughout the Early and Middle Roman period (Wilson & Grigoropoulos 2009: 275; Orton & Hughes 2013: 225). Imported amphora, such as those discussed in Bourriau & French (2007), can also have reasonably wide date ranges. In such cases, it is the context of the find that can restrict the date range possible for the particular vessel (or more commonly sherd) in question and the presence of other objects or types of ceramics with shorter date ranges. The examination of Egyptian material within the framework of its stratigraphic context was first carried out by Gertrude Caton-Thompson at the prehistoric site of Hemamieh (Brunton & Caton-Thompson 1928: pls. LXII–LXV). Prior to this, objects – and particularly ceramics – had only been considered within cemetery contexts within Egypt and not within vertical stratigraphy. It is the connection between the object and stratigraphic context that helps the archaeologist to place the objects in sequence and thus determine a tighter potential date range.

2.1.1.1 Using Relative Chronology in Fieldwork

During excavation or survey, the most useful ceramic finds are diagnostics (rims, bases, handles and decorated sherds) which help us reconstruct the vessel form and assign it to a typology. It is important that all material is kept separated by archaeological context, and the ceramicist can then decide, together with the field director, which sherds/vessels should be kept. (See Box 2.1 for a further discussion of this.) A mixture of common sense, experience, and familiarity with a given area, site or period will largely dictate the outcome. Surface layers of a trench or excavation area are often very mixed and might also represent later re-dumping/ movement of archaeological material resulting from processes such as robbing or the actions of *sebbakhin* dismantling ancient mud-brick structures in order to use the material as fertiliser (Bailey 1999). Whilst these mixed layers might relate to a single event or multiple events of later activity, an experienced ceramicist will be able to work quickly through the material to ascertain what is diagnostic, choose samples representative of the key types, and put the other material aside. During this stage, it is less important to record the quantity of material, particularly where there is a large amount of *radim* (re-dumped material, possibly including the spoil of earlier excavations), but it will nonetheless inform as to the wider ceramic chronology for the site or area as a whole. As discussed in Box 2.1, the quantification of ceramics in the field requires a system to be devised and adhered to so that the ceramics can be compared consistently across a site or region. Ceramics are our most plentiful archaeological resource on Neolithic sites and later, and are variably weighed, and/or counted. What is important is that this work is undertaken for a reason, with specific questions in mind, and an aim of ascertaining not purely overall quantity, but quantity of different types and fabrics, to get a clearer understanding of the type of site and its place within the wider region, or world, over time. At the time when seriation was being developed by Petrie, sherds and vessels were simply counted; more recently, it has become more common to weigh sherds to get an impression of the total quantity and to estimate how many vessels there would have originally been, including estimated vessel equivalents (eve) (Orton & Hughes 2013: 21–22). Systematic study of ceramics can help ascertain whether a site was particularly involved in trade and/or exchange with other regions within or outside Egypt at certain periods of time, and may answer what triggered greater or lesser contact over time. Within the individual contexts, with sufficient and dateable material, however, it should be possible to establish the relative sequence of events through Neolithic, Predynastic, Dynastic and later history. We will return to the absolute dating of these events a little later in this chapter.

In the field, such as during ground survey, ceramic typologies provide information regarding site longevity (e.g. Wilson & Grigoropoulos 2009). Surveying enables archaeologists to assess wider changes, developments and movements of peoples across the landscape – both those dictated by nature (e.g. moving water courses, discussed in detail later in Section 2.3: The Environment) and by human intervention (the latter may be in response to the former). Survey and associated macroscopic examination of ceramic types, even where precise contexts cannot be ascertained, can nevertheless provide important dating information and suggestions as to whether the original context was funerary or domestic. During field survey, it is often only possible to make general comments as to the function of some vessels, especially when dealing with sherds that

Box 2.1   Ceramic Recording in the Field
Janine Bourriau & Peter French

We discuss the differing strategies adopted for recording the ceramics from three Egyptian settlement excavations. The first two concern Buto (Tell el-Fara'in in the Delta), excavated initially by the German Institute of Archaeology in Cairo and directed by Thomas von der Way and Dina Faltings (Faltings et al. 2000: 131–79) and most recently by Ulrich Hartung (Hartung et al. 2003: 199–266). The third was at Memphis (Kom Rabia), directed by David Jeffreys for the Egypt Exploration Society (Jeffreys 2006).

Total Recording of Rim Sherds

At Buto, in the late 1990s, excavations within the girdle wall of the major temple continued and extended work done by Egyptian colleagues. The stratigraphy consisted of sloping layers containing sherds but little else. Apart from some Roman material near the surface, the pottery was Late Ptolemaic and remarkably uniform throughout the six-metre depth. Simultaneously, at some distance from the temple, the upper levels of a house were excavated. This coincidentally yielded pottery of the same date.

It was soon evident that, in these two contemporary assemblages, the typological mix was very different. The decision was taken to catalogue each rim sherd, calculate the percentage preserved and total these to an *eve* (estimated vessel equivalent) (Orton et al. 1993: 172–3, Figure 13.2) for each type. Although several thousand rims were involved, the limited chronological range allowed rapid progress. We deduced that the area within the temple was probably the Sacred Lake, which had been used for rubbish disposal by the temple personnel, and had yielded a mixed assemblage with large cooking pots characteristic of communal living or family units, whereas the numerous small bowls and amphorae, and the paucity of other vessels indicated that the 'house' was probably a tavern or drinking club (French 2000: 168–9, 174–5).

Selective Recording

More extensive excavations have taken place annually in a third area of the city, where continuous occupation from the late 8th to the mid-6th century B.C. was preceded by centuries of abandonment and succeeded only by cemeteries. One key context was recorded by the same total rim method (French & Bourriau 2007: 100–19;

French 2009, 128–33), but generally the first millennium sherds were far too numerous for two people to record in detail, even with occasional assistance. We realised that most, deriving (as always in the Delta) from Nile silt vessels, had a limited typological range, much of it endlessly repeated. Once the dating of the three major phases was established and the recurring types and their variations were drawn, we could give the date of each context to the excavators, with the number of rims present (100 being more trustworthy than 1). Otherwise, we sketched the types in each context, recorded in detail the imports and marls, and set aside for study and drawing only well-preserved or unusual diagnostics.

Random and Purposive Sampling

At Memphis in 1984–2000, Janine Bourriau and a team of four ceramicists and students confronted sherds from 650 years, from the mid-13th to the end of the 19th Dynasty. There were too many for total recording (six early weeks yielded 85,000 diagnostics, mostly rims); yet, to establish a similar statistically valid ceramic chronology following 'best practice', it was essential to note each sherd's fabric, shape, method of manufacture, surface treatment and decoration.

It was decided from the outset to employ random sampling, an objective procedure giving each sherd an equal chance of selection (Orton 2000; Bourriau 2010: 7–10). Only rims were sampled, as the best guide to identifying the whole vessel, and only from uncontaminated closed contexts, omitting walls, excavators' baulks, etc. Every reliable context was recorded so that nothing chronologically significant was overlooked. On the advice of statistician Nicholas Fieller of Sheffield University, sample size varied with context size: 1–15 rims — record all; 16–29 — record 15; 30–75 — 20; 76–200 — 25; 201+ — 30. At another site, numbers might differ to suit context sizes.

The process was simple but had to be strictly followed. All rims from a single context were washed, allowed to dry and placed beside a board divided into squares numbered 1–300. Starting from Square 1, one rim was placed on each successive square until all had been placed; for more than 300 rims we returned to Square 1, placing a second, notionally 301, below the first, and so on. We used a random number table to select the rims, ignoring numbers above the highest numbered occupied square. (Free computer programs are now available on the internet.) By rolling a die, we picked a different starting point in the table each time. The selected rims were bagged and labelled.

Knowledge and experience played no role in random sampling, but to ensure no significant diagnostic went unrecorded a supplementary 'purposive' sample, not confined to rims, was extracted from the complete context assemblage. Sherds selected included: complete or well-preserved vessels; non-Egyptian sherds; pot-marks, wear marks or decoration; sherds from contexts important for other projects within the excavation. The 'purposive' sample was similarly bagged and labelled, and kept strictly separate from the random sample at every stage of recording and publication.

With this method of sampling, all random sample sherds should be published, either in full or as statistics, with as many of the purposive sample as seems useful.

Discussion

The ceramicist must understand from the outset the likely scale of the excavation, its research priorities and the role ceramics are to play, and adopt an appropriate collecting or sampling strategy. Until the 1960s, most pottery was selected for publication based on its state of preservation. Not only did this discriminate in favour of closed forms, especially the very small and very large, it also ruled out any subsequent statistical analysis, limiting the value of the work. Today's ceramicists must make clear their basis for selection and include any statistics extracted; future evaluation of their results will depend on this.

Sherds can yield much information. Imports illustrate exchange networks between Egypt and its neighbours (the Levant, Cyprus, the Aegean); in some parts of the country (especially the Delta, Nubia and the oases) minority fabrics or forms identify internal trade patterns; decorated vessels form part of the artistic heritage; manufacturing techniques chronicle the ceramic industries themselves, and the major employers and contributors to the national economy. Yet for the ceramicist, establishing a chronology will probably always be the first priority.

References

Bourriau J. (2010) *Kom Rabia. The New Kingdom Pottery. Survey of Memphis IV.* London: Egypt Exploration Society.

Faltings D., Ballet P., Förster F., French P., Ihde C., Sahlmann H., Thomalksy J., Thumshirn C. & Wodzinska A.(2000) Zweiter Vorbericht über die Arbeiten in Buto von 1996 bis 1999. *Mitteilungen des Deutschen Archäologischen Instituts Abteilung Kairo* 56: 131–179.

French P. (2000) 'Trench C1: The Pottery of the Sacred Lake' and 'Trench B1: The Pottery'. In: Faltings D. et al. (eds) Zweiter Vorbericht über die Arbeiten in Buto von 1996 bis 1999. *Mitteilungen des Deutschen Archäologischen Instituts, Abteilung* Kairo 56: 168–169 & 174–175.

French, P. (2009) Tell el-Fara'in – Buto In: Hartung U. et al. (eds), 9. *Mitteilungen des Deutschen Archäologischen Instituts, Abteilung Kairo* 63: 128–133.

French, P., & Bourriau J. (2007) Tell el-Fara'in – Buto In: Hartung U. et al. (eds), 9. *Mitteilungen des Deutschen Archäologischen Instituts, Abteilung Kairo* 63: 101–119.

Hartung U., Ballet P., Effland A., French P., Hartmann R., Herbich T., Hoffmann H., Hower-Tilmann E., Kitagawa C., Kopp P., Kreibig W., Lecuyot G., Lösch S., Marouard G., Nerlich A., Pithon M. & Zink A. (2003) Tell el-Fara'in – Buto, 8. Vorbericht. *Mitteilungen des Deutschen Archäologischen Instituts, Abteilung Kairo.* 59: 199–266.

Jeffreys D. (2006) *Kom Rabia: The New Kingdom Settlement (Levels II–V). Survey of Memphis V.* London: Egypt Exploration Society.

Orton C. (2000) *Sampling in Archaeology.* Cambridge: Cambridge University Press.

Orton C., Tyers P. & Vince A. (1993) *Pottery in archaeology.* Cambridge: Cambridge University Press.

are completely out of their original contexts (Wilson & Grigoropoulos 2009: 267). There are, however, a number of observations that can be made in the field or as part of post-excavation analyses that should prove useful for ascertaining the range of functions of ceramic vessels (Orton & Hughes 2013: 20–21; Skibo 2013). While excavating at a single stratified site, archaeologists can plot events through time on both horizontal and vertical planes: with regional survey, it is possible to plot which areas within the landscape appear to have been used at various periods in time. The data collected during ground survey can be integrated with recent and historical map data, with satellite and geophysical survey data, and chronological and environmental data added through hand augering (all of which will be discussed later) to build up a detailed history of the site over time (e.g. Schiestl & Herbich 2013). As well as the ancient history of the site, it is also important to appreciate its recent history, as this can – and has – had a profound effect on what is seen on the surface today, and how it can sometimes be misinterpreted due to recent surface disturbances (Bailey 1999). As explained in Box 2.1, it is essential to begin work with a set strategy for collection and sampling, taking into account the questions that will be asked of the data, as well as recording whether sherds were recovered by different means, e.g. sieving or hand collection. Then we need to ascertain whether the circumstances allow for, or even dictate, analysis to be carried out in the field (such as on field survey) or whether it is more appropriate for selected samples to come back to a storage/study facility for detailed analysis and recording. This latter work might require desalinisation of sherds to remove encrusted salts (Orton & Hughes 2013: 48–64). The aim of drawing is to recreate the profile of the vessel and to reconstruct both interior and exterior detail; marks on the surfaces of the vessel can indicate how the vessel was made – by hand, by a fast wheel, or coiled and finished on a slow wheel, or slab built and finished by hand smoothing , and whether the surface was polished or burnished (Arnold 1993: 15–82). Examination of the fabric of the vessel helps to inform about the origin of the clay, whether local or imported (see Chapter 4: Biography of Objects), Nile silt or marl clay, and what material was added (temper) to change the properties of the clay, such as straw (fine to coarse), sand (varying quantities), crushed shell, crushed limestone, etc. An introduction into ceramic analysis/description is found together with some examples of types in Wodzińska (2009a; 2009b; 2010a; 2010b) and more detailed description can be found in Arnold & Bourriau (1993).The type of fabric of the pottery can also be very important in terms of dating, and fabric types including the Vienna System (Nordström & Bourriau 1993) are discussed in Box 4.1 and elsewhere in Chapter 4: Biography of Objects.

It all sounds logical, but how does it actually work in practice during fieldwork and what does it bring to our understanding of the history of single sites and to our understanding of site development over regions? Relative chronologies are an essential part of Egyptology and archaeology. When we are talking about single sites, or looking at single features or contexts, relative dating is most commonly the means by which specific events can be dated, e.g. a burial, the collapse of a house, or the burning of a site. We look for objects within, above and below these event contexts and by examining these finds we can date events, and then place events in chronological order. One of the key ways to order these events visually is by using a Harris Matrix, as this links to the processes of site formation (Harris 1979). Investigation of the frequency of certain artefact types across a site or region can aid the understanding of changing patterns of site use. For example, the changing position of river branches can affect a site's ability to remain within larger networks of exchange. Such artefact types can date the period over which such changes took place. Analysis of ceramic fabrics and shapes reveals the origins of the clay and the objects themselves (see Chapter 4: Biography of Objects). This can even be a source of information regarding ethnicity (see Chapter 3: Biography of People) and provide nuance regarding trading relations over time.

When objects are used to date contexts, we need to be certain that the context is sealed and not disturbed by later activity. Disturbance by re-building or robbery may cause earlier objects to move into later contexts and later objects into earlier contexts. If the context is found to be undisturbed, the contents of the context could provide dating evidence, such as for a burial, or for the founding of a building when in foundation deposits. Multiple objects may be present, but the date of the event will be fixed at the youngest object present, providing what is called the *terminus post quem* (TPQ, the Latin for 'limit after which') or the oldest object for the *terminus ante quem* (TAQ, 'limit before which'). This may be any type of object, but could include a coin (Orton & Hughes 2013: 225–226; Tassie & Owens 2010: 78–79). For example, if there is a Predynastic burial with ceramic types ranging from Naqada IIIB-C, then the moment of burial could be said may be at least as late as Naqada IIIC. It must be borne in mind that earlier date of that Naqada IIIB vessel, but may be from a little later. Earlier vessels could have been in circulation for a number of years before being buried, or could even be heirlooms (Jeffreys 2003). It is also crucial to remember that sherds may not be in their original place of deposition; they may have been re-deposited into later contexts, and may have worked their way into earlier contexts (Orton & Hughes 2013: 222). As noted earlier, relative dating is not confined to ceramics; other notable early work includes Petrie's Protodynastic Corpus which included relative dating of siltstone cosmetic palettes (Petrie 1921).

Domestic contexts generally present a greater challenge to the archaeologist in terms of stratigraphy and dating, as they commonly have complex stratigraphic sequences representing series of layers comprising floor surfaces, rubbish pits, collapsed walls, and foundations dug through earlier surfaces. With sound excavation methodology, such as the implementation of single context excavation (Barker 1993; Museum of London Archaeology Service 1994), each separate event is given a unique sequential context number, with no two context numbers *ever* being duplicated within a single site (even in different trenches). The cutting of a pit therefore has a different number to the fill of the pit. Each context is recorded separately, and the associated sediment is sieved separately with any ecofacts or artefacts bagged, labelled and recorded as having come from that specific context and registered for future analysis/drawing. During excavation, the relationship between these events is recorded with a running matrix. This is collated to create the final matrix, such as the Harris Matrix noted above (Harris 1979). The matrix, therefore, helps the archaeologist to understand the sequence of cultural and natural events that contributed to the archaeological record as revealed throughout the course of the excavation. It is, however, the examination of the finds and the post-excavation analyses that give relative and absolute dates to these events. The examination of stratigraphy and contextual associations is thus vital.

Ceramics are crucial within relative chronology, but what about communities and individuals in the pre-pottery Neolithic and Palaeolithic periods? Is there any way that stone tools and their contexts can be examined in similar ways to provide evidence as to their chronological sequence? Both relative and absolute dating methods might be applied.

Research was already well underway by the late 1920s, when Sandford & Arkell (1929: vii–viii) investigated the occurrence of lithics embedded within geological formations. They wanted to find a geological date for the first arrival of humans in the Nile Valley and to trace their subsequent history in the region. What was also important to them was the comparison between specific lithic implements found in the Nile Valley and similar types found in Western Europe and establishing whether the implements were of the same geological date. If the European contexts were later, this might suggest a transfer of techniques from south to north over a Sicilian land-bridge (Sandford & Arkell 1929: ix). In the course of their fieldwork, Sandford & Arkell (1929: 28–31) detected prehistoric Lower Palaeolithic artefacts in the Nile gravels of a particular Palaeolithic Nile channel ('the fifth') and, although a range of types were found together, through examination of lithics they distinguished the older and more water-worn Chellean types from the still fresh and sharp younger Acheulean tools. In beaches and river gravels within two different geological strata in Upper Egypt, they located both early and late Mousterian implements, yet failed to find any earlier Mousterian in Lower Egypt (Sandford & Arkell 1929: 45–52). The lithics that were the 'immediate successors of the Mousterian' were detected in silts covering Mousterian gravels from Aswan to Luxor. They named these silts post-Mousterian, and proposed a post-Middle Palaeolithic date (Sandford & Arkell 1929: 63). Their fieldwork led to the identification of four river terraces at specific height ranges above the river Nile, each with associated Lower and Middle Palaeolithic tools (Sandford & Arkell 1929). Such environmentally aware methods of relative dating of lithic implements allowed them to offer a new hypothesis on the relative dating of the Fayum Neolithic cultures that had been recently proposed by Caton-Thompson (1927) and with Gardner (1934). Caton-Thompson & Gardner (1934) believed that the Fayum lake levels fell over time, and, using relative methods, had dated the evidence from the higher locations on the Fayum lake (Fayum A) to a point in time before that from the lower levels of the Fayum lake shore (Fayum B). Sandford & Arkell (1929: 66–72), however, were able to argue convincingly that the lake levels had actually risen again during the course of the Neolithic and that Fayum B, the material culture associated with the lower lake level, actually pre-dated Fayum A. Wendorf & Schild (1976) subsequently proved this to be the case using absolute dating methods (see below). Since the work of Caton-Thompson and Sandford and Arkell, research into lithic industries has advanced greatly, and archaeologists, including Vermeersch (1992), van Peer et al. (2010) and Schild & Wendorf (2010) have created lithic typologies from the Middle to Late Palaeolithic.

2.1.1.2 Systematic Spot Sampling

How is it possible to get reliable dating for sites through analysis of ceramics for dating? And given that there is just so much material, how do we decide what to look at and what not to look at? In Box 2.1 Bourriau describes one sampling rationale in terms of what to sample within the collected ceramic assemblages – be they from excavations or from survey. When in the field on survey we have to decide on a strategy (relative to our research questions) which covers as much ground and/ or as much detail as possible, within the restrictions of the available time and resources that fieldwork allows. This is also important when applying for permission to work in Egypt as it is necessary to state the exact areas of work, with locations marked on 1:50,000 Survey of Egypt maps, and archaeological site or modern village names with global navigation satellite systems (GNSS) coordinates where possible. During field survey, a slightly different strategy for collection/observation of surface material may have to be established for each different location visited, but an adaptable research strategy or design should be drawn up before going into the field (see Tassie & Owens 2010: 38 for a list of important points to consider). Wilson & Grigoropoulos (2009: 2–3) list different possibilities. These include basic/initial area visits as well as site surveys. In the latter, where greater attention is paid to the dimensions of the site and its general location, a sample of pottery is normally collected. At some sites, during follow-up visits, augers are drilled to establish more information about the chronology of the site (including sub-surface material) as well as environmental conditions. This can include proximity to ancient waterways (see Section 2.3: The Environment; Wilson & Grigoropoulos 2009: 2–3). Particularly when a site/suspected site is being visited for survey, or when an attempt is being made to understand the surface distribution of ceramics, the different activity areas, the date of a site, or scatters of material out of the context of a known site and its variable size over time, it is important to adopt a systematic sampling strategy, as opposed to trying to comb every inch of the ground to look for archaeological material (Orton 2000: 67–69; Tassie & Owens 2010: 45–46). The actual strategy will depend on the research questions and the size of the area being surveyed; these could range from total collection of diagnostic material within given transects, total collection of all materials within circles of the same area at regular intervals across a site, or gridded smaller area sampling, with total collection within selected grid squares (Orton 2000: 67–111; Tassie & Owens 2010: 45–46). In prehistoric contexts, if a gridded area collection of lithics is being undertaken, it can be more sensible to adopt a total collection strategy in a selected number of squares. This enables the total toolkit in use and the débitage to be assessed, thereby establishing whether knapping was carried out in the vicinity. Deflation must also be considered; the effects of the environment over many years cause sediments to deflate into thinner layers, resulting in the presence of a mixture of lithics of varied date, as occurs in hearths (Mercier et al. 1999; Wendrich et al. 2010: 1000, 1002).

So although items can be placed into a temporal order, being able to put a specific date or a tight chronological range onto an object, site or event is another issue. How can we accurately and precisely date Egypt and its history?

2.1.2 Absolute Dating

The previous section focused on methods that help order contexts through the relative dating of objects and provide relative dates to sites and findspots across wider regions. How simple is it, however, to assign chronometric (absolute) or calendrical dates to objects, contexts and events? Can we get beyond the ordering of events, and restrictions based on the longevity of ceramics/other finds, to being able to place sites, contexts and specific events into absolute ranges of years? And can all types of objects and materials be dated by chronometric means and thereby help improve the precision of the prehistoric and historic chronology of Egypt? This section explores various methods, with reference to textual and scientific methods of establishing exact points in time. It discusses the possibilities of achieving chronometric dates, but also cautions on possible pitfalls and limitations when selecting material for scientific dating.

It is first important to look at how the Egyptians themselves thought about chronology and the duration of the known past (Redford 1986). There was no single system of dating and no consistent or definitive record of regnal years, but a number of ancient attempts to order the kings and queens do exist (see king list in Appendix I; Hornung et al. 2006a; 2006b; Redford 1986; Ryholt 2006: 26–32; Wilkinson 2000). The most important historical source is Manetho, who working as a priest in the Nile Delta in the early 3rd century BC, wrote the *Aegyptiaca* during the reign of Ptolemy II (Manetho 1940; Shortland 2013: 18–22). The *Aegyptiaca* is only preserved to us in the form of copies, with parts of it also found in the writings of later authors (Manetho 1940; Shortland 2013: 18–22). The earliest list of kings of which we have evidence dates to the 5th Dynasty, the Royal Annals of Memphis, and records events within specific reigns from the 1st through to the 5th Dynasty. It is preserved in several fragments, with the most substantial being the Palermo Stone (Hornung et al. 2006a; Wilkinson 2000). This source attempts to look back in time beyond the first kings of the unified state of Egypt (at around the end of the 4th millennium BC), to what are described as mythical or early historical rulers known as the Followers of Horus (briefly presented in Hornung et al. 2006a). One of the best known examples of a king list is that carved into the walls of the Temple of Seti I at Abydos (see Figure 2.1.; Brand 2000: 162, 373, Pl. 81). The scene depicts Seti I and

Figure 2.1 King list on the wall of the Temple of Seti I at Abydos (Photograph courtesy of Joanne Rowland)

the young Ramesses II making offerings to their royal ancestors. These ancestors extend back in time to a king called Menes, placed at the very beginning of the 1st Dynasty. The Turin King List, also called the Royal Canon of Turin, is one of the most complete lists to survive into the modern era. Given that a text dating to this reign occupies the other side of the papyrus, it dates no earlier than the reign of Ramesses II (Ryholt 2006).

How reliable are these lists, and what information do they give as to who ruled when, and for exactly how long? Firstly, it is very important to consider the contexts of these records and understand that most of these lists were not intended to be accurate and objective historical records of every ruler that had come before; the exception is the Turin King List (Ryholt 2006: 26). Instead, because Egyptian rulers wanted to emphasise their legitimacy through association with the long line of strong and revered rulers of the past, such as King Djoser (3rd Dynasty) or King Sneferu (4th Dynasty), less popular rulers, including Hatshepsut and Akhenaten, may be omitted from such lists, leaving historically inaccurate sources. These sources therefore provide varying degrees of information and accuracy. There is also the issue of co-regencies and periods in Egyptian history where the country is ruled simultaneously from different centres, for example the simultaneous Theban and Heracleopolitan rulers during what is known as the First Intermediate Period, with the country being re-unified by Mentuhotep II of the 11th Dynasty (Grajetzki 2006: 10).

If these sources are not historically accurate, are they still useful? Despite the fact that they lack historical robusticity, the very existence of these lists enables the construction of and challenging of different chronologies. Such sources include the king lists, papyri and other inscriptions that give specific regnal dates for certain events. A relatively recent and comprehensive discussion of chronology, and useful bibliography, can be consulted in Hornung et al. (2006c). The order of Egyptian rulers and the calculation of the length of each reign depends upon how documents have been read, understood, compared and in some instances (e.g. the Turin King List) physically reconstructed (Ryholt 2006: 27). In attempts to link these historical records to exact points in time, Egyptologists have to take such measures as trying to assign a date to astronomical observations of Sothic and lunar dates mentioned in such texts. For example, the Papyrus Berlin 10012 announces the early rising of the star Sothic (Sirius) on the 15th day of the fourth month of spring during the 13th year of the reign of Senusret III. Some 21 lunar dates have been identified for this reign (Dee 2013a; de Jong 2006; Krauss 2006). The matching of dates and astronomical events is, however, not as simple as it seems, since there is often lively debate over precisely which astronomical events are described and from where in Egypt these events were observed (i.e. Aswan, Thebes, Memphis). This can lead to divergent absolute date ranges and, together with such issues as co-regencies, contribute to what are referred to as 'high' and 'low' chronologies, where either older or younger date ranges are preferred for a given reign, Dynasty, or in general (Höflmayer et al. 2013; Shortland 2013: 27). At certain times, the difference between such high and low chronologies can be as little as 25 years. Several debates between high and low dates exist in Egyptian archaeology, including the dating of the Thera Eruption (Baillie 1995: 108–121; Bietak & Höflmayer 2007; Manning et al. 2002). The finding of evidence for new rulers in specific regions, such as the Abydos rulers of the Second Intermediate Period (as found by J. Wegner and team in January 2014, unpublished at time of writing), necessitates further revisions to these chronologies. There is also lively and ongoing debate relating to the dating of rulers, including of the Third Intermediate Period and the late New Kingdom (Dodson 2013; Jansen-Winkeln 2006a; 2006b; Kitchen 1995; Rohl 1995; Shortland 2005; Taylor 2013).

2.1.2.1 Events in Real Time

The first recorded absolute date in Egyptian history (earliest 'day-exact date') is June 20th 688 BC in the Julian calendar, and appears in a document recording the sale of a slave (Depuydt 2006: 468). In 664 BC, Thebes was sacked by the Assyrian ruler Ashurbanipal and Psamtek I was placed in control of Egypt; these events are known through Assyrian records, and are supported by the Egyptian texts (Kitchen 1995; Shortland 2013: 24). Prior to these occurrences, attempts to fix events to absolute dates have focused on lunar and solar observations, and attempts at 'dead reckoning', counting backwards from events with known calendrical dates (Shortland 2013). There are, however, many cases where archaeological contexts provide neither tightly datable objects of material culture (or no objects at all), nor convenient texts providing regnal dates. What can be done in these cases? What ways are there for obtaining reliable absolute dates for such archaeological contexts and events? There are a number of scientific methods available and suitable to help date prehistoric and historic Egypt. These methods allow the calculation of chronometric dates in years before present (bp/BP), which can then be calibrated into years BC (BCE, 'before the common era', is also in common usage) (Bronk Ramsey 2013: 31–32; Manning 2006: 327–334). It is worth noting that, as will be explained below, before present actually refers to years before 1950 AD (Taylor 1987: 5). The most commonly used dating method within historic Egyptian contexts is radiocarbon (^{14}C, also written as 14C or C-14), which provides the possibility of achieving an absolute chronology for Egypt and her neighbours. This, in turn, creates a huge new potential, not only for fixing historical events within Egypt, but also for comparing the absolute dates of events over wide geographical distances and linking these chronologies (and actual events) through what are described as synchronisms or lynchpins (Shortland 2013: 23–24).

What do we have to do in order to get these 'scientific dates'? How are they produced, and what types of materials are needed? These 'dates', or more accurately 'measurements', are most usually achieved through three main methods; they are:

1   Radiocarbon (^{14}C; organic),
2   Dendrochronology ('tree-ring' dating; organic),
3   Luminescence (sediment) dating, which can be subdivided into:
    a   Optically stimulating luminescence (OSL) (inorganic), and,
    b   Thermoluminescence (TL) (inorganic)

There are a range of other methods in use, including Uranium-Thorium dating (U-series), Varve dating and Tephra-chronology. These are not discussed here, but a useful chart showing the age range potentials for various of scientific dating methods can be found in Aitken (1998: Figures p1 & p2; van der Plicht 2004: 3).

2.1.2.2 Radiocarbon Measurements

The most widely used method for producing dates is the radiocarbon (^{14}C) method. This technique was first introduced into archaeology by Willard

F. Libby in the late 1940s (Arnold & Libby 1949), totally revolutionising the possibilities of achieving dates for objects and, by association, events, within archaeology. Some of the very first samples to be dated by the ^{14}C method were known age samples of acacia wood from the Step Pyramid of King Djoser of the 3rd Dynasty at Saqqara (Dee et al. 2012: 869). In practical terms for archaeologists and researchers wishing to produce dates by this method, it requires samples of organic material to be taken. These samples can comprise anything that was once living – plant, animal or human. This gives a huge range of possibilities for 'dating' ancient Egypt. With the most recent of the radiocarbon techniques, using an Accelerator Mass Spectrometer (AMS), it is possible to conduct measurements (from which the chronometric date is then derived) from very small samples of organic material (Bronk Ramsey 2013: 30).

But how does the science of the ^{14}C method actually work? How can a sample of ancient organic material produce an accurate chronometric date? In brief, the radiocarbon method works by measuring how much carbon 14 (^{14}C) remains within a given sample of organic material (Taylor 1987: 2–3). There are two stable types of carbon in all living organisms, carbon 12 and carbon 13, and one unstable or radioactive type: carbon 14 (abbreviated to ^{14}C, 14C, or less commonly C-14; Taylor 1987: 1–2). When a living organism dies, it stops absorbing carbon from the atmosphere, and, although the ^{12}C and ^{13}C levels remain constant within the (now dead) organism, the levels of ^{14}C begin to decay (Taylor 1987: 1–2). Libby established that it is possible to measure the number of decay 'events' that have occurred since the organism died, in other words, how much time has passed since the organic sample ceased to live (Taylor 1987: 2). He calculated that after 5568 years, half the amount of ^{14}C present within dead organic matter has decayed; in the next 5568 years, half of what was left at that point, decays again, and so on (Bronk Ramsey 2013: 30; Taylor 1987: 1–2, Figure 1.1, 9–10). These decay events were called the Libby half-life, and a later revised calculation of it resulted in the figure, known as the Cambridge half-life, being adjusted to 5730 ± 40 years (Godwin 1962; Hughes & Mann 1964; Taylor 1987: 9).

There is another important factor that affects the use of radiocarbon measurements – one which was not realised at the time of the inception of the method. After living matter dies, ^{14}C is not released back into the atmosphere at an equal and steady rate over time, but instead there are environmental fluctuations. Measurements therefore need to be calibrated using a calibration curve (e.g. the *OxCal* online calibration program using the *INTCAL* calibration curve) which represents these environmental fluctuations over time. These calibration curves use data derived from known radiocarbon measurements taken on tree-rings (Suess 1978; Taylor 1987: 19–20; van der Plicht 2004: 1). Given that tree-rings reflect environmental changes year by year, the compilation of extensive continuous tree-ring chronologies allows archaeologists to refer to sequences dating back some 7000 years (Baillie 1995: 9; Taylor 1987: 21). Tree-ring 'dating', known as dendrochronology, will be discussed below in more detail, but its importance the ^{14}C method is considerable. Calibration of radiocarbon measurements against the curve results in dates within the absolute BC/AD timescale, and these dates are expressed as cal BC and cal AD (van der Plicht 2004: 1). Prior to this point dates are expressed as, for example, 1825 ± 25 years, which refers to years BP, but which are not actual calendrical years (van der Plicht 2004: 1). Cal BP, BP (before present) refers to a 'present' of 1950 AD. This marks the rough date when the first ^{14}C results were published, with lower case 'bp' referring to non-calibrated dates (Taylor 1987: 5; van der Plicht 2004: 1). When the dates extend back into the Palaeolithic period of prehistory, they are usually expressed as BP dates and not converted into BC. Tree-ring chronologies have also facilitated high-precision dating through what is known as 'wiggle-matching' (Baillie 1995: 69–72). Radiocarbon measurements on oak tree -rings from Irish and German oak showing remarkably similar sets of results were combined into the international standard for radiocarbon calibration (Baillie 1995: 69; Pearson & Stuiver 1986; Stuiver & Pearson 1986). As an additional result of this work, which was primarily aimed at achieving a precision level of up to ± 20 years for radiocarbon measurements, it was also revealed that there were short-term 'wiggles' in the calibration relationship thought to be caused by the sun (Baillie 1995: 69). This enables archaeologists to sample pieces of wood, or other pieces of organic material within stratified deposits, or consecutive blocks from tree-rings and fit this new series of ^{14}C results onto the calibration curve. By means of the wiggles, the series of new measurements can be aligned exactly to the curve (Baillie 1995: 69, Figure 4.8). The precision of this method depends on the precision of the calibration curve, which itself has come under question, so care needs to be taken (Baillie 1995: 72; Pearson & Stuiver 1993; Stuiver & Pearson 1993).

At some points along the calibration curve there are some rather flat areas which can make high-precision dating problematic. We can turn to the issue of dating the Thera Eruption to provide one such example on the curve in between 1675–1525 cal BC (Baillie 1995: 110, Figure 2; Bronk Ramsey et al. 2004: 326). There remains disagreement between scholars as to exactly when the eruption took place, and the precision available using calibrated radiocarbon measurements cannot be accurate enough to be in favour of one or other argument, leading to dates c.100–150 years apart (Baillie 1995: 108–121, 154–157). It is possible to use ice core data to clarify the situation. A frost-ring exists at 1645 ± 20 BC (1627 BC), thereby favouring the 1628 BC eruption date over the other possibility of 1520 BC (Baillie 1995: 109–121; 154–157; Hammer et al. 2003). There are, however, many other factors to be taken into account. These include environmental issues such as the discharge of sulphur during the eruption (or perhaps the eruption of another volcano entirely) and its record within the ice cores, and the description of storms in the ancient texts which may (or may not) be related to the eruption (Baillie 1995: 109–121; 154–157). Recently, phases before and after the eruption were dated, in addition to samples from the actual destruction layers themselves, thereby allowing for the dating of phases either side of the plateau in the calibration curve. Wiggle-matching was used to increase the precision of the dates (Bronk Ramsey et al. 2004: 326). Short-lived plant remains (16 samples) were used (the value of which are further discussed below), together with bone and charcoal where necessary for *terminus post quem* (Bronk Ramsey et al. 2004: 327). The results were also subjected to Bayesian analysis (see Box 2.2) in the OxCal program. The results of such varied approaches to the radiocarbon method combine to support the earlier date range for the eruption of Thera 1663–1599 BC (95 per cent confidence) based on the radiocarbon evidence (Bronk Ramsey et al. 2004: 337). This example demonstrates the contribution that ^{14}C can make to archaeological debates, but also indicates the considerations necessary with regarding context, types of organic materials and the application of wiggle-matching and statistical analysis. The context and type of materials will be returned to shortly for further and more detailed discussion.

Radiocarbon chronologies have come under criticism in the past for not 'working' in certain environments. An example of this is the 'reservoir effect', relating to the variable ^{14}C concentrations in carbon reservoirs found in terrestrial environments (Dee et al. 2010; Taylor 1987: 33–34). The reservoir effect has been cited as causing problems with Egyptian ^{14}C results in the past, although a recent systematic study of historical material of known date (AD 1700–1900) from Egypt has shown that the possible discrepancy in dates caused by this effect is consistently in the region of 19 years (Dee et al. 2010). Dee et al. (2010: 687) note the cause of the reservoir effect as being "wherever an organism incorporates carbon that is depleted in the radioisotope relative to atmospheric levels"; it is this that causes the dates to be 'offset' and to appear as older. Following analysis of the recent historical samples from Egypt and consideration of the ancient riverine environment, dissimilar in some aspects to today (due to the different yearly cycle prior to the raising of the Aswan High Dam), it was concluded that seasonal variation in plants taking in ^{14}C was likely to be responsible for what is only a small offset from the calibration curve (Dee et al. 2010: 690). This is important as it systematically challenges the application of the reservoir effect to ^{14}C results on Egyptian data, and proves that, from the dataset examined, the offsets are quite minimal (Dee et al. 2010). The reservoir effect is more closely connected with aquatic environments (Dee et al. 2010: 687). Whereas radiocarbon measurements dating back to between 9400 and 11,855 cal BP can be calibrated using tree-ring sequences (Kromer & Spurk 1998), earlier dates require calibration by corals. These have also raised the issue of error due to the 'reservoir effect' and for which a correction is made (van der Plicht 2004: 2–3).

2.1.2.2.1 DIFFERENT ¹⁴C METHODS

As mentioned above, the most recent development within ¹⁴C is Accelerator Mass Spectrometry (AMS), but there are also several other 'conventional' methods: Liquid Scintillation Counting (LSC), Gas Proportional Counting (GPC) and, the first, Solid Carbon Counting (SCC; see Taylor 1987: 82–95 for full descriptions of each). AMS, LSC and GPC are all in use today, dependent upon the laboratory. What are the differences between these methods? Can they achieve equally precise dates, and do they all require the same sample sizes? The different methods have allowed scientists and archaeologists to date ever older samples. The SCC method, developed in the early1950s, permitted dates reaching back to 25,000 years ago, associated with an error margin of 200–300 years; GPC, developed in the mid-1950s, allowed dates as early as 30,000–60,000 years ago to be investigated with errors of as little as 15–20 years and LSC, also developed in the mid-1950s, but only employed widely from the 1960s onwards, was similar (Taylor 1987: 82–90). These methods require larger sample sizes than are currently required by the AMS method, with 10–12g of actual carbon needed for SCC, 7–20g of carbon for GPC. An overall organic sample size of between 20–500g (dependent upon the type of material sampled) is required today by the ¹⁴C laboratory at the Institut Français d'Archéologie Orientale in Cairo (Taylor 1987: 82–90).

One of the massive advantages of the AMS method (a direct counting system method developed from the 1970s onwards and used 'routinely' since 1983; Gowlett & Hedges 1986) is that it is able to detect a far greater proportion of ¹⁴C atoms than had been possible using the previous 'conventional' decay counting systems (Taylor 1987: 90). For archaeologists, as well as other scientists, this means that much smaller sample sizes are needed (Taylor 1987: 90). It is possible, where necessary, to use samples as small as 25–50 mg. This has created huge possibilities for dating Egyptian material, such as, for example, single seeds or fragments of papyri. This means that whereas the application of the ¹⁴C method for certain types of material/objects was unthinkable in the past due to the unsustainability of archaeological resources, or the unjustifiable damage that would have been caused to an object, the AMS method places far less pressure upon archaeological resources and achieves high precision results. This has great repercussions for everyone working with and responsible for archaeological material. Today, it can mean the difference between being able to date an otherwise unknown archaeological context or simply having to say 'we don't know what date it is' because there is only a tiny fragment of charcoal in a prehistoric hearth or couple of fruit seeds in a ceramic vessel found within a tomb. The date limitations have improved since the inception of AMS, which could date back to c.40,000 BP in the mid-1980s, but which today can extend back to closer to 50,000

Box 2.2   Bayesian Statistics in Egyptology
Felix Höflmayer

Within radiocarbon dating, the Bayesian statistical method is used for considering additional (most commonly chronological) information on organic samples in order to increase the precision of radiocarbon determinations. While this method was already used in Aegean prehistory studies in the mid-1990s for assessing the absolute date for the Minoan eruption of Santorini (Thera) in the early Late Bronze Age, in the field of Egyptology the use of Bayesian analysis is relatively new, but nevertheless of increasing importance.

A calibrated radiocarbon determination of a given (preferably short-lived) sample might yield substantial uncertainty ranges or even produce different possible choices for the 'true' calendric age, appearing as separate 'peaks' on the time-line, depending on the specific shape of the radiocarbon calibration curve. Additional information about the samples, e.g. their succession based on archaeological stratigraphy of their find-contexts, can be combined with the respective radiocarbon measurements in a Bayesian model in order to increase the precision of the individual calibrated age ranges. The application of the Bayesian method can be demonstrated by a simple example.

Let us assume two short-lived organic samples (sample A and sample B) from a fictitious excavation. Sample A yields a radiocarbon result of 4040 ± 20 while sample B is radiocarbon dated to 3965 ± 20 (Figure 2.2). Calibration of these two radiocarbon determinations results in possible dates for sample A of between 2580 and 2564 or 2534 and 2493 BC and for sample B of between 2550 and 2537 or 2491 and 2466 BC at 1σ standard deviation. Due to the specific shape of the radiocarbon calibration curve in the third millennium BC, both uncalibrated radiocarbon determinations of 4040 ± 20 and 3965 ± 20 result in two probability peaks on the timeline respectively. Based on radiocarbon evidence alone it is not possible to decide which peak represents the 'real' calendrical age of the respective sample.

Let us further assume that both samples were found in good stratigraphic contexts (like a jar full of charred grain from destruction horizons) and that the stratigraphy of the site shows that sample A comes from a younger phase than sample B. Sample B, therefore, should be older than sample A in absolute time and only the highlighted probability peaks would be in agreement with this assumption (Figure 2.2). This additional information, the so-called 'prior information' (because it refers to information known before measurement of the samples) can be combined with the

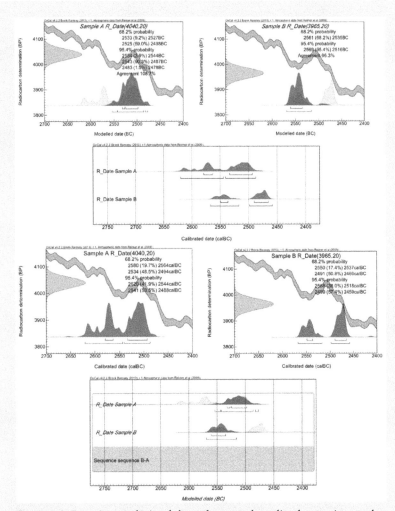

Figure 2.2 Bayesian analysis of dates for two short-lived organic samples, demonstrating the effects of prior information.

radiocarbon determinations of the individual samples using the Bayesian statistical method. The software package most commonly used today for the calibration and Bayesian analysis of radiocarbon dates is OxCal (current version 4.2.2), developed by Christopher Bronk Ramsey of the University of Oxford.

Adding the prior information (sample B being older than sample A) to the calibration process results in a much more precise date for our individual samples. Based on the assumption that sample B is older than sample A, calibrated and modeled age determinations for sample B fall between 2561 and 2535 BC and for sample A between

2533 and 2498 BC at 1σ standard deviation (shown in dark grey on the figures). In the original probability distributions the earlier peak of sample A and the later peak of sample B have been eliminated since they would contradict our prior information that sample B is older than sample A (original probability distributions shown in light grey).

This simple fictitious example shows that adding additional information may be a useful tool for increasing the precision of radiocarbon determinations. It has to be stressed, however, that the posterior probability distribution will – to a certain extent – always reflect the prior information that has been added to the respective models. Therefore, one has to be very cautious in deciding which prior information should be used for creating a Bayesian model, and in publications it should always be stated explicitly what kind of prior information was used for a specific model.

Perhaps the most sophisticated Bayesian modeling on radiocarbon data from Egyptian contexts was done in the framework of the project on 'Radiocarbon Dating and the Egyptian Historical Chronology' that was led by Christopher Bronk Ramsey and Michael W. Dee from the University of Oxford. Within this project, over 200 new high-precision radiocarbon measurements on short-lived samples that could securely be assigned to reigns of specific kings were used to check the reliability of the Egyptian historical chronology. As prior information only the known succession of kings and different models for their respective reign-lengths (plus an additional error of +5/+10 years) have been used. No information on absolute calendar dates aside from the radiocarbon measurements have been used for this model. Three distinct Bayesian models have been created for the Old, Middle and New Kingdoms proving that radiocarbon dating is in good agreement with the Egyptian historical chronology.

BP (Gowlett & Hedges 1986: 63; Higham et al. 2012; Zohary et al. 2012: 17).

These new chronometric dates can be ground-breaking, particularly so in relation to the knowledge of Egyptian prehistory in the pre-pottery phases, but also for synchronising the Egyptian historical chronology with scientific dates where the possibilities and precision is much greater than it was 40 years ago (Bronk Ramsey et al. 2010; Säve-Söderbergh & Olsson 1970). At the time of writing there are no AMS facilities within Egypt (Bronk Ramsey 2013: 30) which means that very small samples cannot be dated in Egypt. This has the additional repercussion that no freshly excavated material can be dated by the AMS method due to the current limitations on export of archaeological material from Egypt.

2.1.2.2.2 SAMPLE SELECTION PROCEDURES

From the above, it is clear that ¹⁴C may have many useful applications, but are there some organic samples that might be better than others, or contexts that stand to be more reliable than others? There are plenty of possibilities for dating Egyptian contexts, prehistoric and historic alike. Samples can derive from museum collections in addition to ongoing excavations, although the latter are subject to the issues raised above due to the lack of currently available AMS facilities in Egypt. The best organic material for dating comes from short-lived plant species. Long-lived species, such as Lebanese cedar, are less reliable as the place of the fragment of tree-ring within the broader sequence is unknown, although they are being used (see Section 2.1.2.3: Dendrochronology). Many materials regularly found on excavations are suitable for radiocarbon dating: fruits and seeds, basketry, papyrus, charcoal, reed matting, grasses, bone collagen and hair. Some of these can only really be sampled using the AMS method as the sample sizes would simply be too small for conventional dating methods (Taylor 1987: 82–90). The question that needs to be asked before proceeding with any scientific dating is simple and yet its importance cannot be overstated: what is the date that we are really looking for? It sounds rather obvious but it does require serious consideration before samples are a) taken and b) submitted to a laboratory for dating; laboratories require submission forms to be completed giving just this type of information. In order to avoid irrelevant dates and the drain on not just financial resources, but more importantly unsustainable archaeological ones, a number of smaller questions need to be answered in order to decide whether the proposed organic sample is likely to produce a reliable date or an unreliable one, and whether that date is intended for the sole purpose of dating the object from which the sample comes, or it is to be used to date an associated context. Two sets of questions relating firstly to the context and secondly to the type of organic material were published by H. T. Waterbolk in 1971 and these 'criteria', or variations thereof, should be answered by the submitter of samples to radiocarbon laboratories; they constitute best practice and are paraphrased below directly from Waterbolk (1971).

The first point to be addressed is the certainty of association between the ¹⁴C sample and a known context – the context for which the date is required (Waterbolk 1971: 15). The following are based directly on the recommendations in the article by Waterbolk (1971: 15–16) in terms of certainty of association:

a   Full certainty: whether the sample actually came from the object for which we are trying to establish the archaeological date; e.g. if a date were required for a papyrus sandal, and the sample was a piece of papyrus taken directly from the sandal, then this is full certainty.

b High probability: where the sample has a 'direct contextual relationship' to the date being sought; e.g. if the date of a burial were under investigation, then food remains found within a sealed jar in an intact tomb would present a high probability of association.

c Probability: where there is no direct relationship between the sample and the context; e.g. if the date of a burial were under investigation and remains of matting were found within a disturbed tomb. The fact that the tomb has been disturbed means that it is probable that there is an association, but there is a chance that the matting could have been introduced at the time of the disturbance.

d Reasonable possibility: where the context of the sample is only suggestive of a known date; e.g. charcoal pieces found below the floor level of a known stratified deposit within a settlement. This association could only really mean that the sample pre-dated the floor, and therefore there is only a reasonable possibility that the charcoal is associated with the floor.

The second point to be established, is the degree to which the chosen ^{14}C sample might have what is called 'inbuilt age'; in other words, how long had the sample existed as raw material or as an object before its deposition (Dee et al. 2012)? The criteria for selection are as follows (Waterbolk 1971: 16–17) in terms of inbuilt age:

1 Negligible: short-lived plant remains; e.g. a piece of reed from a mat, or a seed.

2 Decades: this would include charcoal and also the outer rings of a tree (meaning that the tree was still living and that these outer rings would relate to the time of felling); e.g. a sample could include the outer tree-rings of a type of timber indigenous to Egypt that had been used for the construction of a coffin.

3 Centuries: long-lived wood species such as cedar (imported from the Lebanon); e.g. entire wooden planks used in funerary boats, where the outer tree-rings cannot be determined from the sample.

4 Unknown material: where the sample type is not clear; e.g. earth, ash or sediment found on the floor of a building.

What is important is that, where possible, the sample should be taken from short-lived plant remains. This means that inbuilt age will be minimal and ensure that the material being dated can reasonably be expected to give the date for the event that is being investigated. As such, if a potential sample cannot be classified by the above criteria to belonging to one of the first two categories for both certainty of association and inbuilt age, then it is not a good choice if a high level of precision is required, together with contextual integrity. Samples of wood from long-lived species, for example, could be problematic, with a 100-year-old tree possibly returning dates relating to any point from the date of its felling to the first year of its tree-ring growth (Dee et al. 2012: 873). As mentioned before, the first sample taken on Egyptian material was wood (Arnold & Libby 1949). It was, however, local wood, although the issue of longevity of tree species must still be taken into account.

The best samples to date are those that have been recently excavated. These are samples that have had minimum contamination, for which the archaeologists can state exactly from where they derive, and use to date archaeological sequences. Both in the cases of freshly excavated contexts as well as for samples derived from museums, it is better to have multiple samples than a single sample, and multiple samples taken from a range of strata make it possible to model the series of dates with the method of Bayesian modelling (see Box 2.2; Bronk Ramsey 2006; Dee 2013b: 65–75).

Radiocarbon dating is widely used in archaeology today, but is sampling a straightforward matter in the field or the museum if we follow the guidelines of Waterbolk? Are there some materials, objects or contexts that are better than others? As seen above, there are two best-case scenarios when choosing material for dating by the radiocarbon method: 1) the certainty that the organic material relates precisely to the archaeological context for which a date is sought, and 2) the knowledge that the organic material being sampled is from a short-lived species. To explain it simply, if a seed from a fruit found inside a sealed ceramic jar in a sealed burial is dated, then the date obtained is the date at which the fruit was picked from a tree. It does not automatically mean that this date corresponds exactly to the date of the burial. However, as the objective is to ascertain the date of the burial, this choice of ^{14}C sample might give the best possible indicator of this point in time unless the human skeletal remains themselves are dated directly. Ultimately, the burial is sealed and the jar containing the fruit is sealed, so the context is undisturbed. The seed represents short-lived plant remains (Waterbolk A), and its relationship with the context is highly probable (Waterbolk B; Waterbolk 1971: 15–17). In contrast, an example of a bad choice would be a fruit seed extracted during the sieving of the fill of a disturbed burial. In that instance, although the seed is still a short-lived plant ecofact (Waterbolk A), there is very little evidence to directly relate it to the burial itself. At best it would be a Waterbolk C or D, and therefore not a recommended sample (Waterbolk 1971: 15–17). This explains the difference between good, reliable samples, and samples that are simply not going to provide a reliable date for the context. One further example from a settlement context might be where the archaeologist is interested in obtaining a date for the start of the use of a storage pit or silo. A sample of a grain of wheat or barley found inside this pit could be taken, which would provide a date relating to a single point in time during the history of the use of the silo. Assuming for a moment that this storage pit is lined with a reed basket

(Wendrich et al. 2010: 1000), then a ^{14}C sample of a single piece of this basket would also mark the point in time at which the reed was cut, and a close point in time to the manufacture date of the basket. This date, whilst having a highly probable close relationship to the manufacture of the basket, can only return a date approximating to the time at which the basket went into the pit, and it will only be able to represent a date at some point in the life of the usage of the pit. In short, radiocarbon measurements are an extremely effective way of achieving chronometric dates, but it is the responsibility of the person choosing the sample to make logical choices that will provide a date for an object, a context, or an event (Gowlett & Hedges 1986: 63-71; Waterbolk 1971).

Radiocarbon dating has contributed significantly in various areas of research, as seen earlier, but a further one is in the development of prehistoric chronology. A high proportion of the overall number of radiocarbon measurements carried out on Egyptian material come from prehistoric contexts (Rowland 2013). Both the lack of historical correlates and, in some cases, the lack of any items of material culture have meant that researchers working within prehistory have utilised absolute dating methods to a greater degree than those focused on the historical era. Such significant advances, both in Egypt and in the wider Near East, include the origins of agriculture and the dating of the domestication of plants and animals, as well as the origins and dispersal of cattle-keeping in Africa (Harris 1986: 5; Hassan 2002). The use of AMS has also helped to resolve former problems relating to the dating of plant remains by conventional radiocarbon counting methods. For example, at Wadi Kubbaniya, an exceptionally early date had been considered for barley, amongst other plant remains, of between 18,500–17,000 BP (Harris 1986: 6; Wendorf & Schild 1980; 1984; Wendorf et al. 1979; 1984). This was later found to be incorrect due to re-examination of the plant remains (originally dated through association with charcoal within the stratigraphic context), and found to be intrusive, and re-dated by AMS to 4850 ± 200 BP (Harris 1986: 6–7; Wendorf et al. 1984).

2.1.2.3 Dendrochronology

Dendrochronology is an independent absolute dating method which started to be used widely in American archaeology as early as the 1930s but not in the rest of the world until the 1970s. This was at a point when long chronologies of tree-rings had been assembled by scholars including Baillie, developing from the work initiated by A.E. Douglass (Baillie 1982: 28–35; 1995: 9, 18; Douglass 1919; 1941). Tree-rings are an integral part of the radiocarbon method as it was realised that radiocarbon measurements had to take into account the fluctuating amount of carbon being released into the environment over time. These fluctuations are currently accounted for by calibrating the radiocarbon measurements against a curve based on dendrochronological data (see Kromer 2009; Leavitt & Bannister 2009). But how does dendrochronology work, and how is it useful for archaeologists? Why are tree-rings so important for examining chronology? Dendrochronology has three primary areas of investigation essential to archaeology: chronology, behaviour and the environment (Dean 1996; 2009). Environmental phenomena, and to a lesser extent, human activity, dictate the different growth patterns laid down by trees year by year, making it possible to differentiate between the yearly rings and to understand the impact of the variable climate. An annual growth ring forms throughout the year, directly beneath the bark of the tree: cells are added during the growing season (typically spring), with essentially no growth occurring during the winter (Baillie 1995: 16; Cichocki 2006; Stokes & Smiley 1968). How thick the tree-rings are is dependent on a number of factors, which include how hospitable the local environment was through the period during which a tree grew. For example, in low altitude deserts, such as Egypt, availability of moisture (water) is typically the most limiting factor for tree growth. At high altitudes, temperature is often the most limiting factor for growth. However, other special environmental features during the year (e.g., volcanic eruptions (LaMarche & Hirschboeck 1984) or forest fires (Swetnam 1993)) or whether the tree has been damaged by insects (Swetnam & Lynch 1993) or disease can affect annual growth in individual trees or large populations.

Not all trees are equally suitable for dendrochronology, however, as there at least four distinct requirements that must be met so that ring series can be compared across different datasets (Bannister 1963; Ferguson 1970). The entire process is predicated upon anchoring a tree-ring series (a chronology) to the present, usually via a living tree, thus permitting the assignment of annual calendrical dates for past events and processes. It is essential then to find overlapping sequences where a series of rings in a modern specimen(s) can be tied to progressively older specimens, be they dead trees, historic timbers in use in buildings, or archaeologically recovered wood (see Baillie 1995: 16–18, Figure 1.1). One of the long-lived species that has been used widely in dendrochronology is the bristlecone pine (*Pinus longaeva*) from the USA, where a 7474-year chronology of tree-rings had already been established by 1970 (Ferguson 1970: 237), and now extends to more than 8000 years. Another is the European oak, which, when combined with European pines, extends some 12,400-plus years (Friedrich et al. 2004). These 'long reference' chronologies have been available in some form since the 1980s for application in Old World chronology (Baillie 1995: 11, 17, 18). Trees that grow in tropical and in subtropical regions, however, often provide complications for dendrochronological analyses and, occasionally, are not suitable for tree-ring analyses. This applies to some of the trees that are indigenous to Egypt, such as the sycamore fig which does not have distinct, clear borders between the different rings (Cichocki 2006: 362). The utility of other native Egyptian species for tree-ring research, such as acacia (*Acacia nilotica*), has not been properly evaluated (Creasman 2014; 2015). Long thought to be inadequate for dendrochronology, acacia from neighbouring regions has indeed been demonstrated to be viable (Wils et al. 2010; 2011). In the case of Egypt, the long-lived species that is without doubt useful for dendrochronology is the Lebanese cedar (*Cedrus libani*) which was imported into Egypt for various uses from at least the Old Kingdom onwards, including building timbers for temples, palaces and boats, as well as small objects including coffins (Cichocki 2006; Creasman 2014; 2015; Gale et al. 2000; see also later sections of this volume).

Are samples taken for dendrochronology, as for radiocarbon measurements, or can tree-rings be 'read' *in situ*? Tree-rings can be examined by sampling (including coring into modern living trees) but there are also instances when it is not necessary to do so (Baillie 1995: 25). A cross-section of the wood is necessary in order to examine the rings and to measure how thick they are. This can be achieved either by drilling a core or by cutting a disc (Cichocki 2006: 362). The sample can come equally from a living tree or from timbers in use in buildings, or wood used in any context within the archaeological record (Baillie 1995: 18). In all cases, the surface or section should be cut transverse to the direction of tree-ring growth and needs to be smoothed afterwards (usually by progressively finer grit sand paper or sharp blades) in order for the transverse surface of the rings to become clear (Baillie 1995: 17, 18; Stokes & Smiley 1968). If, by chance, a cross-section is visible, then cleaning may be sufficient for non-destructive examination. This is likely to be required when working with museum objects. Study can then be achieved through photography, video images, or flatbed scanners with high resolution (Baillie 1995: 18–19; Cichocki 2006: 363). X-ray microtomography (micro-CT or µCT) scanning may be a viable non-destructive option for small artefacts in the near future (Grabner et al. 2009). In order to check for circuit uniformity (that the rings are consistent around a specimen) and diversity, several radii on each specimen are usually evaluated. Ideally, specimens should include the bark with the layer of sapwood, as this is needed for establishing the precise felling date of the timber (Baillie 1995: 18–20 discusses the full sampling, preparation and measurement process). Once the specimen has been acquired, the dendrochronologist measures or graphs the individual tree-ring widths and patterns (from the centre outwards). These measurements are then compared with the master chronologies for the genera/species/region (Baillie 1995: 17-19). In this way, when tree rings of unknown date are examined, it should be possible to fix them into a known sequence of tree-rings/master chronologies using 'cross-dating' and employing statistical correlation coefficients (Baillie 1995: 17, 20).

Sometimes, especially early in chronology building efforts for a species or place, there might be gaps at either end of a sequence. These gaps, preventing a single or group of specimens from being anchored to the present, produce what is termed a 'floating' chronology. These specimens are the data against which archaeologists can check the date of their tree-ring samples in order to ascribe a precise calendrical date to their wood. As already seen with the radiocarbon method, the extent to which this can then date an actual archaeological event (e.g., site construction) very much depends on the context and associated data. Depending on the number of tree-ring standards that exist for a time, place or species, it can be possible to have what are described as 'pointer years'. These distinct tree-ring patterns usually appear in 75 per cent or more of the specimens for the group, demonstrating extremely consistent growth trends (Cichocki 2006: 364). Similarly, where there are years where the environment does not support good growth, there might be only very narrow rings or, rarely, even missing rings (rings not found on an individual specimen, but evidenced when a large enough sample size is evaluated). Nevertheless it is vital that these rings are detected as every ring is necessary for calculating chronology; in dendrochronology even "a one-year error is [...] egregious" (Kuniholm 2002: 64).

There are a number of tree-ring databases worldwide, and the project whose foci includes Egypt is 'The Synchronization of Civilizations in the Eastern Mediterranean in the Second Millennium BC' (SCIEM 2000), based at the University of Vienna (Cichocki 2006: 36). Another longstanding effort for the region, for nearly 30 years based at Cornell University (the Aegean Dendrochronology Project), has now expanded to include more than a dozen scientists from the University of Arizona's Laboratory of Tree-Ring Research and Center for Mediterranean Archaeology and the Environment. One key aim of both endeavours is to develop an independent dendrochronological dating method for the ancient Mediterranean world. Thus far, however, the chronologies relevant to Egypt consist of only a few relatively short floating chronologies (500+ years each). Dendrochronology therefore has the potential to date past human events and processes of great importance to archaeologists and Egyptologists, but must be pursued collectively, in consultation with the Egyptian authorities, and with greater vigour for Egypt. Radiocarbon dating will benefit directly from this research for calibration purposes (a regional calibration curve for the Mediterranean and Near East will be essential to improving ^{14}C accuracy), although, as discussed above, short-lived plants, where available, are still the preferred radiocarbon samples.

2.1.2.4 *Sediment-based (or Inorganic) Dating Methods*

So far we have looked at methods using organic materials, methods that work up to c. 50–60,000 BP. However, scientific dates for earlier periods of time are also needed. For contexts/objects containing no organics/insufficient organics, are there alternative methods? If, for example, a radiocarbon sample relates to a period where there is a plateau on the calibration curve, is there another method that can help (Bluszcz 2005: 138)? There are other methods that can help and are called 'sediment-based' dating methods: Optically Stimulated Luminescence (OSL) and Thermoluminescence (TL). OSL and TL dates are expressed as BC or AD dates and can be compared with calibrated radiocarbon measurements which are also traditionally capitalised (Aitken 1985: 32).

2.1.2.4.1 LUMINESCENCE DATING

Luminescence methods function by calculating the amount of time that has passed between the activation of the luminescence dating technique on a sample and the point in time at which the object (or sediment) was previously exposed to light or heat. For ceramics, this equates to the date of manufacture (firing) and where sediments are being dated directly (e.g. samples taken during augering, as discussed in Section 2.3: The Environment), this equates to the amount of time since the sediment was last exposed to light. Luminescence has interested scientists since at least the 1600s (Aitken 1985: 2–3), but luminescence dating methods (also termed 'trapped charge dating methods') have been used by archaeologists only since the late 1960s (Goedicke 2006). Ionising radiation energy is released by natural radioisotopes and becomes trapped within the crystal lattice of the minerals (Bluszcz 2005: 137). Some minerals, including feldspar and quartz (both found frequently in archaeological contexts), can store and then release what is called 'radioactive decay energy' (Goedicke 2006: 356; Sekkina et al. 2003: 94). Energy accumulates, corresponding to the amount of time since the sediment or object containing the sediment was exposed either to heat or to light, so when this dating method is applied, it zeroes the material to the same setting as it was when it was fired (in the case of ceramics) or when it was buried (for sediments) (Bluszcz 2005: 138; Goedicke 2006: 356; Sekkina et al. 2003: 94–95). There are three different types of radioactive decay that have to be examined separately (alpha, beta, and gamma) in order to measure the radiation dose that the object/sample has received; the necessity of measuring all of these leads to a large error margin similar to early ^{14}C dates (Goedicke 2006: 356–357; Sekkina et al. 2003: 94). How much error is there, and what does this mean for the accuracy of the dates that can be achieved? The error can range from c. 7–12 per cent on single samples to as low as 5 per cent if a number of samples from one context are analysed (Goedicke 2006: 358). Electron Spin Resistance (ESR), although not discussed in detail here, is another method that detects trapped electrons. ESR can be used on bone and shell, and does not require that the object being dated was ever heated (Aitken 1985: 4; 1998: Figure p1 & p2).

2.1.2.4.2 CERAMICS AND DATING

As discussed in Section 2.1.1 on relative dating, artefact typologies allow archaeologists to order features/events uncovered during excavation and to examine the stratigraphical relationships between archaeological contexts and the artefacts found within them. The order of events can be reconstructed within a relative time frame. The study of artefact typologies within survey contexts is also valuable in identifying relative concentrations of material of specific dates or date ranges in particular areas over time. This aids understanding of the changing spatial distribution of sites through time, as well as the growth and shrinkage of individual sites over time. Ceramics can help to untangle the order of events as the archaeologists examine them within the vertical and horizontal stratigraphy. But, as has been discussed already, the vessels and sherds do not generally provide absolute dates or precise points in time themselves unless they are known to be used for a very short-lived period of time. There are but a few opportunities for ^{14}C dating on organic inclusions that have survived within ceramics and hence luminescence dating techniques are best applied to ceramics (Orton & Hughes 2013: 18).

2.1.2.4.3 THERMOLUMINESCENCE (TL) AND OBJECT DATING

Prior to the 1940s, thermoluminescence (TL) was used in geology to identify minerals, but due to the development of the photomultiplier in that decade, from the 1950s onwards, TL could be used for both archaeological and geological age determinations (Aitken 1985: 3). The method, as applied to ceramics (since the 1960s) and other fired objects, calculates the time interval between the firing/heating of a ceramic vessel, or other object previously subjected to a minimum temperature of 500°C, and the point in time at which it is subjected to heat or sunlight again (Aitken 1985: 3; Bluszcz 2005: 138; Goedicke 2006: 358). The electrons within the material are ionised by nuclear radiation and some of these are captured at 'traps'; these objects can include ceramics, stones, tiles and bricks, or indeed any object that has been fired or heated to at least c. 500°C, including fired cores used during the lost-wax bronze-casting process or even burnt flint (Aitken 1985: 1, 3, 9; Bluszcz 2005: 137; Ogden 2000: 158). This minimum heating temperature is necessary in order that the minerals will not reflect their original geological age, but rather their age is 'reset' as the date when the vessel/object was fired (Aitken 1985: 9).

For analysis, the material is ground down and heated electrically and rapidly to 500°C (c. 20°C per second, although more recently 10°C per second using infra-red laser beam pulses), with the same rate having to be maintained on each run of a sample (Aitken 1985: 1, 5–6). During the heating process, light is emitted that is TL. This TL comes from the minerals that are within the ceramic. The effect that exposure to radioactive impurities in that material is the creation of a 'weak flux of nuclear radiation' (Aitken 1985: 1, 5–6). It is very important that only the radiation-induced TL is measured, since other TL will serve to artificially increase the age; the nuclear radiation from potassium, thorium, and uranium usually produce the TL, with a smaller amount from rubidium and cosmic radiation (Aitken 1985: 6–7, 10–11). During the dating process, there is a 'plateau test' that must be carried out which will assess the palaeodose within the sample and serves to assess whether only the measurements of the correct type of thermoluminescence have been achieved (Aitken 1985: 7–8, Figure 1.3). This produces a curve which shows the natural and artificial glow curves and the ratio of the two; it is the outcome of this ratio calculation that decides whether the plateau is stable and whether or not it is representative of the natural dose of thermoluminescence and not affected by other age contaminants (Aitken 1985: 7–8, Figure 1.3). The electron traps that have not leaked electrons, which usually have a 'glow-peak' only at 300°C and above, are those that need to be measured (Aitken 1985: 6–7). The electrons between 300°C and 400°C are those that represent an electron lifetime within thousands of years, whereas those above this reflect geological time and millions of years (Aitken 1985: 8).

As with ^{14}C, the radioisotopes also have half-lives, in this case 10^9 years plus, and the radiation flux is constant (unlike ^{14}C) and therefore the amount of thermoluminescence is directly proportional to the time between the first firing of the ceramic vessel, tile or brick, and the point in time at which the archaeologists decide to re-fire the sample for scientific dating purposes (Aitken 1985: 1). The age of the object is the *palaeodose* (which can also be called 'accrued dose', 'accumulated dose', 'archaeodose' or 'total dose') divided by the *annual dose* (Aitken 1985: 9, 61–112). Once this second heating event occurs, it effectively sets the date to zero once more (Aitken 1985: 1). The dates that can be achieved, however, although improved recently, are currently subject to accuracy of 5–10 per cent (at the 68 per cent confidence level), which means that although ceramic vessels can be assigned to broad periods of time, the method is not suitable if fine-tuning of vessel dates is intended (Aitken 1985: 30–31; Bourriau et al. 2000: 144; Bluszcz 2005: 137–138; Orton & Hughes 2013: 225).

Whereas radiocarbon measurements are expressed either as, for example, 2512 ± 25 years, or when calibrated 790–542 cal BC, thermoluminescence dates are as follows, '1070 BC (± 100, ± 220, OxTL 143e)' which is an average date, in this case for seven samples, and ± 220 years is 'the overall predicted error limit' (Aitken 1985: 31). The further back in time, the less accurate thermoluminescence dates become: for younger contexts it is suitable to use this method within the past 2000 years, but beyond that it is less applicable. It is useful for Palaeolithic contexts that cannot be dated by ^{14}C (Aitken 1985: 31).

Is there just one way to carry out TL dating? There are a number of different methods that have been developed and are described in detail by Aitken (1985: 12–30). The 'Quartz Inclusion Technique' has been used for dating ceramics and focuses only on the quartz inclusions found within ceramics; the outer layer of these is removed to ensure that no clay is included (Fleming 1970; Aitken 1985: 17–18). After extraction from the ceramic sherd, quartz grains of 90–120 μm are obtained through sieving and magnetic separation, with any calcite and feldspar grains removed chemically, and with the same process removing the outer layer of the quartz grains (Aitken 1985: 18). All procedures are carried out under red light. The quartz grains are measured for natural thermoluminescence, and the same sample measured for artificial thermoluminescence (they are exposed to radiation) and the results compared. This does not give a very accurate date, so the *additive dose method* must be used (Aitken 1985: 18–19). This involves batches of 5mg of quartz grains from the overall sample, with a number measured only for natural thermoluminescence and others for both natural and added thermoluminescence, with all results plotted against the glow curve temperatures for the palaeodose plateau test (Aitken 1985: 19–20). As already mentioned, in order to calculate the age of the sample, the palaeodose must be divided by the annual dose and, therefore, the latter also needs to be calculated. For a description, see Aitken (1985: 21). Beta and gamma doses need to be measured: beta doses have a very restricted range of just a few millimetres and therefore a sample size of c. 10 cm is sufficient, but gamma doses come largely from the sediment surrounding a ceramic fragment to a radius of 30 cm (Aitken 1985: 22–23). For these reasons, whereas beta doses can be introduced into the phosphor from the powdered sample within laboratory conditions, for gamma doses, due to the larger area of radiation, a capsule containing the phosphor is taken into the field and lowered 30 cm into an auger hole relating to the archaeological context of retrieval and left for several months and then returned to the laboratory for measurement of the gamma dose (Aitken 1985: 22–23).

A second technique is the 'Fine-Grain Technique' developed by Zimmerman in 1971, and involves actual fine grains rather than grains crushed to a small size during powdering of the ceramic for analysis (Aitken 1985: 24–25). The advantages of this method are that it does not need quartz crystals and that the fine grains will have absorbed alpha particles (and therefore have the radioactivity of the clay of the vessel itself). This also means, however, that there is a possibility of contamination where minerals are not known (Aitken 1985: 25). In contrast to the former method, grain sizes of 1–8 μm are collected through washing and floating in acetone (Aitken 1985: 25).

Can objects from museum collections be dated using this method? It is possible to date objects that have been excavated some time ago; however, it is not normal practice as it is important to have access to data on the immediate environment of the object at the time of excavation. This includes details as to the local environment in which the object has been sitting for the preceding long period of time. It is normal for the surrounding sediment to be measured for radiation, its humidity measured, and the degree of certainty that the object had not been exposed to light (for hundreds or thousands of years) to be recorded (Aitken 1985: 1; Goedicke 2006: 357). These methods look at both the radioactive content within the actual object and in the sediment surrounding the object, and consider how sensitive the actual sample itself is to 'acquiring' TL (Aitken 1985: 1).

Sekkina et al. (2003) examined ceramic sherds from the Giza region. The sherds under investigation were not museum-based objects, but were excavated for the purposes of analysis, and a sample of the surrounding sediment was kept for measurement of the radioactive elements (Sekkina et al. 2003: 95). These sherds had to be at least 0.30 m away from context boundaries, given that other contexts could be of completely different date and hence their sediments could influence the results of the dating of the sherds (Sekkina et al. 2003: 95). Since the material to be dated is largely quartz or feldspar, the ceramics had to be crushed, in darkroom conditions, in order to release the minerals (Aitken 1985; Sekkina et al. 2003: 95–96 describe the whole process of sample preparation and analysis in detail). The heating of the object produces the vibrations which cause the electrons to be released and these vibrations increase as the heat increases (Aitken 1985: 42). The materials being dated were Old Kingdom ceramic sherds, for which the TL method yielded a date of 4388 ± 166 years. This is still a large error margin in comparison with radiocarbon dates, albeit a date which roughly correlates with the 4th Dynasty (Sekkina et al. 2003: 99).

TL dating has also been used on prehistoric Egyptian contexts, including Hemamieh and the Tarifian at Qurna (Goedicke 2006: 359; Whittle 1975). UC19678, a ceramic sherd from Hemamieh, was dated in 1980 and returned a date of 3775 ± 330 BC. A series of TL measurements were carried out on lithics from the Palaeolithic sequence at the Sodmein Cave, where there were seven sequences relating to the Palaeolithic occupation in the cave and additional Neolithic occupation sequences (Mercier et al. 1999: 1339). With ^{14}C measurements available for contexts within the cave, there was an estimated chronology for the occupation sequences, but the Lower Middle Palaeolithic sequences reached beyond the limits of the ^{14}C method (Mercier et al. 1999: 1341). Unlike OSL, described below, TL measurements are taken directly on objects themselves, and in this instance from the central core of the tool. This core had to be crushed and then sieved in order to extract the samples (Mercier et al. 1999: 1341). The resultant date, taken from an average of TL dates, for the Lower Middle Palaeolithic layers at Sodmein Cave is 118,000 ± 8000, and is associated with the period during which a specific hearth was in use. The dates obtained from Sodmein Cave can be compared with other date estimates also taken by the TL method from Bir Tarfawi in the Eastern Sahara and such sequences of dates brings new understanding to the 'diachronic succession' of various Palaeolithic industries in the wider region (Mercier et al. 1999: 1344). Furthermore, Sodmein Cave is a good example demonstrating how ^{14}C and TL dates can complement each other, with ^{14}C able to offer greater precision, but TL able to reach much farther back in time.

2.1.2.4.4 OPTICALLY STIMULATED LUMINESCENCE (OSL)

As seen above, TL can be applied to Egyptian objects, although it is of limited practical use due to the level of accuracy possible (Goedicke 2006: 358–360). But in what other ways can luminescence dating methods be used within archaeology? The method of OSL is much used by geologists, and is also applicable for archaeologists working in certain contexts. It developed from TL dating, with the aim of achieving a 'better method' for sediment dating than TL (Aitken 1998: 6). OSL can be applied to provide direct ages for sediments, even if these sediments were only ever exposed to sunlight for a short period of time; it can also be used on ceramic samples. Whereas the precision of date range may not be fine enough to contribute to chronological discussions within the Dynastic era, it has many useful applications within quaternary research, where much larger time periods are examined and where precision to within 500–1000 years is informative. OSL thus gives appropriate resolution for monitoring geological change. As with the application of TL for object dating, quartz or feldspar grains are used when dating sedimentary units (Aitken 1998: 1–5). As with TL dating, the date ranges possible extend back to 100,000 years BP, and possibly as early as 500,000 years BP (Aitken 1998: 2; Bluszcz 2005: 137; Goedicke 2006: 359–360).

In Egypt, OSL dating has been of use in Palaeolithic contexts, notably through the work of archaeologists including Pierre Vermeersch, during investigations of Middle Palaeolithic contexts (Vermeersch et al. 1998). One such context is Taramsa Hill, in the Qena region (close to the Temple of Hathor at Dendera), where chert quarrying sites associated with the early, mid and late Middle Palaeolithic have been identified. The time periods were defined on the basis of lithic assemblages (Vermeersch et al. 1998: 475). During a survey undertaken in this area, the human skeletal remains of one individual, a child, were located and excavated. These human remains were suggested to be contemporary with the quarrying activity, given that they were positioned on and surrounded by quarrying debris and tools dating to the mid and late Middle Palaeolithic (Vermeersch et al. 1998: 475, 478). The burial was also found sealed under a layer of extracted chert cobbles that had not been disturbed, and was in turn covered by another layer of Middle Palaeolithic extraction debris (Vermeersch et al. 1998: 479). Another factor adding to the security of the date is the fact that the human remains were sealed by an aridisol that formed in the area between 9500 and 6000 years ago during a humid phase (Vermeersch et al. 1998: 479). Samples for OSL dating were taken from the infill of quarrying extraction pits. This infill covered the lithics and provided a *terminus ante quem* for the lithics below (Vermeersch et al. 1998: 480). Due to the nature of this method and the fact that the OSL dating measures the time elapsed since previous exposure to light, it is crucial that the samples are extracted without exposure to light. Where possible, therefore, it is prudent for these samples to be taken by someone with prior experience. This sub-sampling must be undertaken in darkroom conditions, with the use of only red light in an otherwise completely dark environment. When sampling for OSL, extraction must be carried out using tubes that do not allow any light to permeate to their contents. In the case of Taramsa Hill, 5.0 m long light-proof PVC tubing was cut and hammered into the freshly cleaned sections under investigation. The samples were then cleaned and prepared in darkroom conditions (Aitken 1985; Vermeersch et al. 1998: 480–481). These types of tubes are regularly available in hardware shops and can be purchased in most towns in Egypt. The resultant dates for the late Middle Palaeolithic contexts, which is the period to which the skeletal remains were dated, range from 49,800–80,400 BP, which were averaged to 55,500 ± 3700 BP (Vermeersch et al. 1998: Table 1, 480 & 481). Osteological analysis has not been possible due to the inaccessibility of the burial; however, from comparative analysis with Levantine and North African populations, and through use of the photographic evidence, it is suggested that the skeletal remains are those of an anatomically modern child (Vermeersch et al. 1998: 481). This evidence constitutes one of only two examples of Middle Palaeolithic burials, and 30,000 years later, there are only three Qadan cemeteries (Midant-Reynes 2000: 43; Saxe 1971; Vermeersch et al. 1998: 483). The ability to provide an absolute date to this degree of accuracy is very exciting in this context; it is an excellent example of the importance of examining archaeological contexts in relation to the relative stratigraphy, the artefactual assemblages, and in this case, the possibility of both relatively and absolutely placing this virtually unique human burial within the sequence of events.

2.1.2.4.5 REHYDROXYLATION

A new method is being applied to the dating of ceramics: rehydroxylation (Orton & Hughes 2013: 224–225). This method involves heating a ceramic sample to ascertain how much water it has taken on during its life. This is followed by exposure to water vapour to see how rapidly the ceramic can then absorb this. Meteorological records are vital for establishing the mean temperature during the lifetime of the ceramic (Orton & Hughes

2013: 224). The principle behind the development of this method is that the ceramic recombines with moisture in the environment over time (Orton & Hughes 2013: 224). This method is still being developed at the time of writing, but it has given some preliminary and apparently reliable results (Orton & Hughes 2013: 224–225; Wilson et al. 2009).

2.2   Finding Sites and Buildings (with Kristian Strutt)

How is it possible to find an archaeological site in Egypt and, without excavation, to identify the towns, villages and houses in which people lived and the cemeteries in which they were buried? This section will answer these types of question and deal with the practicalities of using various different scientific methods to locate sites and buildings in space. The importance of the environment, notably the proximity of water channels, as a reason for site foundation is addressed in Section 2.3: The Environment.

2.2.1   Remote Sensing and Egyptian Archaeology

How is it possible to know what is there without excavating the area? Modern archaeology is increasingly concerned with the assessment, recording and preservation of sites to facilitate implementation of cultural heritage management programmes, with recourse to excavation only at endangered sites (rescue archaeology) and to answer specific research questions. Excavation is a destructive process; how is it possible to identify the towns, villages and houses in which people lived, and the cemeteries in which they were buried, without large-scale excavation? And how can this be done within the context of the expansive modern development that serves the needs of the rapidly expanding population of modern Egypt? Thankfully, technological advances over the past century have increasingly allowed archaeologists to detect sub-surface remains without even putting a trowel to the ground. These techniques collectively are referred to under the heading of Remote Sensing. They are increasingly used as a matter of course by archaeological projects to assess sites and regions, as well as to investigate the use of space and function in specific buildings/rooms, and wider landscape processes (Herbich 2012a: 11). Furthermore, various techniques encompassed within Remote Sensing (many of which are discussed here) are also valuable in the detection, interpretation, or preservation and monitoring of archaeological heritage throughout the world (Cowley 2011; Doneus & Briese 2011: 72). The techniques are introduced here, and complementary techniques for site reconstruction of the ancient environment, including drill coring/hand augering, are discussed below and in Section 2.3.

Remote sensing is the term used for any technique that provides an image or data of a material or an area from a distance (Campbell 1996).

Remote sensing encompasses geophysical survey, aerial photography, airborne sensing techniques and satellite imaging (Schmidt 2004). Techniques can be applied to limited areas, assessing the sub-surface remains on single sites, or over extended areas, collecting data on the distribution of archaeological sites across wide geographical areas. These methods all produce data that can feed into a spatial database (e.g. a Geographical Information System – GIS; Conolly & Lake 2006) and all contribute towards our building a history of the distribution of sites across the landscape through time, and aid in spatial analysis examining site growth, and/or decline across different periods. Information relating to the history of the sites can be obtained through ceramic analysis during site survey, as discussed in Section 2.1.1: Relative Chronology and Relative Dating, as well as by absolute dating, including OSL, which can give useful date ranges relating to sites. Remote sensing techniques were not originally designed with archaeology in mind, but they have been adopted and developed to suit its purposes over the course of the 20th and into the 21st century. The techniques were initially developed for political and military purposes, aiding war reconnaissance, for mapping in the early 20th century, with aerial photography and later satellites (in the 1950s) allowing for observation of large tracts of land, including, in some instances, national surveys (Bradford 1957). Although not discussed further here, historical maps, including those produced during wartime reconnaissance, are a valuable resource to the archaeologist, recording a landscape that is often much altered today. These resources provide links to sites and to toponyms that have now fallen out of use. They can be accessed in many national and university libraries and institutions.

As early as the 1920s, archaeologists began to apply remote sensing methods to archaeological fieldwork, firstly with aerial photography in the UK (Crawford 1923; 1929), and then with air photography to survey huge expanses of land in North Africa and the Middle East (Jones 2000; Reeves 1936) and later to map environmental change and the distribution of archaeological sites within the landscape. One of the groups of techniques used widely in archaeology today is geophysical survey. Geophysical survey comprises a series of methods with different capabilities, each appropriate to different terrains and different sedimentary environments, but the choice of technique depends upon factors including the depth of sensing required and the proximity of modern buildings.

2.2.1.1 Aerial Photography

How can we find archaeological sites within the landscape? Whereas some monuments are still visible on the ground, there may be certain aspects, such as their internal configuration, that are harder to discern unless viewed from above. Furthermore, even once monumental structures are sometimes barely visible from the surface (e.g. Herbich &

Spencer 2006). Structures, such as the remains of temples or buildings, can usually be recognised on conventional aerial photographs as long as they are viewed at an appropriate scale and with a suitable viewing angle. Crop and sediment marks, which identify sub-surface structures such as walls, foundations or ditches, are harder to identify with any certainty. Air photographic survey records of numerous important sites have been conducted, including Karnak Temple and the Giza plateau, and the results of such photographic air surveys also became crucial for mapping archaeological sites in less easy to manage desert areas, away from the Nile Valley and Delta.

2.2.1.1.1 MULTISPECTRAL PHOTOGRAPHY AND SATELLITE IMAGERY

Satellites were first sent into space in the 1950s (Campbell 1996). The scope for using multispectral cameras (especially from space) was realised early on as different features can be recognised using different bands of the electromagnetic spectrum. Multispectral cameras look simultaneously across a wide range of wavelengths; some of these may be more sensitive to changes in vegetation, moisture and temperature than either standard cameras or the eye. The development of such technology led to higher resolution satellite imagery, both using the spectrum of visible light and other ranges in the electromagnetic spectrum (Table 2.1), including near infrared and thermal ranges (Allsop 1992: 122ff; Parcak 2009: 41–80, Table 3.1, Figure 3.1; Scollar et al. 1990). Red and infrared images are less affected by atmospheric water vapour than other wavelengths and the infrared image responds to aspects of the sub-surface of the ground (Donoghue 2001). Although water vapour is not usually an issue in Egypt, these images can be of particular use for identifying certain sub-surface structures and relict river channels.

Table 2.1 Principal divisions of the electromagnetic spectrum

Division	Range of wavelength
Gamma rays	0.03nm
X rays	0.03–300nm
Ultraviolet radiation	0.30–0.38μm
Visible light	0.38–0.72μm
Near infrared radiation	0.72–1.30μm
Mid infrared radiation	1.30–3.00μm
Far infrared radiation	7.00–1,000μm (1mm)
Microwave radiation	1–300mm
Radio waves	≥300mm

The archaeological and environmental applications of satellite imagery have been recognised, and this technology has been used for mapping environmental change, the distribution of archaeological sites in landscape studies and investigation into the diversity of site types. Improvements over the past five years have included very high resolution (VHR) visible light spectrum data provided both by IKONOS and Quickbird (Lasaponara & Masini 2006; 2007), and the cheaper and more widely available satellite photographic products on the internet such as Google Earth and the WorldWind platform developed by NASA (Beck 2006; Parcak 2009: 41–80). Satellite images are usually georeferenced, which means that they can be tied to precise points on the ground (Parcak 2009: 89–90).

In Egypt, although there has been a tendency for remote sensing applications to be focused on the environmental circumstances relating to modern developments, the study of Egyptian landscapes has also benefited hugely from remote sensing techniques. Projects that have used this technology include studies of Palaeolithic settlement and landscape morphology, such as those undertaken in the southern extent of the Western Desert by the United States Geological Survey (USGS) and shuttle imaging radar (SIR-A) (McHugh et al. 1988) that detected former river systems and Palaeolithic artefacts within alluvial sediments which the researchers thought might have been early occupation sites (*McCauley* et al. 1982). The Egyptian Geological Survey and Mining Authority (EGSMA) have also demonstrated the geomorphological and archaeological applications of satellite image interpretation (McHugh et al. 1988). The EGSMA were successful in locating a palaeo-drainage system, and with it a number of Acheulian complexes, the investigation of which could then begin on the ground. Remote sensing techniques aid in identifying and locating sites which otherwise would probably not be found, and following this, the various processes of ground truthing may begin. In the case of the discovery of the Acheulian complexes, after the location had been confirmed, it was then possible to conduct field survey and chronometric dating of the deposits on the ground. All of these processes contributed to a reconstruction of the palaeo-environment of the study area (McHugh et al. 1988). This type of integrated strategy, involving both remotely sensed data and field survey, has been used also in establishing the development of geomorphological and archaeological sequences in Upper Egypt at both Karnak and Luxor (Bunbury et al. 2008). In this case, the data obtained through aerial photography and satellite imagery was used to identify the palaeochannels and islands within the meander belt of the Nile, and then in the field it was possible to conduct an auger coring survey (see Box 2.6) in order to establish a sequence for the different deposits beneath the ground.

How and why might we integrate these methods? What benefits might this give? To enable integration of various sources, and to examine the data within the real world, the various sources need to be *georeferenced* and converted into digital format for use in a GIS (Conolly & Lake 2006: 72–78; Parcak 2009: 88–90). Early work in Egypt, both during and after the First World War, exposed the difficulties of being able to georeference aerial photographs in expanses of largely featureless desert (Collier 2002). This can also be the case across the Nile Valley and cultivation areas, given that photographs suffer from distortion. In more recent years, aerial photographs are generally ortho-rectified and the distortions removed. This makes it possible to overlay these photographs onto satellite images and maps; all of these can be combined or viewed as separate layers within a GIS as *raster* images (Conolly & Lake 2006: 27–29). It is, however, possible to georeference both older aerial photographs and maps, thereby enabling them to be fixed to specific points on the ground (Conolly & Lake 2006: 17–24; Parcak 2009: 89–90). *Vector* images, by contrast, are a collection of polygons, lines and points, all of which relate to exact x and y coordinates (Conolly & Lake 2006: 17–24). These can be real-world coordinates or even arbitrary site grid coordinates. Plans or maps that have been digitised, including by Computer Aided Design (CAD), produce vector objects. These images can be integrated within a GIS and related attribute data files can be joined to them.

Integrated strategies of site identification, such as the use of remote sensing, GIS and algorithms, have been used to locate sites and structures in the Fayum and at Saqqara by the HORUS Project (Di Iorio et al. 2010). Adding orthography and photogrammetry from low-level aerial photographs, combined with field-based GPS (Global Positioning System) and topographic surveys, is a relatively cost-effective way of delineating sites. The benefits and accuracy include allowing the colour, fabric and nature of a site to be recorded, as can be seen from current work in the Fayum, such as at the Ptolemaic and Roman site of Soknopaiou Nesos. A high level of detail has been acquired there using a camera placed in a low-level balloon (Bitelli et al. 2003: 56ff). In Abusir, Saqqara and Dahshur, archaeologists obtained high-resolution satellite images (QuickBird) to gain a better understanding of both the topography and development of Abusir; this imagery was subsequently made available to the community of archaeologists working in the area and to the Ministry of Antiquities (Bárta & Brůna 2005). A similar approach was used by the American Research Center in Egypt (ARCE) at Thebes. On this occasion the orbit of the satellite was specifically set to image a specific area, and this was examined together with surface survey data, 3-D modelling and geophysical survey data and then linked to a database with information on all archaeological features (Bárta & Brůna 2005: 3). This particular imagery is very high resolution; however, it must be remembered that cheaper and free imagery is readily available (see Box 2.3) in the form of images online (and downloadable) at Google Earth and the declassified CORONA satellite images available through University of Arkansas (http://corona.cast.uark.edu/index.html). The latter have the advantage of dating back to the late 1960s and provide a view of a less inhabited and developed landscape in Egypt, as well as recent environmental changes (Parcak 2009: 52–57). Notably, the Aswan High Dam, which substantially changed the behaviour of the Nile, was completed in 1970. Research at Deir el-Barsha highlights the advantages of using CORONA images taken in 1968 since local activities in subsequent years have masked natural features within the landscape, including alluvial fans and the deposition of sand and rubble, which helps explain why the ancient cemeteries were placed where they were. Satellite imagery is likewise used in the monitoring of cultural heritage (Grøn et al. 2004) by locating archaeological sites otherwise 'hidden' in agricultural fields and for mapping illicit excavation activities. This includes developments in the use of spectral imagery to locate small-scale features within the landscape (Grøn et al. 2011).

In addition to satellite imagery there is an ever-increasing number of methods for recording archaeological landscapes and specific features from above, involving both expensive and non-expensive equipment, and specialist to non-specialist prior knowledge. Such methods were originally developed for a variety of purposes, from mining prospection to military, but have been adapted for use within archaeology when they have become cost-effective, or when the possibilities of data acquisition from the companies concerned have existed.

Box 2.3   Satellite Archaeology for Egyptology
Sarah Parcak

The subfield of satellite archaeology, or satellite remote sensing, has great potential for Egyptian archaeology. It can be defined as the analysis and use of satellite images or space-based photographs to visualise and assess the Earth's surface for mapping past anthropogenic and natural changes. The science of remote sensing has only been applied globally by archaeologists since the early 1980s and in Egypt (more intensively) in the past 15 years. A methodology for the science of satellite archaeology is still being developed, and it changes constantly with the development of satellite technology. Satellite archaeology, ultimately, is a valuable tool for broad scale survey and more focused excavation, as it can save time and money for archaeological projects, but requires ground work for the verification of any findings. Egypt's sites are threatened by urbanisation, looting, and groundwater pollution, and, at present, satellite archaeology is the only way easily to map and monitor these sites.

Satellite archaeology allows Egyptologists to see simultaneously in the visual, infrared, and thermal parts of the light spectrum, and is called "multispectral" remote sensing. Every type of landscape feature in the world, including soil types, vegetation, geology, and man-made features, has a specific signature that multispectral satellites can identify. As human beings, we can only see in the visual part of the light spectrum. Past features that are either buried or covered over by modern vegetation, sand, and structures can be made visible through the analysis of multispectral satellite images. Satellites also allow archaeologists to model landscapes in 3D, making past courses of the Nile or monument relationships more clear.

Currently, commercially available satellites have a 0.4–90 m resolution, making it possible to see most ancient Egyptian sites and their associated features. Satellites at present allow scientists to see 0.2–1 m beneath the surface in the floodplain and desert edges, while RADAR imagery increases that depth to 10 m in the Sahara. Future advances in technology will allow even deeper ground penetration. The cost of imagery, once prohibitive, is no longer an issue: satellite images are either free online via NASA, or cost (prior to educational discount) between $10.00 USD and $30.00 USD per km. sq. (with a minimum order area of 25 km. sq.).

Satellite archaeology is most valuable for Egyptology via archaeological survey, including both intersite and intrasite survey. While topographic survey has played a valuable role in Egyptology, starting with the Napoleonic expedition over 200 years ago, it is made difficult today by the very nature of modern Egyptian landscapes. Many villages and cities are built on top of ancient sites, while numerous tells are covered over either partially or completely by cemeteries or fields. Broad transect survey, possible across much of the rest of the ancient Near East, is made virtually impossible by multiple and differing periods of flooding in fields. Access to sites is also difficult due to canals and unreliable roadways, yet satellite maps can help us decide how best to reach sites on smaller roads. Satellite archaeology allows archaeologists to pinpoint the exact locations of sites and features to examine on the ground, thereby maximising the use of available time to survey each site and collect material culture for analysis.

While archaeologists have used satellite archaeology since the early 1980s, its use was more limited in Egypt until the early 2000s. This not surprising, as the field of aerial archaeology in general has not played a strong role in archaeological excavation and survey in Egypt since the 1930s, despite Royal Air Force photos of numerous Nile Valley monuments. These remain invaluable today due to major landscape changes following the construction of the Aswan High Dam. The most famous and early application of satellite

Figure 2.3 Quickbird 432-RGB image, pansharpened with high pass edge detection and a 95% clip linear stretch, which reveals part of the plan of the southern Roman Period town at Tell Timai

archaeology took place in the Western Desert, with the discovery of the so-called "RADAR Rivers" using SIR-A NASA imagery (McHugh et al. 1988). Later use of RADAR imagery led to the discovery of mortuary remains in Dahshur (Yoshimura et al. 2006).

There are many possibilities and great potential for satellite archaeology in Egypt. Starting in 2002, satellite archaeology led to the discovery of 44 previously unknown sites in the east Delta, and 70 sites in Middle Egypt via the analysis of Landsat, ASTER, Corona and Quickbird satellite imagery (Parcak 2007; 2008; 2009). Tells throughout Egypt, even those largely covered over by modern towns, have their own distinct spectral signature. During periods of greater moisture (such as during winter), it is possible to differentiate between modern towns and ancient tells through the use of multispectral analysis. Varied remote sensing techniques, which include supervised and unsupervised classification, thermal analysis and principal components analysis, have all aided in site detection. This can be seen in the image of Tell Timai (Figure 2.3). Site detection from space is ultimately affected by atmospheric conditions, weather patterns, local soil conditions and field drainage, which all need to be taken into account when doing satellite archaeology in Egypt. Multiple intensive survey seasons in each region have confirmed both the location of the sites and their dates. Ongoing remote sensing project research across the Delta and Nile Valley has revealed hundreds of additional sites.

Additional ongoing research using satellite archaeology is testing the potential of space-based subsurface survey. This could stand to change the field of Egyptology radically: if archaeologists can map the subsurface remains of a tell from space, they can choose precise areas for excavation, thus saving the cost of magnetometry and GPR work. At this moment, however, satellites cannot see as deeply as GPR. Also, while archaeologists have identified and tested techniques to locate surface sites, the techniques to locate specific features on sites are being refined for each geographic region of Egypt. These techniques can only be tested with excavation, coring, and survey. Ultimately, the question of "what is there left to find" in Egypt becomes relevant. Calculations based on known Delta region excavation and survey suggest that archaeologists have uncovered less than 1 per cent of all Egyptian remains. Satellite archaeology thus has an important role to play in the mapping of Egyptian sites and features, in order to map Egypt's sites before they are gone forever.

References

McHugh W., Breed C., Schaber G., McCauley J. & Szabo B. (1988) Acheulian sites along the "Radar Rivers," Southern Egyptian Sahara. *Journal of Field Archaeology* 15: 361–379.

Parcak S. (2007) Satellite remote sensing methods for monitoring archaeological tell sites under threat in the Middle East. *Journal of Field Archaeology* 42: 61–83.

Parcak S. (2008) Survey in Egyptology. In: Wilkinson R. (ed.) *Egyptology Today*. Cambridge: Cambridge University Press. pp. 41–65.

Parcak S. (2009) *Satellite Remote Sensing for Archaeology*. London: Routledge.

Yoshimura S., Kondo J., Hasegawa S., Sakata T., Etaya M., Nakagawa T. & Nishimoto S. (2006) A preliminary report of the general survey at Dahshur North, Egypt. *Annual Report of the Collegium Mediterranistarum Mediterraneus* 20: 3–24.

2.2.1.1.2 AERIAL PHOTOGRAPHIC METHODS

Kite Aerial Photography (KAP) is one of the methods used, fixingcameras to airborne devices and raiseing above the landscape, and hence, with KAP, photographs can be taken from heights of c. 75–110 m. The camera can be positioned c. 10–30 m below the kite, and is held within a cradle. Pulleys help to keep the camera in position, which not only avoid distortion in the photographs, but also allow for photographs to be taken from whichever angle is required; this can be very useful for highlighting certain features/ shadows which might not be so clear when taken horizontally from above in bright sunshine. It can be advantageous, as with aerial photography in general, to take photographs earlier in the morning or later in the afternoon when the shadows are longer (Peeters & Peeters 2011: 154). Images captured from directly above, however, are very important in the recording of archaeological features and structures during excavation, as this is not always possible from photographic ladders or even towers. Another method is to attach a camera to a helium balloon, as has been used at Amarna and Antinoupolis (Kemp 2010; Pintaudi 2008; Stevens 2012a).

Unmanned Aerial Vehicles (UAVs) (drones) are also used in imaging. These are devices controlled by onboard computers or by remote-controlled units on land, and have cameras (and sometimes laser scanners) attached for imaging. Imaging using UAVs can achieve a very high level of accuracy (c. 2–5 cm). This accuracy is achieved by recording the GPS readings of five fixed points within each square km covered by the UAV so that the images can be subsequently rectified against the GPS readings. This umbrella term encompasses a wide variety of devices of various capabilities and various sizes that can be flown at various heights. Methods currently used in archaeology which are included under this umbrella are the 'Oktokopter', or 'Microcopter', which can record features, sites and wider landscapes. These are all devices that can be flown c. 30–40 m above the surface, and which have cameras attached to record high -resolution digital images. It can be possible to tilt the camera while airborne in order to capture images from a variety of angles, and this means that it is not only possible to image accessible, but also more inaccessible archaeological sites, such as those in rocky or coastal areas and those submerged by water at certain times of day or year. Images can also be taken of the same surfaces at various time intervals, thereby enabling the monitoring of erosion and the implementation of suitable conservation strategies. Such methods also facilitate the planning of sites. This is particularly important if excavation is being carried out on a very limited timescale, such as on rescue projects, as the site and excavation plans can be drawn directly from the photographic images.

2.2.1.1.3 PHOTOGRAMMETRY

Photogrammetry can be used to create Digital Elevation Models (DEM) by capturing overlapping digital photographic images taken with a 20–40 per cent overlap (Koutsoudis et al. 2013: 4452). These DEM images are then processed to create 'orthophotographs' by matching the pixels between the digital images (Shaw & Corns 2011: 7). DEMs permit the creation of models of the topography of the wider landscape but do not have sufficiently high resolution to provide models of small and/or complex sites (Shaw & Corns 2011: 78).

In archaeology, a related method of photogrammetry has been used in recent years called Structure from Motion (SfM) imaging (Koutsoudis et al. 2013: 4451; Powlesland 2014; Torres et al. 2012). This can be used both for the imaging of individual objects and of features within sites and wider landscapes. Targets can be placed on structures and surveyed for metric evaluation of the quality of the model. It is possible, however, to take photographic images of features without targets and to process these images using freely downloadable (or purchased – professional) software to knit the individual photographs together into single 3-D images. Agisoft Photoscan is one of the most commonly used packages in archaeology at the time of writing. As long as the camera is calibrated in advance (to counter optical distortion) and the images are captured in the correct way, the images can be re-processed over time using more advanced software as new packages become available (Verhoeven 2011).

Photogrammetric methods can be used for capturing individual objects (e.g. Koutsoudis et al. 2013) as well as larger features, both on the surface and in underwater contexts. To ensure that the size of the original object or feature can be incorporated into the model, targets and a scale must be placed within a single field of vision. The scale might be a 1 m scale for an archaeological structure, down to smaller cm scales for small features or objects. The use of the scale and targets enable the size of the original object to be known and included within the digital model during image processing. A series of photographs are then taken from each surface of the object which are then all uploaded into the processing software. Imaging and image manipulation technology is developing and changing quickly. Future-proofing, where possible, should be attempted when using these methods. See, for example, Box 4.9 discussing Reflectance Transformation Imaging.

Photogrammetric methods can also be implemented when imaging via UAV devices; these methods are very efficient ways of capturing images of sites during and after excavations, as well as of specific contexts and objects (Koutsoudis et al. 2013; Torres et al. 2012). One problem that can be faced when trying to detect and image sites within broader landscapes is that the archaeology can be obscured by trees, other vegetation and/or modern buildings. There are methods that can be used to help counteract such problems, such as using computer programs to remove vegetation from satellite images. Another option is to use Airborne Laser Scanning (ALS).

2.2.1.1.4   LASER SCANNING

There are two main types of laser scanning: terrestrial laser scanning (TLS) and airborne laser scanning (ALS; also known as Light Detection And Ranging [*light and radar*] – LiDAR; Parcak 2009: 76–79). Both methods produce large point clouds of 3-D data (where each point has an x, y and z coordinate), and these point clouds can be used to generate digital terrain models (DTMs) or be rendered and modelled accordingly. These can be very large datasets, so it is important to consider how and where such digital data will be stored.

LiDAR is a method of laser scanning first developed in the 1960s and used widely in archaeological contexts since the 2000s (English Heritage 2010). It measures the height of the ground surface and can be applied to reconstruct natural or man-made features across the landscape through the creation of DTMs (Devereux et al. 2008; Doneus & Briese 2011). In terms of resolution, a good DTM will require a single measurement per square metre of land (Doneus & Briese 2011: 60), although data now exists with a resolution of 0.5 m or even 0.25 m. Although DTMs can be acquired cost-free, these are often supplied without knowledge of the time and date of the imaging, the weather conditions at the time, or other potentially useful metadata (Doneus & Briese 2011: 59).

How does ALS work and what can it achieve that cannot be obtained through satellite images or aerial photographs? This method really comes into its own when features are obscured by vegetation, particularly trees (Doneus & Briese 2011: 59). There are conventional scans using ALS (discrete echo scanners) and full-waveform recording systems, with the latter being most advantageous when trying to combat dense vegetation cover (Doneus & Briese 2011: 62–63). The method works on quite a different principle to those already described. It is not a photographic method, but instead uses laser ultraviolet, visible light and/or infrared pulses (with the most usual being infrared). The pulses emitted from the aircraft build up a number of echoes, and GPS data is also collected so that the ALS data can be related to real-world coordinates (Doneus & Briese 2011: 60). The pulses do hit trees, but also continue to the ground (Doneus & Briese 2011: 60–61). Fixed to a manned aircraft (either plane or helicopter), the laser pulses to the ground 20,000–100,000 times per second and, although the vegetation and tree cover reflects these pulses back up, the pulses try to continue to the ground so the ground cover can be penetrated (Shaw & Corns 2011: 79). It is not only tree cover that can obscure the archaeology, but also modern structures. In order to alleviate these problems, the data can be processed to remove point clouds that relate to trees, vegetation and modern buildings. It is therefore important when processing the data to take into consideration the type of context, landscape and season (Doneus & Briese 2011: 65–66; Shaw & Corns 2011: 79). In addition, it is possible to use aerial photographs (as discussed above) together with the LiDAR data, to produce digital orthographs (Shaw & Corns 2011: 79). LiDAR is also helpful for imaging structures at intervals over time, thereby helping in the monitoring of landscapes for erosion. This permits testing and monitoring of any conservation measures implemented (Shaw & Corns 2011: 82).

Although airborne laser scanning is usually fixed to a manned aircraft, it is also possible to mount the laser onto a UAV, such as a microcopter. GPS points taken on the ground can be tied to known points from the images taken when the aircraft or UAV flies over the ground through post-processing of the data with real -world coordinates. Where used with a UAV, it is also possible to program the device to return to land if the battery is running out. Some non-standard LiDAR systems (including the FLI-MAP 400 helicopter -based system) use a Class I laser scanner (150 KHz), with two GPS receivers, which are dependent on the availability of local fixed GPS base stations. These types of geo-referenced data can be used in GIS, and a Spatial Data Infrastructure (SDI) also allows all elements of geographic data including metadata to be collated and synthesised, and then permits the data to be accessed via the worldwide web (Shaw & Corns 2011, 85). This can take the form of a database accessible via a map interface, such as the CORONA datasets at the University of Arkansas (http://corona.cast.uark.edu/index.html), or archived data from archaeological projects, with metadata, that have been saved in particular stable formats for digital curation with a repository such as the Archaeology Data Service in the UK (ADS; http://archaeologydataservice.ac.uk/).

2.2.2 *Geophysical Survey Techniques*

'How big is the site? What are its dimensions? How far below the ground is the archaeology, and how deep does it go?' These questions can be subdivided into two types: questions dealing with issues of horizontal dimensions such as size and shape, and questions dealing with issues of depth. Certain geophysical methods are appropriate for one or other of these questions, with only two methods producing horizontal plan and depth below surface and dealing with 3-D data (Ground Penetrating Radar – GPR and Electrical Resistance Tomography – ERT; Conyers 2012: 19–20; Imai et al. 1987). All the methods allow the archaeologist to obtain information on the extent of sub-surface remains, be it the horizontal or vertical extent, but the results will be affected by the type of sediment, the level of contrast between the archaeology and the surrounding sediment and also proximity to the water table. Geophysical methods therefore better inform decisions as to which areas to investigate further either through excavation with tranches, or augering, as discussed in Section 2.3: The Environment. It enables the archaeologist to know specifically where to look to help answer targeted research questions relating to specific buildings on a site, to investigate particular environmental features, or to find sub-surface evidence relating to a site no longer visible on the surface. The methods cannot reveal how old a feature is; neither can they tell exactly what materials and sediments are beneath the ground. It is through experience of working within a given environment and ultimately through ground-truthing that educated interpretations of geophysical survey data become increasingly possible.

Geophysical survey techniques have been developed over many years, and have been used widely in Egypt since the 1990s. Work has been carried out by surveyors from Egyptian universities, sometimes in collaboration with foreign missions and/or the Ministry of Antiquities, and at times with other surveyors coming from foreign institutions (Ghazala et al. 2005; Herbich & Peeters 2006; Ibrahim et al. 2002). Its first application in the Nile Valley was in the 1960s by Albert Hesse at the Second Cataract Fort at Mirgissa in Lower Nubia, where results had to be plotted manually; only in the 1980s were measurements fully automatic (Herbich 2012a: 11; Vercoutter 1970: 51–121). One early collaboration developed between Ain Shams University, the then Egyptian Antiquities Organisation (EAO) of Antiquities (currently the Ministry of Antiquities), and a physicist from a Californian institute. They started work in 1974, applying acoustic sounding, resistivity and magnetometry, in conjunction with aerial photography, and thermal-infrared imagery obtained through the National Academy of Scientific Research and Technology Remote Sensing Center (Moussa et al. 1977). The National Science Foundation-funded team worked in various environments in the Nile Delta and the Nile valley (Moussa et al. 1977), most recently undertaking gamma-ray spectroscopy (Moussa 2001). The developments in computer technology, as well as refinements in the geophysical survey equipment itself, means that various methods have been employed frequently in Egypt since the 1990s. On sites such as Qantir (ancient Piramesse) in the Delta, magnetic survey by Helmut Becker and Joerg Fassbinder has produced more data as to the settlement layout than could have been obtained from many years of excavation (Herbich 2012a: 11; 2012b: 383; Pusch 1999). As a result of pressures on funding for excavation and more recent agendas being focused on minimal excavation to answer specific questions, as well as as issues of preservation and conservation, geophysical survey techniques are increasingly used to obtain a broad overview of a site, thereby allowing for minimal targeted work to proceed if required (Conyers & Goodman 1997: 11–12).

One of the key names in geophysical survey in Egypt, who worked with and developed different types of geophysical survey methods was Ian Mathieson of the Saqqara Geophysical Survey Project (Mathieson 2013; Mathieson et al. 1995; Price 2009). The early work of the project combined aerial photographs, historical maps and geophysical survey by Mathieson and Jon Dittmer, adding valuable new information to previous plans of the area between the Gisr el-Mudir and the old Abusir Lake. This included locating the northern wall and gate of the Serapeum enclosure (Mathieson et al. 1995) and mud-brick structures facing the Serapeum (Leahy & Mathieson 2002; Mathieson 2013). One of the main purposes of the project was to examine which geophysical survey methods could perform in terrain where archaeological remains are deeply overlain by sand, and in the case of Saqqara, at a very complex site used over a very long period of time (Leahy & Mathieson 2002: 14). Geophysical surveyors working in Egypt have applied multiple geophysical methods in different conditions throughout to produce incredibly detailed and informative plans of archaeological features (see Box 2.4; Bunbury et al. 2009; Forstner-Müller et al. 2010; Graham et al. 2014; Herbich & Spencer 2006; Rowland & Strutt 2011; Strutt et al. 2012; 2014).

Techniques such as magnetometer survey and ground penetrating radar, given appropriate environmental conditions (discussed below), can establish the horizontal extent of archaeological features beneath the surface. They can provide high -resolution images of sub-surface anomalies and the results have the potential to outline roads, walls, buildings and pits (e.g. Herbich & Spencer 2006). Once manipulated through computer programs, magnetometer survey data produces images showing sub-surface anomalies that can sometimes be detected and identified as potential walls, roads, buildings, or graves. With magnetometry, the data cannot be separated to inform as to the stratigraphic layering of the features, and to what depth they extend (see Figure 2.4). By contrast, ground penetrating radar data (GPR) is manipulated to produce a horizontal plan made up from a series of vertical profiles, and it is possible to show a number of horizontal plans at given depths, e.g. 0.30 m, 0.60 m, 0.90 m or 0.50 m, 1.00 m and so on. In some situations, GPR allows the depth of walls to be traced, and even to detect where floors appear and end. Electrical Resistivity Tomography (ERT) also provides a series of vertical profiles of data that can then be combined to produce a series of horizontal plans at different depths. These techniques can give the precise depths of features or other anomalies and can detect which features underlie which. Geophysical survey data can be used in conjunction with satellite imagery, aerial photographs and techniques such as drill coring to build up 3-D reconstructions of site topography and geomorphology. These issues are discussed further in Section 2.2.3.

So, which geophysical technique will provide the most information about the site? The answer to this question depends upon several issues, and different techniques are often applied to single sites or areas to assess what might produce the most significant results. One further key question exists: what does the archaeologist want to find out about the sub-surface features? Is the key aim to gain a general idea of the plan of sub-surface features, or is it to gain an in-depth knowledge of stratigraphical relationships (which can be tested through excavation or drilling, see Section 2.3: The Environment), or is it both? It might be that a combination of methods is necessary to provide the required information.

Geophysical survey techniques use different methods to locate anomalies in the contrasting properties of the sediments/features beneath the ground, and are affected by variations in the physical properties of the sediment. These include, for instance, the sediment's ability to conduct electricity and the contrast, or lack thereof, between the surrounding sediment and the features within it. For example, if there is a simple pit burial with few or no grave goods, then the grave might not be detected using magnetic survey. However, if the grave is a mud-brick lined tomb, with large numbers of ceramic vessels, then the chances of detection increase. This is the case at Tell el-Farkha, where graves with high numbers of large ceramic jars exhibited a higher magnetic intensity (Herbich 2012b: 388–389) and at Quesna, where Roman fired ceramic coffins at the Necropolis were located using magnetometry (Rowland & Strutt 2011). In contrast, a grave pit with nothing inside can feasibly be detected by GPR due to the changes in the stratigraphy and slumping of the sediment in the area of the pit (Conyers 2012: 137–139). Modern features, however, may cause anomalies within the results. For example, a modern drainage channel beneath the surface might show as a linear feature on all types of geophysical survey, whereas in a built-up environment, modern buildings and metal pipes all cause anomalies due to their magnetic properties and are visible alongside the archaeological features (Rowland & Strutt 2011). In addition, modern narrow streets which run across ancient mounds (*koms*) are frequently unsuitable for magnetic survey as this method benefits from surveying areas of at least 0.5 ha in area. In such cases, where koms are heavily built over in modern times, with much interference from metal pipes and other structures containing metal, one of the other types of geophysical survey may be more appropriate, such as GPR or ERT (Conyers 2012; Graham et al. 2012a; 2012b). The context impacts greatly on the most appropriate choice of method, as in some situations, small scale survey may obtain better results than resistivity, such as undertaken by Herbich at Pelusium (Herbich pers comm)

How do these methods actually work? To understand this it is necessary to recognise the different physical parameters that geophysical survey techniques measure, and the practical and theoretical considerations that allow us to understand why certain methods are preferable in particular given circumstances.

2.2.2.1 Magnetometry

Magnetic prospection of sediments is based on the measurement of differences in the magnitude and directionality of the Earth's magnetic field at points over a specific area. The iron content of a sediment principally provides the basis for its magnetic properties. When a trench has been dug for the construction of a building, different types of material may be deposited in it, thereby potentially changing the resultant magnetic properties and leading to differences between the trench and the surrounding sediment. When walls are constructed of mud-brick and/or stone they also have different magnetic properties relative to the surrounding sediment. Although even small changes in the magnetic field can be identified and thus sub-surface features detected, one of the key benefits of magnetometry is its ability to identify kilns and hearths (as the past heating events at these produce massive local distortions in the Earth's magnetic field). Presence of magnetite, maghaemite and haematite iron oxides all affect the magnetic properties of sediments (Aspinall et al. 2008: 23–24). Although the variations in the Earth's magnetic field associated with archaeological features are weak, human action does change the magnetic properties in the ground. Examples include activities involving heat/burning, through bacterial action in domestic waste as well as where ceramics or brick or metal fragments are lying within sediment (Aspinall et al. 2008: 21, 24–25).

There are three basic types of magnetometer available to the archaeologist: 1) proton magnetometers, 2) fluxgate gradiometers, and 3) optically pumped magnetometers (such as alkali vapour magnetometers or caesium magnetometers). These three types also fit into two main groups: scalar (proton magnetometer and alkali vapour type) and vector (fluxgate) (Aspinall et al. 2008: 29). How do they differ and when is one type of magnetometer more suitable than another? A brief summary of the different types of equipment available is presented below (see Aspinall et al. 2008: 34–56 for details on all types of magnetometers).

Fluxgate gradiometers are the most commonly used instruments in magnetic survey in archaeology in Egypt (Aspinall et al. 2008: 33–41). They are based around a highly permeable nickel iron alloy core (Scollar et al. 1990: 456) which is magnetised by the Earth's magnetic field, together with an alternating field applied via a primary winding. Due to the fluxgate's directional method of functioning, a single fluxgate cannot be used on its own as it cannot be held at a constant angle to the Earth's magnetic field. Gradiometers therefore have two fluxgates positioned vertically to one another on a rigid staff (Scollar et al. 1990: 456). This reduces the effects of instrument orientation on readings. Fluxgate gradiometers are sensitive to 0.5 nT or below depending on the instrument. The range of prospection depends on magnetic properties of searched features and the contrast with the values of the surrounding area (signal ratio). The nature of archaeological deposits, such as mud-brick, means that such features are rarely detected deeper than 1m below the modern ground surface. The very robust nature of this type of instrument, its portability and the ease with which it can be transported by air has led to it being used commonly in archaeological survey, but see also Herbich (2014)

Box 2.4   Magnetic Survey of the Great Temple Enclosure in Tell el-Balamun

Tomasz Herbich

The great temple enclosure in Tell el-Balamun was excavated in 1991–2008 by a British Museum expedition following sporadic digging in the early 20th century. The research resulted in the tracing of the plans of the temple of Amun (erected in the New Kingdom, rebuilt in the Late Period) and the temples of Psamtik I and Nectanebo I, as well as the layout of a fort in the southern corner of the enclosure (all of Late Dynastic date). Ptolemaic architecture was discovered in the northeastern part of the complex and Late Period burials were located north of the Amun temple. The 30th Dynasty outer enclosure wall was traced thanks to surface vestiges of the foundations; the size of the enclosure was determined at 450 by 400 m. The less well preserved inner wall of the 26th Dynasty was mapped following excavations (Spencer 1996; 1999; 2003).

Geophysical methods of prospection were suggested for the investigation of the area inside the enclosure wall as its sheer size was a deterrent to extensive excavations. The magnetic method, which registers changes of intensity of the Earth's magnetic field, was selected following an analysis of the known remains and the building materials once in use at the site. The method has been proven in the past ten years to be extremely effective on tell sites in the Nile Delta (Herbich 2003; Hartung & Herbich 2004; Hale & Wilson 2003; Spencer 2004). Sun-dried mud brick was the principal building material at Tell el-Balamun; baked brick is rare, being limited to Roman-age structures. Stone must have been used in monumental structures, but later salvaging of building material from the abandoned architecture has effectively removed most traces of it. Sand occurs in the spaces between temple foundations. An iron-oxide presence in the Nile silt from which mud bricks were produced (as was the case of Balamun lying in the Nile Delta) gives the bricks sufficiently high magnetic susceptibility to make them easily traceable by magnetic mapping. Fired clay (bricks, furnaces, hearths) also have higher magnetic properties. The susceptibility of stone depends on its type: magmatic rock (granite and basalt in the case of Balamun) has high magnetic susceptibility, whereas sedimentary rock (sandstone and limestone) and sand only very slight (Aspinall et al. 2008: 21–28).

The magnetic mapping of 21 ha inside the enclosure provided a clear overall image of features, both already excavated and previously unrecorded, in areas that had apparently been well-researched,

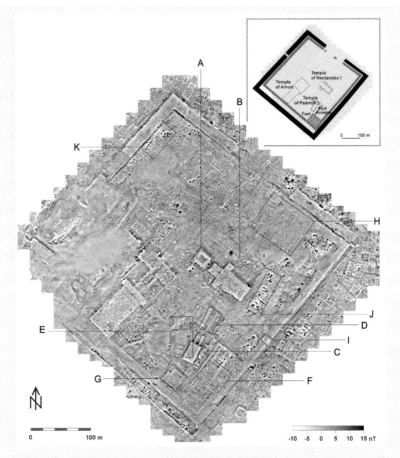

Figure 2.4 Magnetic survey map of Tell el-Balamun (locations denoted with letters are discussed in box text)

such as the surroundings of the temple and fort, as well as in the northwestern part of the enclosure which had hitherto escaped archaeological investigation (Herbich & Spencer 2006; 2007; 2008; 2009; Herbich 2009). The features most in evidence on the magnetic map are the sand fillings of temple foundations, remains of enclosure walls and traces of industrial activities (Figure 2.4).

A square structure (marked A) was detected in front of the Nectanebo temple. The anomaly corresponding to it has values markedly lower than the surroundings, indicating that the building materials in this case were devoid of magnetic properties. Excavations showed that it was a bark-station, built of limestone blocks, apparently the sole surviving stone building in Balamun. At the rear end of the Nectanebo temple, prospection revealed remnants of a mud-brick wall with bastions on the exterior (B). Judging by the magnetic map, this wall is cut by the back of the temple foundation, with parts surviving on either side.

The magnetic map clearly reflects the outline of a narrow colonnade foundation in front of the Amun temple naos (known from excavation, Spencer 1999: 15–16). Prospection revealed a similar colonnade (C) behind the entrance pylon in the temple of Psamtik. It also appeared, judging by the two parallel anomalies of lower magnetic value registered in front of the pylon (D), that the approach to the temple was more extensive than previously anticipated. There seem to have been structures lining the sides of a processional way leading to the temple.

The less than distinct T-shaped structure oriented directly to the north (E), was found to correspond with the outlines of a foundation-pit for an unknown temple (Spencer 2009: 45–49). This mud-brick sanctuary had been destroyed so extensively that its tracing by traditional methods of archaeological excavation would not have been possible.

In places as, for example, in the Fort Annexe, the magnetic map revealed evidence of structures on multiple levels. The southern part of this building had proved very difficult to trace in excavation (Spencer 1996: 59–62). The magnetic map showed a later structure of rectangular shape, the long axis aligned NW–SE, superimposed on this part of the building (F). It turned out to be much easier to detect the presence of the later structure on the magnetic map than through actual excavation. The prospection also brought to light a previously unnoticed structure (G) adjoining the NW wall of the fort. In the area where digging uncovered Ptolemaic vestiges (H), the magnetic map provided data for the reconstruction of the street grid in this district.

Geophysical mapping improved the known layout of the enclosure walls. The magnetic image of the northeastern and northwestern sections of the 30th Dynasty wall reflects its method of construction as separate, alternately projecting and recessed, panels of brickwork. The detailed image of particular panels on the magnetic map permits their size to be established. In places where erosion had obliterated the wall entirely (the southeastern section), the map evinces the presence of casemate-type buildings probably from the Saite Period. Since the inner wall shows less clearly on the surface than the outer wall, the magnetic survey has provided valuable information on the alignment and thickness of this wall at various points. A gate has also been identified in the southeastern section of the wall from the 26th Dynasty (I).

Data from the magnetic survey have precisely pinpointed areas of industrial activity inside the enclosure. For example, concentrations of furnaces are visible at the back of the Nectanebo temple (J) and in the northern corner of the enclosure (K). An industrial complex was also observed sandwiched between the southwestern sections of the 26th and 30th Dynasty enclosure walls, proving that the two walls coexisted at least for some period.

The survey was run as a joint project of the British Museum and the Polish Center of Mediterranean Archaeology of the University of Warsaw, between 2005 and 2008.

References

Aspinall A., Gaffney C. & Schmidt A. (2008) *Magnetometry for archaeologists.* Lanham: Altamira Press.

Hale D. & Wilson P. (2003) Geomagnetic surveys at Sais, Sa el-Hagar, western Delta, Egypt. *Archaeologia Polona* 41: 185–188.

Hartung U. & Herbich T. (2004) Geophysical investigation at Buto (Tell el-Farain). *Egyptian Archaeology* 24: 14–17.

Herbich T. (2003) Archaeological geophysics in Egypt: the Polish contribution. *Archaeologia Polona* 41: 13–55.

Herbich T. (2009) Magnetic survey of the Late Period grate temple enclosure in Tell-el-Balamun, Egypt. *Archeosciences. Revue d'Archéométrie* 33 (suppl.): 77–79.

Herbich T. & Spencer AJ. (2006) Geophysical survey at Tell el-Balamun. *Egyptian Archaeology* 29: 16–19.

Herbich T. & Spencer AJ. (2007) Tell el-Balamun: Geophysical and archaeological survey, 2005. *Polish Archaeology in the Mediterranean* 17: 117–123.

Herbich T. & Spencer AJ. (2008) Tell el-Balamun. The geophysical and archaeological survey, 2006. *Polish Archaeology in the Mediterranean* 18: 101–111.

Herbich T. & Spencer AJ. (2009) The magnetic survey. In: Spencer AJ. (ed.) *Excavation at Tell el-Balamun 2003–2008.* pp. 104–109. http://www.britishmuseum.org/research/research_projects/excavation_in_egypt/report_in_detail.aspx

Spencer AJ. (1996) *Excavations at Tell el-Balamun 1991–1994.* London: British Museum Press.

Spencer AJ. (1999) *Excavations at Tell el-Balamun 1995–1998.* London: British Museum Press.

Spencer AJ. (2003) *Excavations at Tell el-Balamun 1999–2001.* London: British Museum Press.

Spencer AJ. (2009) *Excavations at Tell el-Balamun 2003–2008.* http://www.britishmuseum.org/research/research_projects/excavation_in_egypt/report_in_detail.aspx

Spencer N. (2004) The temples of Kom Firin. *Egyptian Archaeology* 24: 38–40.

Optically pumped magnetometers function on the principle of monochromatic light passing through a magnetic field in an appropriate material and the break-up of this light into several lines of the spectrum based upon the field intensity (Scollar et al. 1990: 462). Contemporary instruments use caesium 133 as the material, although rubidium was also utilised in the past. Within the instrument, the material is heated slightly to vaporise it, with the cell illuminated by a lamp, passing through a lens and a polariser onto a photodetector, with the pumped light orientating the atoms of the cell with the external magnetic field (Scollar et al. 1990: 464).

Free precession proton magnetometers utilise a pair of coils immersed in a liquid to cancel out external magnetic interference. The amplitude of the proton signal is primarily dependent on the polarising current, the Earth's magnetic field and the local gradient.

The results of surveys with fluxgate gradiometers successfully detect positive archaeological features including brick walls, hearths, kilns and disturbed building materials, as well as showing more ephemeral changes in sediment, thereby helping the archaeologist to locate foundation trenches for buildings, pits and ditches (Herbich 2012a; Herbich & Spencer 2006). That these negative features can also be detected makes this type of survey vitally important because of the frequency with which building stones were removed for re-use, either locally or at greater distance. This occurred both in antiquity and in historical and more recent times. Similarly, the detection of pits may reveal simple, lined or brick -built graves, rubbish pits, and storage pits. The accuracy and definition of the results are, however, extremely dependent on the geology of the particular area, and whether the archaeological remains derive from the same materials as found in the sediments (e.g. mud-brick) or are contrasting (e.g. stone). It must be remembered that this method will take readings on all magnetic fields. This means that if there is a dense layer of ceramic sherds across the survey area, for example, then data resulting from this ceramic layer might well mask the clarity of other sub-surface features. Likewise, if sites have been subjected to action by *sebbakhin,* such as digging for mud-brick for fertiliser, or by robbing activity, then these later features will also be detected by the magnetic survey, hence adding some distortion to the results (Bailey 1999).

Magnetic survey is, and has been, widely used in Egypt, in both the Nile Valley and the Delta (Herbich 2012b: 383); it works well on level floodplains and through alluvial sediments, as illustrated through survey work at Qantir (Abdallatif et al. 2003), at Tell Toukh El-Qaramous (Ghazala et al. 2003), and at Tell el-Dabᶜa (Bietak et al. 2007). Survey at the archaeological site of Schedia (Bergmann & Heinzelmann 2003; 2004) to the east of Kafr el-Dawar in the Western delta has shown how large areas can be surveyed using this method. Magnetometry works well as a stand-alone technique for archaeological field survey, on a local and regional scale, but can also play an integral role in sites that have been chosen for excavation. A good example is the work carried out at Myos Hormos on the Red Sea Coast (Peacock & Blue 2006: 34ff).

When a magnetic survey is conducted, a grid system is used (for example using a pre-existing site grid), but the grid should be angled at a tangent to the archaeology. For example, should north-south, east-west structures be expected, the grid should then be oriented northeast-southwest. The surveyor, who should not be wearing any clothing or jewellery containing metal, nor carrying anything metallic (keys, spectacles or metal eyelets in shoes), walks in traverses across the grids, always in the same direction (Herbich & Spencer 2006: 18). This is 'parallel' mode and is preferable to the (cheaper) zigzag mode as it avoids the data loss that can occur using the latter system. A (non-metallic) tape is placed at the starting and finishing end of each grid square (typically 30 × 30 m or 40 × 40 m) and if readings are being taken at 0.25 m intervals, then strings are laid in the walking direction at 1 m, 2 m, and moved at 1 m increments as the grid square is completed (Herbich & Spencer 2006: 18). The surveyor walks first to the left of the string, returns, and then walks to right of the string. Two strings are normally on the ground at any one time and these are moved along by 1 m after every traverse is completed (Figure 2.4). If, however, readings are being taken only at 0.50 m intervals, then the strings are placed only at odd numbers, 1 m, 3 m, and the same procedure followed as for 0.25 m intervals, but with only half the density of traverses being walked (Aspinall et al. 2008: 33; Rowland & Strutt 2011: 332).

The use of the technique is well illustrated with recent survey work undertaken at Deir el-Barsha. At this site, a magnetometer survey was conducted over the area of a necropolis consisting of mud-brick structures cut into a cone of sediment deposited by a wadi (Herbich & Peeters 2006: 14). The survey results demonstrate the clarity of the response to the mud-brick burial shafts located close to the modern ground surface. These showed clearly against the finer fluvial sediment of the surrounding sediment. They also indicate the advantages in undertaking a higher resolution survey at a site where a good response to buried archaeological features could be expected. The survey was carried out on a series of 0.5 m traverses with measurements taken at 0.25 m intervals, providing higher resolution in the resulting survey data. Traverses taken at such close intervals provide high-resolution survey data, thereby having greater possibilities for data processing, and ultimately enabling the plan of the results to deliver much more information.

2.2.2.2 *Methods of Measuring Earth Resistance using Electrical Current*

What happens when magnetic survey is not suitable, when evaluation of deeply buried features is needed, or where the archaeology is within a modern village? What methods will work in these instances? In such cases, one of the methods based upon the conductivity of electricity can be used. There are multiple methods which rely on the conduction of electricity to detect sub-surface features. The main methods are earth resistance survey and electrical resistivity tomography (ERT), including vertical electrical soundings (VES). For the first, the resistance in a given volume of earth is measured in ohms (W), whereas the second gives apparent resistivity values in ohm-metres (Ωm).

2.2.2.2.1 EARTH RESISTANCE SURVEY

In what contexts, and for what specific purposes, is it advisable to use earth resistance survey? The methodology for earth resistance survey is based upon the ability of sediments to conduct an electrical current (Clark 1996: 36; Schmidt 2013)). Although all materials allow an electrical current to pass through them to some extent, depending on the composition of the material, the electricity will meet with greater or less resistance. There are extreme cases of conductive and non-conductive material (Scollar et al. 1990), but differences in the structural and chemical composition of sediments mean that they offer varying degrees of resistance to an electrical current (Clark 1996). How does this inform about archaeological features? Although the earth resistance results might not be able to inform as to the exact material type beneath the ground, the results show the range of different apparent resistances and their relative positions beneath the surface. Some archaeological features (such as those comprising stone or brick) have a higher resistance than typical sediments, whereas others (such as ditches which may contain water or organic deposits) tend to have lower resistivity and thus conduct electricity better (Clark 1996: 36).

Whereas magnetic survey involves the surveyor walking in a single direction across multiple grid squares without the machine itself touching the ground (a passive technique), resistivity involves a number of probes being pressed into the earth (an active technique). It functions by passing an electrical current from these probes down into the ground to measure variations in resistance at points over a survey area. Passive methods do not require the induction of a phenomenon in the ground to measure some specific property of the soil (e.g. magnetometry), whereas active methods involve the creation of a process that makes it possible to measure the given property (radar [involves the introduction of an electromagnetic wave], seismology [requires the creation of a wave]) (Gaffney & Gater 2003: 25-26). The application of resistivity survey usually takes one of two forms:

1   The use of a mobile probe array with a set spacing for taking multiple measurements over an area at a fixed depth to produce a plan of

resistivity values. In standard resistivity surveys, a fixed probe array is utilised, either affixed to a frame for the taking of measurements or built into a mobile system such as the French RATEAU (Querrien et al. 2009: 195; Scollar et al. 1990: 344). Measurements are taken on traverses at equal intervals on a grid system, and the data are used to produce a plan of the changes in resistivity values across the survey area.

2   The use of expanding probe arrays used in profiles to create resistivity tomography sections through an area. This method can be used for long transects, such as through village streets, or across long tracts of uncovered land. The use of resistivity in combination with augering will be revisited in Section 2.3: The Environment. This method of survey has been applied with a degree of success at locations in the Nile Delta, e.g. at Kom el-Ahmar (Minuf) (Box 2.5; Rowland & Strutt 2011: 335–338) and at other locations such as Kharga Oasis (Atya et al. 2005; Kamei et al. 2002), Thebes and Karnak (Graham et al. 2012a; 2012b) and particularly well at Tell el-Daba (Herbich & Forstner-Müller 2013).

2.2.2.2.2   ELECTRICAL RESISTIVITY TOMOGRAPHY, INCLUDING VERTICAL ELECTRICAL SOUNDINGS

Electrical resistivity tomography (ERT) is designed to produce an unbroken profile of resistance measurements through an area, and can take readings at much greater depths than magnetometry (Schmidy 2013). ERT can be utilised both in the form of vertical electrical sounding (VES), which focuses on a single point in a similar way to drill coring, and for the surveying of resistivity sections. The latter form has been widely applied at archaeological sites across Europe and North Africa, including at Saqqara (El-Qady et al. 1999). The application of VES is proving of particular significance where drill core samples form a significant part of the field methodology, as the drill core can be placed at exactly the same location as the VES (Ibrahim et al. 2002; Rowland & Strutt 2011). Work around Tell el-Dab'a using VES was able to prove which areas would have been flooded in antiquity, and which were suitable for settlement (Forstner-Müller 2008: 12–13; Herbich 2012a: 12). This demonstrates that it is a method equally useful for environmental data, enabling the reconstruction of the palaeoenvironment, and hence the development of better understanding of the reasons for settlement foundation and movement over time (as discussed further in Section 2.3: The Environment). The results of the VES are subsequently ground-truthed to discover the exact nature of the archaeological features, natural features and sediments identified through the VES. It is possible to take a number of profiles across a chosen area using VES, such as along the streets of an inhabited *kom*, or stretches along the Nile floodplain (Bunbury et al.

Box 2.5   Topographic Survey
Kristian Strutt

The modern ground surface or topography often contains important information on the conditions and nature of an archaeological site, and the potential existence of structures buried beneath the soil (Bowden 1999). Changes in topography can also have a great influence on determining the nature of features in other forms of survey, such as geophysics. Therefore it is vital to produce a detailed and complete topographic survey as part of a field survey. This can involve survey of extant archaeological remains, and their relationship with modern features, or a broader topographic survey recording elevations across a site, sometimes referred to as an earthwork survey. This generally entails the recording of elevations across a grid of certain resolution, for instance 5 or 10 m intervals, but also the recording of points on known breaks of slope, to emphasise archaeological features in the landscape. To fully comprehend the need for and practice of topographic survey it is, however, critical to understand the role of such techniques in establishing survey control over a site or landscape.

Establishing Control

At the simplest level a theodolite or total station survey can be conducted purely through the setting up of a baseline between two points, and surveying using angle and distance measurements to locate other points. However, many sites or landscapes at a local, regional or national scale, require more control over any potential errors derived from the process of surveying. To do this many theodolite or total station surveys use a system of stations in a control survey network, either through a system of triangulation or through the setting out of a survey traverse. With the former, the points of a baseline are used with intersecting angles to build a network of survey stations, with any error calculated and distributed through the network. With the latter, a series of foresights and backsights are used to establish a traverse which is then closed onto the original starting station (a closed loop traverse) with the misclosure or error being calculated and distributed around the traverse (Bannister et al. 1998). Many countries have their own survey network, and Egypt is no exception, with the Survey of Egypt network established in the late 19th and early 20th centuries (Lyons 1908). Survey teams may choose to utilise these systems; however, with the age of some equipment systems and the removal of old survey points from the landscape, many projects choose instead to establish their own control (Graham et al. 2012; 2014). The use of differential or Real Time Kinetic GPS has, to a large degree, removed the need for such systems of topographic control, with GPS survey being dependent on base stations and the location of satellites.

Site Survey

Once a control has been established the archaeologist can utilise survey instruments to carry out topographic survey. With either GPS or total station, or older techniques such as theodolite or plane table surveying (Bowden 1999), the aim is to measure accurately a series of points in terms of their location (2D), and potentially their elevation (3D). These points may indicate spot elevations across the survey area, or points in the line of a wall or structure that relate to one another. With plane table, theodolite and total station survey a series of angles and distances are measured, giving a horizontal angle, sometimes a vertical angle, and a sloping distance. These values are used to calculate a coordinate location and elevation relative to the position of the survey instrument. For GPS survey the position of a survey point is based on the relative location of satellites, and the relative position of the GPS base station and the rover are used to take measurements.

For survey of possible earthworks the instrument is used to take a series of elevation points on a grid at a set approximate interval, for instance every 4 m, 6 m or 10 m depending on the area being surveyed and the nature of the features. However, more crucial to a successful topographic survey is the recording of breaks of slope in the topography of the site, the points at which the gradient in a slope

Figure 2.5 Topographic survey being undertaken (Photograph courtesy of Kristian Strutt and Joanne Rowland)

changes, indicating the form and nature of features in the topography. For this a series of points along the top and bottom breaks of slope need to be surveyed. In traditional survey these were recorded using a plane table, and were inked up with hachures showing the position of the top and bottom breaks of slope, with the head of each hachure at the top and the tail of each hachure at the base, and the steepness of the slope indicated by the spacing of the hachures. A final plan of extant features and changes in the topography can be produced in this way. With modern survey equipment much of the data visualisation is now done on computer, with the production of DEMs (Digital Elevation Models) of topography with other features, such as buildings, superimposed.

Topographic Survey and Remote Sensing

The effectiveness of topographic survey needs to be considered in relation to many of the other techniques mentioned in the text, as topographic survey provides excellent complementary data for other forms of survey. The topographic uses of remotely sensed data are, however, of primary concern. In many cases the need for spot elevation survey has been superseded by remote topographic data collection techniques, in particular LiDAR data. Where LiDAR data can be sourced at 1 m or 0.5 m intervals then the survey of spot heights is unnecessary. For the purposes of archaeological interpretation, however, the recording of breaks of slope in a terrestrial survey is still preferred as the best method for assessing and mapping archaeological remains.

References

Bannister A., Raymond S. & Baker R. (1998) *Surveying. Seventh Edition.* Harlow: Longman.

Bowden M. (1999) *Unravelling the Landscape: An Inquisitive Approach to Archaeology.* Stroud: Tempus.

Graham A., Strutt KD., Hunter MA., Jones S., Masson A., Millet M. & Pennington BT. (2012) Theban Harbours and Waterscapes Survey, 2012. *Journal of Egyptian Archaeology* 98: 27–42.

Graham A., Strutt KD., Emery VL., Jones S., & Barker DB. (2013) Theban Harbours and Waterscapes Survey, 2013. *Journal of Egyptian Archaeology* 99: 35–52.

Graham A., Strutt KD., Hunter MA., Pennington BT., Toonen WHJ. & Barker DS. (2014) Theban Harbours and Waterscapes Survey, *Journal of Egyptian Archaeology* 100: 41–53.

Lyons HG. (1908) *The Cadastral Survey of Egypt 1892–1907.* Cairo: National Printing Department.

2009). For VES, the resistivity array is expanded over a single location to provide a column of changing resistivity values. The variations from each location are then interpolated to provide a schematic profile of the changing deposits across the landscape, as at Sa el-Hagar in the Nile Delta (Ghazala et al. 2005: 123ff). The use of VES in conjunction with ground-truthing can help to reconstruct the formation processes of an ancient mound (*kom*).

Electrical Resistivity Tomography (ERT) has been used successfully in a number of urban locations for detecting large-scale buried structural remains (Bunbury et al. 2009; Graham et al. 2012a; 2012b; Shaaban & Shaaban 2001; Strutt et al. 2012; 2014; Sultan 2004). ERT uses multiple probes in an expanding array, with the probes moved further apart each time a profile is traversed. It is not necessary to only move in one direction along a survey line or 'traverse'. With ERT, for example, a street could be selected as a profile; the first readings are taken moving north to south, with readings also taken on the return over the same line, from south to north. The traverse is then moved. With each traverse, the distance between the probes ('the probe array') is expanded to increase depth, allowing the variations in high and low resistance measurements to be used in the production of a model of the data (El-Gamili et al. 1999; Loke & Barker 1996a; 1996b; Noel 1992). The number of traverses is directly proportional to the depth of reading required (Rowland & Strutt 2011: 335). Due to the increased spacing, the results will be truncated at either end, as seen in Figure 2.6. It is common to use four probes for electrical profiling (Gaffney & Gater 1993; Gaffney et al. 1991): two current and two potential probes. Survey can be undertaken using a number of differently named probe arrays, such as twin probe, Wenner, Double-Dipole, Schlumberger and Square arrays. One application of ERT within a heavily built-up environment was the survey in Islamic Cairo, which aimed at locating structures associated with the mausoleum of Al-Ghouri (Sultan 2004). It has also found applications in the fields of geomorphology and geoarchaeology (Abbas et al. 2004; Bunbury et al. 2009), either in relation to engineering work and the rapid development of parts of Egypt such as Sinai, or as a part of an extensive archaeological research programme. Other successful applications of ERT include investigations of ancient *kom* sites that are heavily built over with modern structures (Gaber et al. 1999; Rowland & Strutt 2011). In such locations, magnetic

Figure 2.6 ERT transverses demonstrating truncation due to increased spacing

survey of narrow streets lined with modern buildings is impossible as too much interference occurs from the buildings and associated utilities for successful magnetic survey.

2.2.2.3 Ground Penetrating Radar

Two groups of methods have so far been examined; the first, magnetic survey, produces a horizontal plan of results, while the second, methods of earth resistance and ERT, produce horizontal and vertical profiles respectively. Ground penetrating radar (GPR) can be compared with the latter as it collects data as a series of profiles that can then be processed to give horizontal plans of the survey area. GPR has been used increasingly widely since the 1980s and there have been major advances in terms of data collection and processing (Conyers 2012: 13). The method relies on the propagation of an electromagnetic radar wave through the sediment in order to search for changes in sediment composition and the presence of structures. GPR measures the time taken in nanoseconds (ns) for the radar wave to be sent and for the reflected wave to return. The propagation of the signal is dependent on the relative dielectric permittivity (RDP) of the buried material (Conyers & Goodman 1997). As with other methods, GPR detects sediments, structures and objects with different properties beneath the surface, but it is the interpretation of these readings that presents the greatest challenge (Conyers 2012: 13–15).

The method is based upon the radar being moved along a traverse and then back again in the opposite direction to collect a series of profiles. As with the other methods, the traverses are taken within a survey grid. The GPR cart or sled is pushed or drawn initially in one direction along one side of a tape measure, before returning along the other side of the measure, etc. It is effective over large areas and, depending upon the surface, the radar sled or cart can cover ground quite quickly. The radar box, which takes the actual readings, can be mounted in various ways. These include a SmartCart (Rowland & Strutt 2011), to which the radar box is mounted and the cart is pushed along (see Figure 2.7), or the radar box can be housed within a small casing with wheels so that the radar box sits just a few centimetres above the ground and is pulled along (Conyers & Goodman 1997: 24). The radar box can also be placed inside a housing on a sled and pulled across the ground (Conyers 2012: Figures 2 & 3; Conyers & Goodman 1997: 24). The method is particularly helpful where there is good differentiation between the sub-surface features and sediments. It has proved effective in Egypt within very sandy contexts that are not close to the water table (see Rowland & Strutt 2011). In lower lying areas, however, within alluvial silt environments, it has not proved to be so effective (see Conyers 2012: 17–18; Conyers & Goodman 1997: 197–200, Table 6 for description of the feasibility of GPR in certain environments). Because of the way in which the radar box

Figure 2.7 A SmartCart with 500 MHz attenna in operation at Gurob, being pushed by Rais Omer Farouk el-Quftawi (Photograph courtesy of Kristian Strutt and Joanne Rowland)

is pulled, or dragged, it can also be difficult to use where the topography is challenging, although the results can nonetheless still be good; its use on very steep slopes, however, where the elevation of the terrain changes dramatically along a transect is more problematic, such as encountered during the Tausret Temple Survey at the embankment to the modern road (Conyers & Goodman 1997: 2–64, 201; Creasman & Sassen 2011: 152–153). GPR is suitable for application in urban environments where deep stratigraphic deposits are present. The deep deposits in the city of Alexandria, and the modern design and construction of the city streets, meant that the use of low frequency (50 Mhz and 80 Mhz) GPR antenna was critical to the discovery of archaeological remains below the modern city (Hesse et al. 2002).

The technique has a wide variety of applications for archaeological sites. It is commonly used to target specific areas of interest where features have already been located using other geophysical survey techniques (i.e. magnetometry or resistivity). GPR, however, brings an added dimension to both of these techniques. Not only do the results provide both horizontal and vertical plans, but they also provide amplitude slices at specific depths beneath the surface, which can be subjected to 3-D mapping (Conyers 2012: 19; Conyers & Goodman 1997: 165–166, plates 4–8). As a result, it is possible to detect firstly the roof of a structure, then to detect the individual walls of the structure beneath the roof, and eventually the floor level, with potentially either another structure or virgin sand below. It is possible to mount a viewing screen, e.g. where a cart is being used, which means that the profile being traversed below is visible immediately on the screen. This can show, for example, hyperbola created by buried objects or features, during the course of the transect (Conyers 2012: Figures 2–4; Conyers & Goodman 1997: 30). This can be of great practical benefit if a quick test is required over a certain area before systematic survey begins.

The frequency of the antenna dictates the depth to which the signal will propagate, with high frequency antennae giving shallower propagation, and low frequency providing greater depth. This is, however, a general rule of thumb. The relative dielectric permittivity (RDP) of the material being surveyed will dictate the velocity of the radio wave, and the amount of energy expended in propagating through the material, and therefore the depth at which the signal of the GPR will attenuate or become too weak to be able to give a recognisable reflection.

2.2.2.4 Seismic Survey

Some geophysical survey methods are more appropriate than others for detecting features that are deeply buried, but is there a method that can be applied to deal with very deep structures, or geological features, that are heavily overlain with buildings, or so deep that other methods cannot detect them. This is seismic survey, a technique that can detect features down to 15–30 m. The method has neither been widely adopted in Egypt, nor widely used in archaeological geophysics. This is because the method, in general, is limited by the nature of thearchaeological deposits and the spatial resolution of the obtained results. Seismic survey functions on the principle of measuring the elasticity of buried deposits through the passing of a low frequency signal through the ground, and measuring the time it takes for the signal to arrive at a series of deliberately placed receivers, termed geophones (Kearey et al. 2002: 49; Telford et al. 1990: 186). The signal is normally generated either by a sledgehammer striking a metal plate or by a weight dropping onto a metal plate, with a sensor measuring the frequency of the impact.

The resulting seismic energy radiates out from the transmission point, and the time taken for it to reach the geophones is measured (Kearey et al. 2002: 49). The arrival time and strength of the seismic signal is recorded by a plotter with a trace line drawn for each geophone. The most notable use of this technique has been by Hesse and colleagues (2002) in Alexandria in locating parts of the ancient city, and the seismic profiling of mastaba at Saqqara (Metwaly et al. 2005). The accuracy and nature of seismic survey lends itself well both to locating variations in geological and geomorphological facies, and also locating substantial and deeply buried archaeological structures.

2.2.3 3-D Site Reconstruction

3-D reconstructions can be made of objects, sites and landscapes. 3-D laser scanners are used, as are other photographic methods including Reflectance Transformation Imaging (RTI) which is discussed in Box 4.9 (Orton & Hughes 2013: 198–200).

Digital Elevation Models (DEMs) can be made for sites, as they are today, and the sub-surface remains can be assessed to a great depth, but can we really get a good impression of what the sites looked like in the past? Can 3-D reconstructions be accurate and, moreover, useful? New analytical techniques integrate common functionalities of 3-D modelling software with GIS, permitting viewshed analysis to be undertaken on objects of any form and shape modelled in 3-D. This is currently most successful for extant and partially preserved historic and prehistoric built structures, and can be used widely from considering building interiors to townscapes and landscapes (Paliou et al. 2011). Although not yet fully employed in Egypt, the approach has been used to explore visual perception issues associated with the urban remains of Late Bronze Age Akrotiri in Greece (Paliou et al. 2011).

2.2.4 Considerations

"So how do we decide what to use and in what order?" A series of potential questions are presented in the Appendix. These or similar questions must be asked in order to decide which methods can be applied to general Egyptian contexts and in what order they should be undertaken. These are obviously limited by cost and access to equipment and so this section might be considered as an idealised situation and should be modified in response to actual feasibility.

2.3 The Environment and Palaeoenvironment

Is the environment in Egypt today a good indicator of that of the past? Was it as hot and dry in the past as it is now? How did the Nile behave before the Aswan High dam was built? How and when did people start to practice agriculture, and what type of plants did they grow? What types of animals were hunted in the deserts and how do we know how people used animals within domesticated contexts? There is a myriad of questions that can be asked about Egypt's ancient environment and, as will be seen throughout this section, there are different answers for these questions depending upon the time period being looked at, or the exact geographical location. Egypt is a large and diverse country in terms of its climate, its resources, and the possibilities for irrigation and transport offered by the Nile. From Aswan in the south to the Mediterranean coast in the north, it is a distance of over 850km as the crow flies, and the different opportunities and challenges for mobile and settled communities living in different geographic regions during prehistoric and historic times are stark. When trying to understand the environment of ancient Egypt, it is perhaps helpful not to focus too much on the symbolism of 'The Two Lands of Upper and Lower Egypt', or even the 'Red Land' (the desert) and the 'Black Land' (the fertile band along the Nile Valley and the Delta), but rather to think about the actual effects that diverse environments had on the people who lived in Egypt over the millennia.

There is a huge geographic and regional perspective to consider, as well as the temporal scale. The preceding section on absolute and relative chronologies (Section 2.1: Time) has already introduced methods for working with time and establishing specific points on the temporal scale, but here the intention is to start exploring the extent to which environmental change and corresponding technological development occurred in Egypt, examining research questions and examples associated with the prehistoric and the dynastic eras. It is this diversity and fluidity over time and space that *is* Egypt and which created the possibilities and challenges faced by the people within her borders. Likewise, it is this Egypt today that presents the challenges of a changing environment, rising water tables, increasing humidity, as well as expanding agriculture and housing to meet the demands of the modern population. These issues present great challenges for the preservation and protection of Egypt's cultural heritage. This diversity must be remembered, especially when looking at publications where sites are plotted on static maps that do not take into account either the movement of the main Nile branch/Delta branches over time. Maps are often based on a single (often modern) course of the river Nile flowing along the Nile Valley, bifurcating at approxima the location of modern Cairo and running northwards to the Mediterranean coast as two branches (as in recent maps). Some volumes, notably Baines & Málek (1980), do present a series of very useful maps that include ancient environmental features and the differences over time. Although the task of creating accurate maps of ancient Egypt, taking into account the shape of the ancient coastline, the higher number of Nile river branches, and the fluctuating position of the main Nile, is a difficult one (Said 1993: 85), there has been much research conducted since the 1970s aiming to prove and disprove various environmental hypotheses (ancient Delta shorelines are shown in Said 1981: Figure 53). In order to stand a good chance of understanding the actions of people of diverse ethnic backgrounds who populated Egypt throughout its past and who are responsible for its diverse archaeological record, recourse to the ancient environment is essential. This section looks in detail at various ways in which the ancient environment can be reconstructed at various points in time, focusing upon the Nile and its waterways, and the flora and fauna that were available in wild and domestic contexts.

2.3.1   Environmental Reconstruction

Depending upon the period of time under investigation, the question 'what was Egypt's climate like in the past?' will bring different responses. The most extreme shifts in climate were seen during the Pleistocene, and these events affected Palaeolithic groups (c.700,000–7180 BP), as these dramatic fluctuations drastically changed the environment. The focus of occupation in northeast Africa changed as the climate transformed the environment from desert to savannah and back again, in addition to radically changing the water level in the Nile basin (Vermeersch 2006). As a result of the desiccation of the Sahara at 7350 BP there was a movement of peoples from the desert because it was too arid for groups to survive year round (Bubenzer & Riemer 2007). From this time onwards, up until the present day, although Egypt has maintained a roughly similar overall climate, there have been many short-term fluctuations that have favourably as well as unfavourably affected the population.

There are several disparate and distinct types of environmental evidence that can be examined to reconstruct the ancient environment or the palaeoenvironment. Using these distinct data sets, comparisons and analyses can be made to investigate whether there is evidence for environmental or climatic change, either -sudden or gradual. It is possible to use botanical (plant remains), zooarchaeological (faunal remains), human skeletal remains (as discussed in Chapter 3), as well as sediments (described below), to develop environmental reconstructions.

2.3.1.2   The Case of the Nile Delta

As mentioned in Section 2.2: Finding Sites, one of the challenges to archaeologists is the reconstruction of landscapes that are changing rapidly due to the fast rate of population expansion and the accompanied increase in construction and agriculture (Said 1981: 1). This affects not only the archaeology but also the geology of the region (Said 1981: 1). The Nile Delta, one of the most heavily affected regions, has benefitted from the upsurge in environmental research in Egypt since the 1970s, although research into the river and its origins, both within and outside of Egypt's modern borders, extends back to 1896 (Said 1981: 8). It was also only in 1976 that the term 'geoarchaeology' was introduced; more recently a volume has been published that examines how geoarchaeology can help to provide information on the actions of humans in the past (Goldberg & Macphail 2006). Before discussing the scientific methods applicable to the study of the ancient environment however, it is imperative to remember the importance of combining ancient Egyptian textual sources (as discussed in Bietak 1975: 117–139) with additional written sources and maps (e.g. Herodotus II; Strabo 17.1), and these together with more recent historical accounts and maps. Such documents can then be combined with scientific methods. Bietak examined the palaeogeography of the Delta, taking into account the impact of the ancient landscape on the location of settlements within the region, including the location of sand *geziras*, and the reconstruction and dating of the ancient Nile branches (Bietak 1975: Figures 23–25, 27). Stanley (and colleagues) (e.g. Stanley & Warne 1993a; see list of contributors in Rowland & Hamdan 2012: 12; core locations up until 1981 are listed in Said 1981: Figure 2 & Appendix A) have extracted the largest collection of cores across the northern Nile Delta as part of the Smithsonian Institute's Nile Delta Project (see location of cores in Stanley & Warne 1993b: Figure 1, 436), and Wunderlich has worked in both the eastern and western Delta, using an approach based on satellite data and a motor-driven gouge auger in combination with electrical soundings (Andres & Wunderlich 1992; Wunderlich 1993). Many samples have been extracted in the Delta in the course of prospection for water and oil (Said 1981: 93 & Figures 1–3; 1993: 32). The collation of data obtained since the 1970s provides information on the location and size of the sand *geziras* that formed in the Pleistocene period. Today these ancient *geziras* mainly lie beneath the accumulated silts of the Delta. These sand hills provided some of the most suitable locations for settlements and also cemeteries as they remained above the level of the annual inundation. During the flood part of the year, much of the Delta could have been under water and traversable only by boat. Such programmes of coring and hand augering have improved our understanding of the formation and location of the *geziras*, including the rare examples which still rise above the surrounding landscape (Rowland & Hamdan 2012). In turn, this research has enabled scholars to produce maps that provide a much clearer idea of the ancient Delta landscape, including the multiple river branches that existed in antiquity (Bietak 1975: Figure 23; Butzer 1976: Figure 4; 2002; Hassan 1997: Figure 4; Said 1981: 80 and Figure 52; 1993: 32). Deep coring has also investigated and delineated both the modern and ancient coastline (Stanley & Warne 1993a; 1993b). Furthermore, along the length of the Egyptian Nile Valley, coring and hand augering has been carried out by geologists and archaeologists in order to answer a variety of inter-disciplinary questions (see well and core data in Said 1981: Figures 1 & 2, Appendix A). Examples relating to this work are discussed below.

2.3.2 Obtaining Environmental Data

How do we know what the climate was like in the past? Was it vegetated or desert? These types of questions can be answered using a variety of different scientific techniques. Aspects of the local environment can be deduced from studies of the faunal remains or from plant macro and micro remains. These will be discussed later (see Section 2.3.3). Probably the most common mechanism by which environmental data is derived in Egyptian archaeology is through the analysis of sediment samples obtained from drill cores and augers.

2.3.2.1 Sediment Extraction in Research

Augering and coring can be used in conjunction with other site prospection methods, e.g. geophysical survey techniques and surface survey, in order to ground-truth and assess the remaining sub-surface deposits, including the chronological profile of the site. It can also be used alongside the results of transects of Electrical Resistance Tomography (ERT), for example (as described in Section 2.2.1: Remote Sensing and Egyptian Archaeology). Bunbury et al. (2009) used a combination of ERT and hand augering (see Box 2.6) to investigate the possibility that Kom el-Farahy in the Edfu floodplain used to be an island in the Nile. The ERT profile confirmed this suspicion and the location of the *kom*, and this was positively tested using the hand auger, with datable ceramics present in the auger samples. The ceramics themselves can give important information as to the chronology of the ancient site, and the sediments can indicate the migration of the river away (eastwards) from Kom el-Farahy (Bunbury et al. 2009).

Cores can be used to help detect the presence of archaeological material beneath the surface, ground-truthing the results of geophysical survey. In addition, they can provide important data as to the surrounding environment at the time that these sites were in use, as well as changes in the environment subsequent to the site's use or even prior to its usage. This helps build up a clearer hypothesis as to why a site might have been founded, why it flourished and the subsequent reasons for its abandonment. Due to the yearly inundation before the raising of the Aswan High Dam, the archaeology is often deeply buried by alluvial silts deposited during the flood. As a result, techniques such as augering are required, followed by test trenching, if necessary and/or possible, to access the archaeology below and within these deposits. On average, in the Nile Valley there are 9.8 m of silts above the archaeological layers, and in the Delta c. 12 m; it can however be significantly greater (Hassan 1997: 63). Outside the valley and Delta, however, in the desert regions there is not such an issue. Drill coring, however, is still employed by archaeological teams in order to recreate ancient local environments.

Although the Delta suffers from a very high water table today, and often the thickest deposits of alluvial silts, it also has huge potential for bringing new information to our understanding of various phases of Egyptian prehistory and history. Since the 1970s, increasing work has been carried out in the region; this work is notable for bringing much to the knowledge of the ancient environment. Archaeological fieldwork can be more labour intensive as de-watering equipment is sometimes required to cater for the problems of the high ground water, but the research potential is great. A good example of the use of varied methods to reconstruct the ancient environment within one region of the Delta is the work carried out in and around Sais in the western Nile Delta (Wilson 2006). Through collaborative inter-disciplinary work carried out with the University of Mansoura using VES and hand augering, a prehistoric settlement was found to be located on a river levee on the inside of a major river channel, with marshes and sand *gezira* detected in the wider area (Wilson 2006: 75, 77, Figure 1, 102–3, Figures 13–18, 106). It is important that an augering programme is carried out systematically and as objectively as possible; for example, around Sais, the placement of auger locations occurred in transects with roughly equal spacing (Wilson 2006: 76, Figures 14 and 15). This helps ensure even coverage over the area, and facilitates comparison of the sedimentary data between the auger locations over space (e.g. Wilson 2006: 77 Figure 16). If further detail is needed, then additional cores can be placed in order to fill in this detail.

During the augering survey at Sais, prehistoric ceramic sherds were brought to the surface. In one core, these were at a depth of 7.0 m below the surface (Wilson 2006: 80; Wilson & Gilbert 2003: 65). Excavation, using de-watering equipment, confirmed stratigraphic deposits through the Neolithic and then after a c. 300 year hiatus, the development of the Predynastic (Wilson 2006: 75, 83). Given the proximity to Merimde Beni Salama to the southwest, this substantially alters the view of the Neolithic in the Delta, especially with regards inter-site contact (Wilson 2006). The results from the excavation suggest a change in the environment between the early Neolithic and the Predynastic, with the action of a river channel causing the later material to wash down onto the earlier layers (Wilson 2006: 89; Wilson & Gilbert 2003: 71). Visual examination of the layers of sediment between the Neolithic and Predynastic phases indicated that there was a very arid period followed by the return of the flood waters and subsequent re-settlement (Wilson 2006: 101). Examination of the sediments can clarify whether the surface had dried out and been exposed to air for a long time through their orangey-brown colouration, and conversely, when thick black mud is present, as at Sais, that there was an anaerobic deep water channel (Wilson 2006: 103).

The data from these cores is just one example of how to recreate the different local environmental conditions at a given site, conditions which impacted, in some cases severely, upon the settlers. But how can one be certain about the dating of specific events and sedimentary layers found (in particular, in cores and augers, but also in excavation)? Sometimes examination of the sediments in which ceramic sherds are found, the assemblage within the sample and the examination of the abrasion or rounding of the sherds themselves, can help clarify whether the sherds have a direct contextual relationship to the sediments, and whether relative dating methods can be employed (Graham et al. 2012a: 31–32). It is, however, not uncommon for sherds to be drawn into lower layers when using a hand auger, hence the association between sedimentary deposits and artefacts is not always straightforward. This can be especially common if the ceramic fragment is on the very outside of the auger sample. When using a continuous coring method, as detailed below, if there are organic remains (e.g. charred seeds) within sediment, it might also be possible to extract these for ¹⁴C measurement (this would not be advisable using a hand auger). Through the combination of such methods, valuable plant and faunal remains data can be collected. These can add new data to the questions of how, when and where these species were introduced into Egypt, as well as recognising which species can thrive in which environmental conditions.

2.3.2.2 *Methods for Extracting Sediment*

Various types of equipment are used for hand augering and coring, but what is the difference between them – and what equipment is better suited to which investigations? The different types of methods are explained briefly here, thereby contributing advice as to what might be appropriate for specific research questions or a timeframe for work.

The types of equipment available can be divided into three main types: interrupted non-continuous hand augers, continuous coring machines (manually powered) and continuous coring machines (motor operated). Each have their advantages and disadvantages (as listed, for example, in Tassie & Owens 2010: 47), and a combination of continuous coring (manual or motor driven) with hand augering can produce very satisfactory results.

When deciding which 'tool' might be most appropriate for which job, we can ask a number of questions relating to the type of investigation being undertaken, the scale of the work, and even the funds available. Such questions include:

1   Is a continuous record of the stratigraphy needed, or are interrupted samples from certain depths preferable?
2   Is it important that the samples remains closed on extraction and uncontaminated for laboratory analyses (including extraction of ¹⁴C/ OSL samples)?
3   Is it necessary/preferable to examine the sediments, artefacts and ecofacts from the samples in the field, to maximise our understanding of the contexts and to modify the work programme as required?
4   Are there sufficient funds to hire equipment/cover the salary of operators for continuous coring?
5   Is it possible to collect all the samples within a short period of time (if hiring equipment), or is the collection of data over a longer period of time preferable?

These are just a few of the questions that will affect choice. Ultimately, given that the types of equipment not only produce the samples in different sizes/forms, but also are more or less suitable for analyses in the field or the laboratory, and vary greatly in terms of cost, it is usually a reasonably easy choice to make. Some additional brief practical information is provided below on each of the different types of equipment (with additional detail in Box 2.6), information which will help clarify the applicability of different types of equipment for different research agendas, followed by details as to how to record samples, and what types of analysis can be performed on them.

Does this aid in the decision as to what type of equipment might work best? It is important to bear in mind that what might work best might comprise a combination of different methods of sediment retrieval. It might be wise to retrieve a larger number of hand auger samples initially, to get an impression of the local conditions in a cost-effective and expedient way. After the hand auger samples have been extracted, and where it might then prove helpful to read continuous core stratigraphy at selected locations, a limited number of continuous cores could be extracted with manual or motor operated equipment, thereby facilitating laboratory-based analysis. Alternatively, it is also possible to extract the continuous cores initially, examine these in the laboratory, and then 'fill in' the detail with a number of hand auger samples in between.

2.3.2.2.1 HAND AUGERING

As described in Box 2.6, this is the most widely used and most cost-effective sediment sampling method used in archaeology in Egypt today. It has been successfully applied to examine wide areas, often in transects at regular intervals (e.g. every 10 m, or 25 m or 50 m etc.), to track differences and changes in the sub-surface material. It can also be used at specific locations to answer specific questions, such as around the edges of an occupation mound (*kom*). The method is very simple. The equipment consists of a handle, extension poles of 1 m in length, and a selection of drill heads, which can be changed depending upon the type of sediment being extracted (e.g. predominantly sandy, silty or clayey). Once the location is chosen and the immediate area cleared, the total length from the underside of the horizontal drill handle to the point of the drill head is measured and recorded on the logging form being used. A note is usually also made of the GPS position of the core for plotting, together with a written visual observation of the location. The precise location enables the point to be placed within a GIS, and be directly related to geophysical survey data in the field. To start the extraction, the auger is rotated in a clockwise direction, pushing down into the ground until the required depth is reached; this could be every 0.10 m or 0.20 m depending on the data required. In order to determine the current depth of each auger sample, a measurement is taken from the base of the handle to the surface and then subtracted from the total length of the auger. The result is the depth below the surface. Once the required depth is reached, this exact depth is recorded and the auger is pulled up directly (not twisted up). The sample is then removed from the head and is placed on a clean plastic plate, with the number of the sample and depth recorded. Annotations are usually in a form such as KAHM01-1, 0.00–0.20 m, meaning that the auger location is the abbreviation for the site, here Kom el-Ahmar, it is the first auger location, hence '01' and the '1' indicates that it is the first sample to be extracted, together with the depth range of the extraction. The excess sediment around the handle and the head is scraped clean, at which point the sample can be photographed, with colour photo scale and details of the sample number included in the picture. The sample is removed and the description of the sediments and associated artefacts and ecofacts logged (e.g. see Goldberg & Macphail 2006: 316–334).

2.3.2.2.2   DRILL CORING

When deeper samples and continuous cores are required (10+ m), a number of different coring rigs can be employed, variably using percussion or rotational force, in order to push an empty casing of 1 m (or other length) into the ground. After each new depth has been reached, the machine removes the core from the ground either complete within its case, or, in the case of the 'Vibracom' or 'Terrarig' types of machines, the core is extracted and pushed directly out of the machine with the sediment initially exposed to the elements. If laboratory analyses are being carried out it is advisable to reseal the core immediately or, if enclosed, to keep it within the casing until the work is to be carried out. Permission can be sought from the Ministry of Antiquities to transfer samples to off-site laboratories. The cores that are extracted without casing have the advantage that the sediment and any artefacts or ecofacts can be documented directly; in addition, changes in comparison to higher levels in the bore hole or with other cores drilled in the vicinity can be observed immediately and the drilling strategy modified in the field if required, in a similar way as when using the hand auger.

2.3.2.2.3   MANUAL DEEP CORING RIGS

Other drilling equipment, including the 'Handle Casing' drilling machine, is operated by manual force and uses a combination of rotary turning and percussion. With this equipment, when the core is extracted from the ground it is contained within a metal cylinder comprising two longitudinal parts. It is possible to lay the casing horizontally on a clean surface and then tap it along the line of one of the joins in the casing so that it falls open. Depending upon the type of sediment, it either remains as a complete core or divides. At this point, if the core is opened, a field photograph is normally taken. Data on the field photograph normally include a scale, a note of which end is top and bottom, and a complete label including the depth below surface. If subsequent analysis requires laboratory conditions, to avoid contamination from airborne materials, the core can be wrapped as it is, or transferred (preferably not outdoors) into a length of plastic drain piping, labelled as to the core number, length, and the top and bottom end of the core, and wrapped for transport. It is also possible, at this stage, to examine the core and record details of any artefacts and ecofacts. If cores are being extracted for OSL dating, it is vital that opaque core liners are used during their extraction so that no light can get into the cores, and that they are immediately sealed (see Section 2.1.2: Absolute Dating). Any lithic materials within the cores can be examined for relative dating alongside the chronometric dating.

2.3.2.2.4   RECORDING CORES AND SEDIMENTS

In practice, with suitable training in operation and logging, the hand auger is a piece of equipment that is fairly simple to operate. However, the presence of someone with experience of logging sediments is advisable initially. The precise details/methods of recording are influenced by the specialism of the individuals working in the field; for example, methods will vary depending on whether the person recording is a geologist, a soil scientist or an archaeologist (as discussed in Goldberg & Macphail 2006: 321–322). There are certain recording criteria that should be followed (see Goldberg & Macphail 2006: 323–328, examples of recording given in Tables 15.6–7). These include the recording of the colour (or colours) of the sediment according to a Munsell colour chart (usually recorded when moist), a consistent system of describing the texture of the sediment, water content, structure, the degree of compaction or friability of the sediment, the actual composition and the sizes of the internal elements, their frequency and distribution, whether plant roots are present, and information on any artefact fragments as well as ecofacts (e.g. snail shells) noted in the core description. Reconstruction profiles can be drawn for each core in association with photographs taken of the cores as they are lifted. The method is slightly different when continuous cores are described. This often occurs in the laboratory, rather than in the field, and, as such, it is useful to note (and mark) the major divisions between different units within a single core, photograph the cores, and make the description. If not undertaken in the field, cores should be recorded as soon as possible after extraction so that they are still damp, since if left for a while the sediments can dry out, form hard clumps and change colour.

On site, sediment samples are also taken routinely during excavation and described for profiles within excavation areas. Samples can be cut out of sections (possible methods of doing so are described in Goldberg & Macphail 2006: 328–333), or extracted during excavation, so that each different context number may have an associated sediment sample. It is important that all samples are well wrapped and/or double bagged and labelled at the time of extraction, with photographs taken of the location from which the sample was extracted (Goldberg & Macphail 2006: 331). Examples of possible forms are easily and widely available, such as the Environmental Sediment Sample Registers, Sediment Sampling Forms and Environmental Processing forms in Tassie & Owens (2010: 196, 273, 274).

The exact treatment of the samples and the types of materials used during the sampling process – both for extraction and storage – depends on the type of analysis planned (e.g. for ^{14}C, samples should not be touched by bare hands and should not be wrapped in organic material; Goldberg & Macphail 2006: 328–333). Goldberg & Macphail (2006: 333) suggest two very prudent approaches when considering sampling during seasons/site visits: 1) always assume that it will not be possible to re-visit the site, and make sure that the sample collection approach is thorough, based on this premise, and 2) take many samples, regardless of whether or not analysis can be afforded at the time, or even when uncertain as to what analyses could ever be undertaken – samples can be stored for a long time until the questions and the funds are available. Samples can, of course, be tested using a range of analytical techniques to answer varied questions. These require a range of specialists, and it is important that archaeologists remain flexible and open to unplanned analyses that might become possible and/or relevant during planned tests (Goldberg & Macphail 2006: 333).

2.3.2.3 The Location of Waterways

In Section 2.2: Finding Sites, the importance of understanding the ancient landscape in Egypt has been stressed and one integral part of this is information relating to the locations of ancient waterways. The environment fluctuated during prehistoric and historic Egypt, but to what extent were the Egyptian Nile Valley and Delta fundamentally different in terms of how they looked and functioned? Egypt was and is dependent on the Nile for its agriculture – but how was it different in the past? What used to be an annual cycle, with phases of drought, flood (ranging from low, regular to high), subsidence and harvest, has been totally changed through the building of the dams at Aswan, most significantly through the raising of the Aswan High Dam. As recently as the end of the 19th century, there were catastrophic floods that swept entire villages away (Willcocks 1904: 70–72; Wilson 2006: 8). One such example, the village of Antuhi in the Delta, was identified on maps until the end of the 19th century, but is absent from modern maps. The villagers in the neighbouring village of Misjid el-Khidra believe it could be the village that was swept away by very high floods in the late 19th century (Rowland 2009: 38).

Box 2.6 Migrating Nile: Augering in Egypt
Angus Graham & Judith Bunbury

The landscape archaeology of Egypt entered the spotlight with Butzer's (1976) seminal work on the hydrology of Egypt. Since then augering has been used to explore the positions of ancient lakes, waterways and other areas into which sediment has been deposited, including dumps, canals and wadis, to determine how they have changed with time. We now know that the Nile is in constant flux and that aggradation of the Nile floodplain and the migration of the Nile through the natural process of meandering are observed throughout Egypt. Typical mean rates of aggradation are around 1 mm/year (cf. Ball 1939; Butzer 1959; Hassan 1997; Lyons 1906; Said 1993), while lateral migration of the river has a long-term average of around 2 m/year (Hillier et al. 2007; Lutley & Bunbury 2008).

Augering projects that have taken a regional view of past landscapes and waterways include the Amsterdam University Survey Expedition's work in Sharqiya province in the Eastern Delta (de Wit & van Stralen 1988; Sewuster & van Wesemael 1987) and the Survey of Memphis led by David Jeffreys. The 140 boreholes in the Memphite floodplain make it one of the best understood areas in Egypt (Jeffreys 1985; Jeffreys & Bunbury 2005; Jeffreys & Tavares 1994). Other projects focus on specific sites and their surroundings, e.g. at Buto (von der Way 1984; Wunderlich 1993), Tell el-Dab'a and Piramesse/Qantir (Dorner 1994; 1999; Tronchère et al. 2008, 2012) and Sais (Wilson 2006). In the Nile Valley, augering around Amarna (Parcak 2006), Coptos (Pantalacci 2009), Karnak (Bunbury et al. 2008; Graham 2010; Graham & Bunbury 2004; 2005), Edfu (Bunbury et al. 2009) and Hierakonpolis (Bunbury & Graham 2008) has been used to develop an understanding of the landscape architecture.

Hand-augering is a relatively intensive method of obtaining data and is most effective when used in conjunction with other methods of exploration. The selection of coring site locations can be strategically undertaken when maps, air photographs and satellite images are used to evaluate the landscape before augering commences. Cartographic survey can be extrapolated to provide a geometry but often no precise time-scale for past landscape change. A surface survey can be further extended by incorporation of archaeological data from monuments in the region, which can normally be assumed to have remained on mostly dry land since the time of their construction. Geophysical methods, such as Electrical Resistivity Tomography, can also be used to add a component of subsurface stratigraphy and extend observations into three dimensions.

The main advantages of using a hand auger are that it is light and easily transported, cost-effective, can be operated in restricted spaces including within archaeological excavations, and is minimally invasive (Rapp & Hill 2006). The Eijkelkamp or Dutch auger is designed to sample sediment to a depth of 6 m, but, depending on the type of sediment, can be used safely up to a depth of around 10 m. Retrieving medium to coarse sands from below the water table, however, remains problematic. The issue of downhole contamination can be eliminated by continuous casing or can be quantified by detailed study of the materials retrieved when they form a statistically large population (often some hundreds of items). For example, a single artefact may be conspicuous within a largely homogenous assemblage by its different abrasion patterns or chronological parameters.

Alternative methods of sampling sediments include coring and drilling. The former can recover a continuous column of sediment using a hollow cylinder, and provides a relatively undisturbed section. There is, however, some distortion from compaction and it is inappropriate for cultural deposits thicker than 3 m (Rapp & Hill 2006; Schuldenrein 1991; Stein 1986). Both the auger (screw bit for soft sediments) and the drill (diamond bit for harder sediments) use rotary or percussion action to penetrate the ground so the core is discontinuous (Stein 1991), but details such as laminations and mud rip-ups can still be seen. Although bioturbation of the sediments is common, these intrusions are generally on the millimetre to centimetre scale. In the case of river sediments, any intrusions do not disturb the overall stratigraphy in which beds are formed on a decimetre to metre scale, and flood and overbank deposits are on a centimetre to decimetre scale. Exceptions to this are fine flood laminae, which are often preserved but may be blurred by bioturbation, and the plough layer, which can affect the upper metre or two of the sediments destroying sedimentary structure.

In many instances, artefacts recovered by augering are not in the archaeological context of a room or floor deposit, but in the equally meaningful context of a marsh, river or other water-body that can reveal the prevailing environment of the site. Sedimentological records of the grain size, sorting, rounding, clast size and type, Munsell colour, percentage of macroscopic organic remains and other features of the sediment can all be made in the field as the material is retrieved. These attributes enable interpretation of the environment; in the case of a watery environment, a fast flowing river produces

Figure 2.8 Augering alongside Marie Millet's deep sondage SW of the Sacred Lake at Karnak to enable comparison of augered and excavated material and extend the record of sediments below the watertable (Photograph courtesy of Angus Graham)

sandy sediments or, at the other end of the scale, anoxic (dark and smelly) mud is produced in still water. At the same time artefacts (and arteclasts) contained in the sediment reveal the anthropogenic environment in which deposition occurred.

Wet-sieving all the sediment to 2 mm in diameter (phi −1) retrieves all clasts larger than sand-sized particles. (Larger clasts >4 mm diameter may be separated from those smaller for ease of handling and study.) The clasts can then be weighed to calculate the percentage of each sample (individual body of extracted sediment in auger head) that is comprised of clasts. They are then sorted by type into groups/species using tweezers. Ceramic fragments, chippings of building stone related to known construction, faience and other artefacts may all enable both *post quem* and *ante quem* chronological parameters to be placed upon successive deposits. Study of new types of object that join the assemblage through time, an analysis described by geologists as 'sequence stratigraphy', can also be used. In addition abrasion and rounding of sherds provide additional information about the depositional processes. Rhizocretions and other organic remains add further evidence to the interpretation of the environment. Combining these sedimentological results with the archaeology of a site makes it possible to construct a sequence of successive former landscapes as the sedimentary processes develop.

References

Ball J. (1939) *Contributions to the Geography of Egypt*. Cairo: Government Press, Bulâq.

Bunbury, JM. & Graham A. (2008) There's nothing boring about a borehole. *Nekhen News* 20: 22–23.

Bunbury JM., Graham A. & Hunter M. (2008) Stratigraphic landscape analysis: charting the Holocene movements of the Nile at Karnak through ancient Egyptian time. *Geoarchaeology* 23: 351–373.

Bunbury JM., Graham A. & Strutt KD. (2009) Kom el-Farahy: a New Kingdom island in an evolving Edfu floodplain. *British Museum Studies in Ancient Egypt and Sudan* 14: 1–23.

Butzer KW. (1959) Environment and Human Ecology in Egypt during Predynastic and Early Dynastic Times. *Bulletin de la Société de Géographie d'Égypte* 32: 43–87.

Butzer KW. (1976) *Early Hydraulic Civilization in Egypt: a study in cultural ecology*. Chicago and London: University of Chicago Press.

de Wit HE. & van Stralen L. (1988) *Geo-archaeology and ancient distributaries in the eastern Nile delta. Results of the 1987 AUSE survey in Sharqiya, Egypt. Reports of the Laboratory of Physical Geography and Soil Science* 34. Amsterdam: University of Amsterdam.

Dorner J. (1994) Die Rekonstruktion einer pharaonischen Flusslandschaft. *Mitteilungen der Anthropologischen Gesellschaft in Wien*, 123/124: 401–406.

Dorner J. (1999) Die Topographie von Piramesse – Vorbericht. *Ägypten und Levante* 9: 77–83.

Graham A. (2010) Ancient landscapes around the Opet temple, Karnak. *Egyptian Archaeology* 36: 25–28.

Graham A. & Bunbury JM. (2004) Pottery from the alluvial environments at Karnak North. *Bulletin de Liaison du Groupe International d'Étude de la Céramique Égyptienne* 22: 55–59.

Graham A. & Bunbury JM. (2005) The ancient landscapes and waterscapes of Karnak. *Egyptian Archaeology* 27: 17–19.

Hassan FA. (1997) The dynamics of a riverine civilization: a geoarchaeological perspective on the Nile Valley, Egypt. *World Archaeology* 29: 51–74.

Hillier JK., Bunbury JM. & Graham A. (2007) Monuments on a Migrating Nile. *Journal of Archaeological Science* 34: 1011–1015.

Jeffreys DG. (1985) *The Survey of Memphis I: The Archaeological Report*. London: Egypt Exploration Society.

Jeffreys DG. & Bunbury JM. (2005) Memphis, 2004. *Journal of Egyptian Archaeology* 91: 8–12.

Jeffreys DG. & Tavares A. (1994) The Historic Landscape of Early Dynastic Memphis. *Mitteilungen des Deutschen Archäologischen Instituts, Abteilung Kairo* 50: 143–173.

Lutley K. & Bunbury JM. (2008) The Nile on the move. *Egyptian Archaeology* 32: 3–5.

Lyons HG. (1906) *The Physiography of the River Nile and its Basin.* Cairo: National Printing Department.

Pantalacci L. (2009) Coptos. In: Pantalacci L. & Denoix S. (eds) *Travaux de l'Institut français d'archéologie orientale 2008–2009.* Cairo: IFAO. pp. 565–567. http://www.ifao.egnet.net/ifao/recherche/rapports-activites/

Parcak S. (2006) The Middle Egypt Survey Project, 2004–06. *Journal of Egyptian Archaeology* 92: 57–61.

Rapp GR. & Hill CL. (2006) *Geoarchaeology: the earth-science approach to archaeological interpretation,* 2nd edn. New Haven, CT & London: Yale University Press.

Said R. (1993) *The River Nile: Geology, Hydrology and Utilization.* Oxford: Pergamon Press.

Schuldenrein J. (1991) Coring and the identity of Cultural-Resource environments: A Comment on Stein. *American Antiquity* 56: 131–137.

Sewuster RJE. & van Wesemael B. (1987) *Tracing ancient river courses in the eastern Nile delta: a geo-archaeological survey in the Sharqiya province, Egypt.* Rapporten van het Fysisch Geografisch en Bodemkundig Laboratorium. Amsterdam: University of Amsterdam.

Stein JK. (1986) Coring Archaeological Sites. *American Antiquity* 51: 505–527.

Stein JK. (1991) Coring in CRM and Archaeology: A Reminder. *American Antiquity* 56: 138–142.

Tronchère H., Salomon F., Callot Y., Goiran J.-P., Schmitt L., Forstner-Müller I. & Bietak M. (2008) Geoarchaeology of Avaris: first results. *Ägypten und Levante* 18: 327–339.

Tronchère H., Goiran J-P., Schmitt L., Preusser F., Bietak M., Forstner-Müller I. & Callot Y. (2012) Geoarchaeology of an ancient fluvial harbour: Avaris and the Pelusiac branch (Nile River, Egypt). *Géomorphologie: relief, processus, environnement* 1: 23-36.

von der Way T. (1984) Untersuchungen des Deutschen Archäologischen Instituts Kairo im nördlichen Delta zwischen Disûq und Tida', *Mitteilungen des Deutschen Archäologischen Instituts, Abteilung Kairo* 40: 297–328.

Wilson P. (2006) *The Survey of Saïs (Sa el-Hagar) 1997–2002.* Egypt Exploration Society Excavation Memoir 77. London: Egypt Exploration Society.

Wunderlich J. (1993) The natural conditions for Pre- and Early Dynastic settlement in the Western Nile Delta around Tell el-Fara'in, Buto. In: Krzyzaniak L., Kobusiewicz M. & Alexander J. (eds) *Environmental Change and Human Culture in the Nile basin and northern Africa until the second millennium B.C.* Studies in African Archaeology 4. Poznán: Poznán Archaeological Museum. pp. 259–266.

2.3.2.3.1 USING REMOTE SENSING TO LOCATE WATERWAYS AS AN INTEGRAL
             WAY OF FINDING SITES

As noted at the start of Section 2.2.1.1.1: Multispectral Photography
and Satellite Imagery, satellite images can help locate ancient waterways
(McHugh et al. 1988; Parcak 2009: 27) thereby making it a very useful
tool within regional survey. Multispectral imaging can detect differences
in moisture levels in vegetation, thus identifying potential water channels.
It is possible to get useful information from free satellite imagery, including
Google Earth and Corona (Parcak 2009), which can help identify shifting
river courses. A bend in the Rosetta branch of the modern Nile River can
be seen just to the north of the Neolithic site of Merimde Beni Salama.
This type of imagery permits us to see clearly how, at this specific location,
the large bend in the river has moved west over time and this movement
is reflected in the new bands/patterns of fields that have taken advantage
of the freed agriculture land. Figure 2.9, a black and white image dating
from November 9th 1968, shows very clearly the white curve to the east
of the then modern Rosetta branch and the advancement of fields in
bands to the west as the land has become available.

   A combination of the analysis of maps (from various periods), satellite
imagery and site visits was undertaken at the apex of the Delta, resulting
in a computer simulation of Nile movements over time (Lutley and
Bunbury 2008). Comparing Nile positions on maps of various dates,
and calculating rates of migration, it was suggested that the Nile actually
shifted up to 9 km per 1000 years in the Giza region (Lutley & Bunbury
2008: 3), whereas the analysis in the Luxor region suggested an average of
only 2 km per 2000 years (Hillier et al. 2007). This research demonstrates

Figure 2.9 Corona satellite image of the Nile river near Merimde Beni Salama,
taken on November 9th 1968

how much detail it is possible to extract remotely using satellite data, aerial photographs and maps, and how it is possible to gain additional information in the field by using geophysical survey techniques (including GPR and ERT) to test anomalies beneath the ground associated with ground-truthing using hand augering. For example, the precise locations where the geophysical survey data suggests certain features are present, such as former levees, *koms*, riverbanks, etc. can be tested (e.g. Bunbury et al. 2009; Hillier et al. 2007). Topographic data is observable from some of the satellite images, but is also collected in the field using a Differential GPS (DGPS). This provides exact positions which can then be tied in with the locations of the geophysical survey profiles, the auger positions, and related to the satellite imagery. In addition, the map data can be geo-referenced. As a result, a fully 3-D topographic model can be developed of an area, a site, a *kom*, or even a landscape, complete with actual sub-surface detail supplied by means of the coring data and the geophysical survey data. One example where the river has shifted is at the site of Elkab, south of Luxor. Here, the southwest corner of the 'Great Walls' appears to have been eroded away due to the shift in the river branch northwards.

2.3.3 Animals and Plants

As noted earlier, much data on past environments can be derived from both macroscopic and microscopic analyses of plant and animal remains. However, the quality and/or recovery rate of both faunal and botanical remains depends upon when, where and how the remains were excavated (for example whether sieving was used), whether they were recorded in detail, whether their archaeological context was secure, and whether they have even survived to the present day. Light faunal or plant remains in archaeological contexts may be intrusive, such as resulting from windborne contamination (as noted above in Sections 2.3.2.1 & 2.3.2.2 regarding sedimentary cores). The actual preservation of both faunal and botanical remains depends upon the local environment. Some organic remains that survive very well in dry desert conditions may survive sporadically at best in damper environments such as in the Delta region.

The processing and use of plants and animals as foodstuffs is discussed in detail in Section 3.4: Diet and Subsistence, but their recovery, documentation and analysis is also essential for our understanding of the ancient environment and the resources available to human groups at various points of time in the past. As Fahmy noted (2004: 711), plants tell us about "nutrition, economy, agricultural production and the environment in the past", and the term *palaeo-ethnobotany* has been applied, since the 1950s, to "the study of human–plant relationships in the past". The discussion here, therefore, largely focuses on the recovery and identification of plant remains and their importance within environmental reconstruction. Examples of potentially suitable recording forms for botanical sampling and analysis can be found in Tassie & Owens (2010).

2.3.3.1 *Plant Remains*

How can the study of plant remains really help to piece together the history of a site? Plant remains can help in many different ways, but, depending upon the local environment of the archaeological site, their preservation may range from exceptional to very poor. As noted above, some plant remains that survive very well in dry desert conditions may survive poorly in damper environments such as the Delta. Certain plant species thrive in certain conditions and not in others, so their survival contributes towards our understanding of the past climate and local environment, as well as changes that occurred over time. This is crucial for a better understanding of the plant-based resources available in the past. These resources affect the materials from which people could weave textiles (see Section 3.5: Clothing & Adornment), their nutritional choices, and even what types of structures they could build. If imported timber was not an option, construction forms might have been restricted due to the types of timber available from indigenous species of tree within Egypt (Gale et al. 2000; see also Section 3.1.1: Burial Rites and Rituals).

What is the best way to recover and to record plant remains in the archaeological record? As already seen above in relation to several different methods, it is essential to retain the contextual information for any type of find, be it a ceramic sherd, or a seed. Plant remains are preserved either by desiccation, mineralisation, waterlogging or carbonisation. As noted by Smith (2003: 25), most methodological advances have been concerned primarily with carbonised remains (Brown et al. 1993; 2009; Fuller et al. 2012; Jones et al. 2008; van der Veen 1985; van der Veen & Fieller 1982). Fahmy (2004) provides a review of the various recovery, sampling and identification methods employed within ancient Egyptian settlement contexts, noting the importance of archaeobotanists being involved in the planning stages of the project. What decisions need to be taken in terms of what kinds of samples to collect for which kinds of plant material? These might comprise charred or desiccated macro remains, pollen, diatoms or phytoliths (Emery-Barbier 2008; Flower et al. 2013; Leroy et al. 2011; Madella & Zurro 2007; Miller et al. 2013). This is both a difficult and very important question, and relates directly to the project research questions. The strategy should be decided in advance of fieldwork, and usually prior to the application for permission to conduct fieldwork in order to ensure that appropriate specialists can work with the material onsite.

Generally, bulk samples for the recovery of macro-remains should be taken from any sealed archaeological features e.g. pits, ditches, hearths, ovens, floors. For most large contexts, given the sheer quantity of material, it is essential to opt for a sampling strategy to recover representative samples. During most archaeological excavations, where possible, c. 2–10 litres of sediment samples per context are routinely taken (Fahmy 2004: 714; Van der Veen 2001). Van der Veen (2001; Fahmy 2004: 715) notes that in sites with large quantities of desiccated remains preserved (usually in domestic dumps), after collection, dry sieving with a 0.5 cm mesh should be carried out with later water flotation; these are regarded as 'small samples'. Large samples are 20 litres, and are normally dry sieved through a 2 mm mesh, with only large botanical remains picked out; hand-picked remains are those collected as work proceeds on a daily basis. It is also noted that they might (as with the large samples) include the rare species that small samples may miss (Fahmy 2004: 715; Van der Veen 2001). In other situations where the context is known to be closely associated with plant remains, such as a brewery or bakery or the area around a hearth, then the density of sampling is greatly increased (see Box 2.7; Smith 2003: 26–27), and total collection can be implemented if, for example, a floor level is being excavated. Dependent on the type of plant remains being retrieved, and their condition, then either flotation or dry sieving might prove to be the most suitable recovery method (see Smith 2003: 27 for a description of the process). Dry sieving is employed using decreasing sieve sizes (to 1.0 mm and 0.5 mm meshes, using geological sieves) in order to pick out smaller macroremains, which Fahmy (2004: 715) notes is a very suitable method in arid regions.

Water flotation is conducted to recover charred plant remains and other macroremains. At many settlement sites in Egypt this is the only type of plant remains preserved, and thus flotation is normally conducted for all samples. This can result in large numbers of carbonised and some desiccated remains; a description of the process is given in Fahmy (2004: 716). Electrostatic separation has also been implemented by Thanheiser (1995; Fahmy 2004: 716), thereby avoiding the need for water and thus preventing the damage that could be caused to fragile plant remains. For all recovery methods, it is critical that the volume of the soil samples is measured and recorded prior to processing, in order for certain statistical analyses to be conducted. In addition to the recovery methods described above, there are some exceptions. When, for example, food storage pits are being excavated, samples are normally collected at least at every 5 cm of depth, unless the context divisions are finer (in which case they are kept separated by context). It is also important that the same quantity of sediment be sampled within each level and/or context, to allow for an understanding of whether certain plant species are increasing or decreasing in quantity over time (Fahmy personal communication).

Short-lived plant remains can also be accurately dated using the AMS ^{14}C method. This already contributes to an understanding of the timing of the introduction of domesticated species not only into Egypt, but across south-west Asia and Africa (Zohary et al. 2012: vii, 17). For example, important work in the Wadi Kubbaniya, as noted above, demonstrates the exceptional preservation of plant species within the arid conditions of Egypt, including the remains of tubers identified as *Cyperus rotundus* and of *Scirpus maritimus/tuberosus,* and AMS ^{14}C dated to 19,000–17,000 BP (Fahmy 2005: 292).

After recovery, how are the different plant species actually identified? Botanists have been working to identify plant remains in archaeological contexts in Egypt since the late 1800s (see Fahmy 2004: 711–714; Hepper 1990, 4–6; Malleson in press, for details), although Arab botanists began work much earlier in the 11th and 13th centuries, investigating the properties of plants for pharmacological purposes (Täckholm & Täckholm 1941: 4). Vivi and Gunnar Täckholm together with Mohammed Drar were responsible for setting up the Cairo herbarium (Hepper 1990: 5) and for the publication of the four-volume Flora of Egypt (Täckholm & Täckholm 1941; Täckholm & Drar 1950; 1954; 1969). Reference volumes such as these, alongside reference collections of modern specimens, are necessary for the identification of seeds, grains and plant parts from samples recovered from excavation (Fahmy 2004: 717; Hepper 1990: 6). Other important works of reference are the four-volume *Flora of Egypt* by Boulos (1999; 2000; 2002; 2005), Cope & Hosni (1991), Thulin (1993; 1995) and Zohary (1973; Zohary & Hopf 1988; 1994; 2000; Zohary et al. 2012). The majority of desiccated and charred remains are identified based on their morphology, under low-powered binocular microscopes (usually 5–30 × magnification). Some particularly well-preserved specimens, including flowers, fruits, seeds and leaves from tombs (Hepper 1990: 6; Täckholm & Täckholm 1941: 4–5) may not need examination under a microscope. Textiles, basketry and wood, as well as pollen and phytoliths, need to be examined under high-powered microscopes, including scanning electron microscopy (SEM). This allows for examination of the cellular structure of the lower epidermis of the plant, sometimes using dyes to emphasise tissues, which allows for samples from groups of species to be separated (see Hepper 1990: 6). SEM can be used to identify plant materials used in textile production, for which the exact plant species cannot always be identified accurately using light microscopic analysis alone. It can also be useful in identifying seeds of some species.

Once the samples have been processed and all the remains have been identified, a number of analyses can take place, and questions can be addressed using statistical analyses. Statistical analyses include multivariate and non-multivariate analyses, as well as correspondence analysis (see Fletcher & Lock 2005: 139–153; Shennan 1997: 308–360) such as that described by Fahmy (2001; 2004: 718; Abd el-Ghani & Fahmy 2001). Fahmy (2004: 718–722, Table 2) examines plant remains by context and association with specific features associated with food production at the Hidden Valley at Farafra Oasis, and the analysis also includes information on the amount of sediment that was actually sampled per area. The analysis helps to look at changing, or consistent, patterns of site use over time, and also the changing plant species over time.

Certain plant species thrive in certain soil conditions and not in others, so the presence or a lack of plants which favour e.g. wet/waterlogged/dry/ arid/saline soils, contributes towards our understanding of the past climate and local environment, as well as changes that occurred over time. This is crucial for a better understanding of the plant-based resources available in the past. These resources affect the materials from which people could weave textiles (see Section 3.5: Clothing and Adornment), their nutritional choices, and even what types of structures they could build if imported timber was not an option, due to types of timber available from indigenous species of tree within Egypt (Gale et al. 2000; see also Section 3.1.1: Burial Rites and Rituals).Trade and exchange relations between groups in different regions can be detected by the presence of non-indigenous species. The present of hitherto unknown species may also give valuable information as to the date by which a specific species started to be cultivated in Egypt (Zohary et al. 2012). This is very important, particularly in the study of Egyptian prehistory, as was referred to above in the section on AMS ^{14}C measurements. This is, for example, notable in the case of the plant remains from the Wadi Kubbaniya, as these were intrusive to the early contexts to which they had originally been considered associated (Wendorf & Schild 1980; 1984; Wendorf et al. 1979; 1984), thereby stressing again the importance of secure contexts.

The abundance of certain species, such as of woods, can be helpful in ascertaining the level of tree cover in a region in antiquity; however, wood was frequently reused and the finds could originate from earlier contexts (Dee et al. 2012: 872–874). Likewise, the presence of charcoal, in itself important in the understanding of ancient Egyptian pyrotechnology (see discussions in Sections 4.4.3: Metals & 4.4.4: Glass), may be derived from reused wood sources (Dee et al. 2012: 874). The relative quantity of wood charcoal found within domestic contexts can indicate aspects of environment and economy – the choice to burn wood as opposed to dung or cereal processing waste indicates either a greater availability of wood resources, or possibly higher status individuals utilising this scarce resource (Murray 2009). Fahmy (2004: 716–717) notes the recovery of desiccated plant remains from c. 5500 BC due to the increasingly dry climate, but also that there is a higher incidence of carbonised material at Neolithic settlements in association with food preparation. One of the biggest issues addressed via the study of plant remains is agriculture.

As a final note, it is important to stress that even if contextual integrity is maintained in combination with the implementation of a careful strategy for the recovery of plant remains during archaeological investigation, it is still unlikely that the complete spectrum of plant species used at a given site will be visible (Smith 2003: 34). If a site is more water-logged or

Box 2.7   Archaeobotany of the Giza Plateau
Mary Anne Murray & Claire Malleson

The remains of Old Kingdom settlement on the Giza plateau have been excavated by Ancient Egypt Research Associates (AERA, directed by Mark Lehner) since 1988. The work has focused on two different areas situated either side of a modern cemetery to the south of the triad of the great pyramids of Khufu, Khafre and Menkaure. The so-called 'Heit el-Ghurab' site (HeG) dates to the late 4th Dynasty, and is characterised by a major complex of formally laid out buildings likely to be some type of communal accommodation. To the north, associated with the mortuary complexes of Menkaure and Khentawes (KKT), there are smaller complexes of domestic buildings dating to the 5–6th Dynasties (Lehner et al. 2004–11).

Throughout their excavations AERA have maintained an extensive program of archaeobotanical analysis. Samples are taken from individual archaeological features – hearths, floors, pits and dumps – and processed by flotation, which recovers plant remains with the use of water. At both the sites of HeG and KKT, the plant assemblages have been preserved by charring. The samples consist primarily of the remains of cereal processing such as weeds, chaff and straw which were used as fuel alongside wood and/or wood charcoal in hearths and bakeries. The process of cleaning cereals in preparation for the production of bread and beer, the two staple foods of the ancient Egypt, produced large quantities of cereal chaff and the weed seeds associated with the crop. These items were gradually removed through threshing, winnowing, pounding, and sieving and finally hand sorting to obtain a cleaned grain product (Murray 2000). Most of these were daily routine activities in every Egyptian household and the cereal processing 'waste' produced was a valuable commodity as a prime fuel, and is commonly found in ancient Egyptian settlement remains – hearths and ovens, domestic dumps where the ash was disposed of, and in rooms, courtyards and walkways where it became trampled and scattered or windblown.

The study of these plant assemblages provides details of not only the environment and ecology of the cereal fields, but also valuable insights into the daily life in an ancient town, including food consumption, cereal husbandry and the processing and details of the economy (Murray 2009).

The compositions of charred plant assemblages are dependent on a large number of pre- and post-depositional variables, such as plant characteristics (e.g. leafy vegetables vs. robust fruit stones; number of seeds produced; oily vs. non-oily seeds); the taxa brought onto a site, either deliberately or unintentionally (e.g. a harvested cereal crop and its associated weeds; seeds preserved in animal dung fuel); the processing of the plants (e.g. the stages to remove weeds, cereal chaff and straw); the charring process (i.e. the likelihood of plants coming into contact with fire – whether foods required cooking or not; the use of wood, weeds and chaff or dung as fuel); deposition (e.g. disposal contexts vs. storage contexts); the nature of the post-deposition environment (e.g. dilution, mixing and loss of composition due to later occupations, animals, plants, etc.); excavation and sampling (e.g. the selection of only 'rich' looking contexts); and the recovery process itself (by bucket or machine flotation, wet or dry sieving, or by hand selection). The alteration of a botanical assemblage, through loss or fragmentation, for example, occurs at every stage of these processes (e.g. Jones 1991).

At Giza, due to the location of the HeG town on the low floodplain adjacent to the desert edge, the archaeological remains have been exposed to both the annual Nile flood (until the 1960s) and the fluctuating water table. The alternating wet–dry conditions are detrimental to the preservation of organic remains; no plant remains have been preserved by desiccation (or drying), as is often common in Egypt, and even charred remains are usually poorly preserved. In contrast, at the nearby site of KKT, plant remains are also preserved by charring but because the site lies just above the water table and flood plain, the remains are in much better condition, and have provided a wealth of detail about activities within KKT, particularly in the domestic dwelling of 'House E'. The disparity in the preservation between these two Old Kingdom sites and therefore the amount of information we can obtain is notable (Murray & el-Gendy, submitted; Malleson, in preparation).

One measure of the relative level of preservation of plant macro-remains on a settlement is the number of items found in each litre of soil sampled, providing the samples come from comparable contexts and that the remains are preserved in the same manner (i.e. desiccated or charred). This is a valid method of comparison here since all the plant remains excavated in HeG and KKT (House E) are preserved in the same manner (charring) and derive from similar types of archaeological contexts (i.e. domestic/household debris primarily from the use of plants as fuel). At present, the average density of charred plant remains (items per litre or IPL) from HeG overall is 8.2 IPL, while in KKT (House E) the average is 389 IPL – 47.4 times richer.

Another factor to consider is the number of different taxa present in the assemblage, i.e. plant types or species, since the diversity of plant taxa present is some reflection of the variety of plants brought into a settlement. At both Giza sites, the majority of the plant remains derive from the use of cereal crop by-products as fuel; the seeds found can almost all be considered to be from the plants which grew as weeds in the cereal fields. In HeG we find 13–43 different plant taxa on site, while in KKT House E, at least 54 different plant types have been found thus far (Malleson, in preparation; Murray, submitted). As a result of the poor taphonomic conditions at HeG, many of the seeds are missing diagnostic characteristics. We are able to identify most taxa only to genus level, e.g. *Phalaris* sp. – canary-grass type, *Rumex* sp. – dock/sorrel, or indeed family. At House E in KKT, however, many of the seeds are in nearly perfect condition with diagnostic features present and it is possible to identify the species of these common taxa, e.g. *Phalaris paradoxa* – hooded canary grass, *Rumex dentatus* – dentated dock.

This variable preservation resulted in fewer firm details regarding the plant economy of HeG than that of KKT House E. At both sites, it is clear that the farmers cultivating these crops allowed their fields to become infested with wild grasses, clovers, reeds and many wild taxa, which they then harvested along with the cereals. At KKT, we also know that the crop was harvested low to the ground (i.e. collecting many low-growing plants) and then only partially threshed, removing just the bulk of the straw. All of the cereals, together with the chaff and weeds were stored in granaries in the house for further processing as needed. showing that not only were the by-products highly valued as fuel, but the resulting ash also held great value to the ancient Egyptians (Malleson 2014).

Samples from beneath the granary in House E indicate that a mix of ash collected from the nearby kitchen area and freshly burned straw was laid down as a foundation, possibly to deter insects. The assemblage at KKT House E has enabled us to suggest that the abundance of weeds was tolerated (and stored) because they bulked out the cereal processing by-products providing valuable additional household fuel. Perhaps more importantly, by allowing clovers to grow in cereal fields, the residents had a supply of enriched animal feed without the need for separate fodder cultivation, thereby integrating arable and livestock agriculture into one efficient farming system.

The exceptional preservation, relative to the HeG botanical record, has allowed us to identify a broader range of species, adding details to our knowledge of Old Kingdom plant use and has highlighted the importance of differential preservation even between neighbouring sites.

References

Jones GEM. (1991) Numerical analysis in archaeobotany. In: van Zeist W., Wasylikova K. & Behre KE. (eds) *Progress in Old World Palaeoethnobotany*. Rotterdam: AA Balkema. pp. 63–80.

Lehner M. et al. (2004–11) *Giza Occasional Publications* 1–5. See also *AERAgram* 1–14 www.aeraweb.org

Malleson C. (2014) Ancient Egyptian Insecticide. *AERAgram* 14: 2. Boston: Ancient Egypt Research Associates Press (AERA).

Malleson C. (in preparation) *Archaeobotanical investigations in House E*. Giza Occasional Publications. Boston: Ancient Egypt Research Associates Press (AERA).

Murray MA. (2000) Cereal production and processing. In: Nicholson P. & Straw I. (eds) *Ancient Egyptian Materials and Technology*. Cambridge: Cambridge University Press. pp. 505–536.

Murray MA. (2009) Questions of continuity: Fodder and fuel use in Bronze Age Egypt. In: Fairbairn A. & Weiss E. (eds) *From Foragers to Farmers*. Oxford: Oxbow Press. pp. 254–267.

Murray MA. (submitted) *The plant remains from the Royal Administrative Building*. Giza Occasional Publications. Boston: Ancient Egypt Research Associates Press (AERA).

Murray MA. & el-Gendy R. (submitted) The ancient plants from Building E at Khentkawes Town North. In: *Proceedings of the Analytical and Publication Field School*. Boston: Ancient Egypt Research Associates Press (AERA).

desiccated, plant preservation is richer than for carbonised assemblages. The normal implication of this is that sampling can be undertaken using smaller quantities, both by volume and/or weight (Smith 2003: 34). It is also current practice to separate out the differentially preserved archaeobotanical remains, and analyse them independently, despite the fact that this may not say anything about how they were used by people in the past (Smith 2003). It is hoped that, in the future, more analytical methods will be developed to enable greater synthesis of these different forms of archaeobotanical data The different preservation of these forms of preserved plant remains may have little or no reflection upon their past use and/or human activity.

2.3.3.2 Faunal Remains

Although this section considers animals primarily in terms of their use for the development of models of palaeoclimate and palaeoenvironment, zooarchaeological and iconographical studies of fauna can provide evidence of diet, domestication and religious practice. Aspects of these are considered in Sections 3.4: Diet and Subsistence & 3.5.5: Mummified Animals. Ikram (2000) provides an explanation of detailed scientific analyses and descriptions of animal slaughtering techniques and meat preservation methods which can aid in such reconstructions.

Animal species are primarily identified using macroscopic methods. Species commonly identified visually include dogs (*Canis*), sheep and goats (*Ovis* and *Capra*), and cattle (*Bos*). In some sites, such as Badarian, Maadi and A-group sites, species that were originally identified macroscopically as gazelles have more recently been re-identified as goat (Boessneck 1988; Flores 2003). Identification of more complex or unusual species requires comparison with reference standards and reference collections, such as the zooarchaeological collections of major museums including the Natural History Museum in London. In addition, some published reference texts (also called bone atlases) provide detailed guides for macroscopic identification (e.g. France 2009; Hillson 1992; 2005; von den Driesch 1976; Walker 1985). Such data are then normally compared with catalogues and inventories of fauna found, either zooarchaeologically, ethnographically or iconographically, in Egypt. Examples of such treatises include Houlihan (1986) for birds, Brewer & Friedman (1989) for fish, Osborn & Osbornová (1998) for mammals, and, more generally, Brewer et al. (1994) for domesticates (including plants).

In many latitudes small vertebrates provide good indicators as to environmental conditions, either in terms of large-scale climate or small-scale vegetation (O'Connor 2000: 123). In order to comprehend the data however, it is essential that the spatial and temporal catchment of the deposit studied is understood. In certain contexts, such as latrine pits, particular species might be attracted to the area and thus appear overly represented. For example, mice or other small rodents may be attracted to storage pits containing seeds and grains, thereby apparently being overrepresented relative to the rest of the site. Other small animals might be attracted to a latrine pit due to the relative abundance at such a location of their preferred diet of invertebrates. Furthermore, some animals, particularly predatory birds such as barn owls, can lead to the development of new bone accumulations deriving from ejected pellets of undigested food material (O'Connor 2000: 125). Small animals tend to be particularly good for environmental reconstruction as they are not commonly consumed by humans as food. Furthermore, small mammals are thought to be more susceptible than large mammals to alterations in their local environment (Luff 1984). Even when small mammals were clearly consumed as a food, some environmental information can still be determined as some small animals prefer particular climates or environments. For example, jackrabbits are generally more common in open vegetation whereas the desert cottontail rabbit prefers low scrub vegetation and hence differences in the frequency of these two species in an assemblage may reflect differences in vegetation cover (O'Connor 2000: 130). Similarly, in Egypt, long-eared hedgehogs are commonly found in olive groves and cultivated areas, desert hedgehogs prefer semi-barren desert whereas Ethiopian hedgehogs prefer vegetated valleys and dense stands of vegetation (Osborn & Osbornová 1998: 19–21). Large mammals, as recovered through hunting, also provide this type of information.

Domestication of most species occurred before the Predynastic period, with some domestic forms apparently deriving primarily from the local wild animals. African fossil ancestors do not exist for dogs, sheep, goats and horses and so these species, in either wild or domestic form, must have derived from neighbouring areas (Clutton-Brock 2012: 50). Early domesticates included sheep and goats, cattle, pigs, donkeys and poultry although it is possible that domestication was attempted for other species including hyena and antelope (Ikram 2000). What is certain is that both tomb decoration and faunal remains from the Old Kingdom onwards suggest that these unusual wild animals were locally available, and hence provide evidence of the local climate and palaeoenvironment. Certainly, until the Old Kingdom, the moister climate supported a diverse fauna in the Nile Valley and the Red Sea Hills (Butzer 1976; Ikram 2000). Differentiation and delineation of the difference between a hunted and a domesticate assemblage can be undertaken through analysis of age at death of the individual animals as the mortality patterns differ between these groups. Furthermore, wild and domestic can also be distinguished on the basis of body size, and hence gross macroscopic measurement is commonly employed (O'Connor 2000; von den Driesch 1976).

The faunal assemblages from sacred animal necropoleis throughout the country provide another important source of information relating to animals within the ancient Egyptian environment. Although the analysis of these faunal assemblages has not always been at the forefront of the research, recent publications (including Nicholson et al. 2013: 7; in press; Rowland et al. 2013) are doing much to inform us about the range of species and quantities of animals and birds involved in the operation of the sacred animal necropoleis. Due to developments in recording methodology, recent excavations and analyses are allowing for these remains to be considered within their original contexts, although the disturbance caused within animal necropoleis due to robbery is problematic. The work on the Late Period–Ptolemaic falcon necropolis at Quesna (Rowland & Ikram 2013; Rowland et al. 2013) has highlighted the sheer quantity of birds deposited within the galleries, the variety of species of both

Box 2.8   Interaction Between Man and Animals in the Prehistoric Nile Valley

Wim Van Neer & Veerle Linseele

The type of animal species found on an archaeological site, and the proportions in which their remains occur, permit detailed inferences to be made on the former relationship between man and animal. These may be of nutritional, utilitarian, ritual or religious natures. Animal bones can be found in a wide variety of depositional contexts, but refuse deposits corresponding to the left-overs of butchering, food preparation or consumption are usually the most common finds. Since the late 1970s, there has been a growing awareness among archaeologists working in Egypt that careful retrieval of small animal remains, preferably through sieving, is necessary in order to document the full range of animal species exploited in the past (Wendorf et al. 1980).

Most archaeozoological information available for Egypt comes from sites dating to the Late Palaeolithic and onwards, and there is more data from the Nile Valley than from the adjacent desert areas. The temporal and spatial distribution of desert sites depends of course on the climatic conditions, and, by their mere presence, these sites are already an indication of former conditions that were better than today (Kuper & Kröpelin 2006).

When the faunal record of the Nile Valley is considered through time, it appears that subsistence was always based on a wide range of food resources and that fish and wild birds were important providers of animal protein. Both fishing and fowling were seasonal activities, involving the capture of migrating water birds during winter, and the harvesting of fish in conjunction with changing water levels throughout the year. When considering the fish from numerous archaeofaunas from the Late Palaeolithic to the Dynastic period, it becomes clear that fishing strategies practised along the Egyptian Nile changed diachronically. Until the end of the Late Palaeolithic, fishing was restricted to the capture of a limited number of species (mainly Nile catfish and tilapia) that typically spawned in the shallow waters of the inundated flood plain at the beginning of the flood season. Later in the year, when the Nile waters receded and shallow pools were formed on the floodplain, it was possible to harvest massive quantities of small fish of the same species trapped in these residual pools. The open water of the Nile river itself was apparently not exploited until the Epipalaeolithic (around 8000 years ago), when a much wider range of fish species was captured, including large Nile perch that typically live in deeper parts of the Nile (Van Neer 2004).

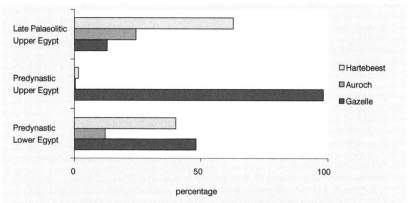

Figure 2.10 Proportions of the major terrestrial game animals in the prehistoric Nile Valley

The graph is based on almost 5000 bone finds from Late Palaeolithic sites in Upper Egypt (E71K12, Wadi Kubbaniya, Kom Ombo and Isna-Idfu area, Abadiya 1 and 3, Shuwikhat, Makhadma 2 and 4), Predynastic sites in Upper Egypt (Hierakonpolis, Adaïma, El Abadiya 2, Armant, Maghar Dendera, El-Mahâsna and Naqada) and Predynastic sites in Lower Egypt (Merimde, Maadi, Buto and El Omari). For references to the individual faunal reports, see Linseele & Van Neer (2009).

Such habitats only became accessible once a better fishing technology developed and the means of navigation (rafts, dug outs or small boats) also improved. Fish became an even more important food resource from then on, being also accessible when the levels of the Nile were low and the floodplain exploitation was over.

When considering the terrestrial wild fauna from settlement sites in the Egyptian Nile Valley dating from the Late Palaeolithic to the New Kingdom period, changes appear in the species spectrum and in the proportion of species (Linseele & Van Neer 2009). Late Palaeolithic hunters focused on three major species: aurochs or wild cattle, hartebeest (a large antelope) and gazelle. These were the main providers of animal meat, whereas hare, porcupine, wild donkey and Barbary sheep were only minor hunted species. Hartebeest and aurochs populations were essentially limited to the Nile Valley proper and it is likely that their densities were never very high because of the narrow floodplain, especially in Upper Egypt. It is in that region that both wild cattle and hartebeest declined markedly during the Predynastic (Figure 2.10), probably as a result of competition with humans and their flocks. Domestic livestock became the major provider of meat and, from then onwards, hunting was a minor activity, almost exclusively concentrating on gazelles that were more abundant in the adjacent semi-desert and desert than in the Nile Valley.

In Lower Egypt, the wild fauna seems to have been affected by human interference too, but apparently later than in the south. The decline in aurochs and hartebeest populations in Lower Egypt only seems have taken place during the Old to New Kingdom periods. This no doubt had to do with the lusher and more humid environment and with the fact that the floodplain was much wider in the Delta area, thereby making it possible for large game to resist human predation pressure for longer.

Typical for Lower Egypt is also the important role that the hunt for hippopotamus played from at least Predynastic times, in comparison with Upper Egypt, where finds of that species are relatively rare. Again, this may be partly a result of the differences in environmental conditions of both regions. Hippos need water as well as grassland to graze on. It is likely that such grasslands were more numerous in the north. Although the intensity of hippo hunting prior to Predynastic times has not been documented thus far for Lower Egypt, a plausible explanation for their abundance on sites of that period could be their competition with humans. It has been suggested that hippos may have been hunted intensively in order to protect fields of cultivated crops (von den Driesch & Boessneck 1985).

The faunal shifts that are observed both in Upper and Lower Egypt can be related to the increasing impact of humans on the environment. This includes not only heavier hunting pressure but also habitat loss due to growing human population densities and, since Neolithic times, competition for land used for domestic stock keeping and for crop cultivation.

References

Kuper R. & Kröpelin S. (2006) Climate-controlled Holocene occupation in the Sahara: motor of Africa's evolution. *Science* 313: 803–807.

Linseele V. & Van Neer W. (2009) Exploitation of desert and other wild game in ancient Egypt: The archaeozoological evidence from the Nile Valley. In: Riemer H., Förster F., Herb M. & Pöllath N. (eds) *Desert animals in the eastern Sahara: Status, economic significance and cultural reflection in antiquity.* Colloquium Africanum 4: 47–78. Cologne: Heinrich Barth Institute.

Van Neer W. (2004) Evolution of prehistoric fishing in the Nile Valley. *Journal of African Archaeology* 2: 251–269.

von den Driesch A. & Boessneck J. (1985) *Die Knochenfunde aus der neolitischen Siedlung von Merimde-Benisalâme an westlichen Nil Delta.* München: Institut für Paläoanatomie, Domestikationsforschung und Geschichte der Tiermedizin.

Wendorf F., Schild R. & Close AE. (eds) (1980) *Loaves and fishes.* New Delhi: Pauls Press.

birds and rodents, and raised several questions in terms of the breeding of these animals. Statistical analyses of the different bones represented, including their lengths, and comparisons between their types, allow for an assessment of the age of the birds and animals used as offerings within the animal cults, as well as the variety of species used as offerings (Rowland et al. 2013: 65–74). The bones at Quesna were analysed by element, family or species to as accurate a taxonomic identification as possible; however, as noted above; access to and comparison with reference collections, in this case one of raptors, is very beneficial (Rowland et al. 2013: 66). The results of such analyses, however, do not necessarily indicate the breadth of the original local faunal assemblages, given that the animals and birds represented in the necropoleis were likely fulfilling a very particular purpose. What also remains problematic to assess is whether the species identified in each necropolis were bred locally, given the difficulty of identifying archaeological remains for mews, and uncertainty as to whether the species were migratory at the time (Rowland et al. 2013: 76).

2.4 Organisation of Human Burial Grounds

As during the early years of archaeological investigation in Egypt, cemeteries remain some of the most frequent sources of archaeological data in Egypt today. The balance of cemetery to settlement data is not as skewed as it was 100 years ago; the fact remains that because of the environmental conditions in Egypt associated with the Nile inundation, cemeteries have had better prospects of preservation and re-location. Cemeteries were located on the highest ground to avoid flood waters. Due to the yearly deposits of silt from these same floods, settlements can be much more deeply buried than cemeteries or even destroyed by the fluctuating position of the Nile and its river branches in the Delta (see Section 2.3: The Environment).

Methodological advancements have allowed for a huge increase in the types of data that can be obtained during and prior to excavation. Since the mid-1990s (see Section 2.2: Finding Sites) it has become increasingly plausible to determine the geographical extent of a cemetery before implementing potentially intrusive investigation strategies, by using geophysical survey. Magnetic survey has proved especially useful, although due to the nature of the method the clearest images relate to tombs with large numbers of ceramic vessels, those with mud-brick or burned brick architecture, and ceramic coffin burials (Herbich 2012b: 388–389). Osteological field methods and post-excavation analyses mean that it is part of usual excavation practice to record data that permit investigation of the health of local populations, their diet, and their demographic distribution. Through scientific approaches (see Chapter 3: Biography of People), it is sometimes possible also to bring new understanding to questions such as 'where did this person come from?' or 'are these people related?' Osteoarchaeological analysis (see Sections 3.1, 3.2, 3.3 and 3.4) provides answers to questions such as: 'how old was this person when they died?', 'do we know what caused them to die?'; and it informs as to the sex of the deceased and on the good, or bad, health that they experienced during their lives.

Cemeteries have long been investigated as sources to inform about social changes over time, and more recently, changing patterns of diet, disease and mortality at different periods, both within and between cemeteries. With refined relative chronologies, and particularly with AMS ^{14}C dates, archaeology is much better equipped to monitor and pinpoint events and changes in communities in order to observe temporal patterns (Chapman 2007). Botanical analyses can go so far as to suggest seasons of death (McAleely 2013) through the presence of specific floral or food remains, and these together with the remains of animal bones inform about the wider environment in which people lived, what they could eat, what foods were consumed as part of the funerary ritual, and what was thought essential for the afterlife. It is through integrated and synthetic consideration of those varied data that analysis of burial grounds can be approached in order to answer a multitude of research questions.

Here in this chapter, the evidence from cemeteries is examined holistically, in order to help consider questions relating to the kind of society within which the deceased lived, as well as questions relating to individuals within the community. Was everyone equal in the community, or did some families/individuals have a higher status than others? Were there regional hierarchies of communities? What types of occupations did people have? Particularly in the case of prehistoric cemeteries, does the evidence support varied or common beliefs in the afterlife? What is known about the rituals surrounding burial? Scales of enquiry vary from examining single burials, to comparing the burials and graves of individuals within one cemetery, through to inter-regional comparisons of cemeteries. It is important to bear in mind that one cemetery does not necessarily equate to one settlement. Ultimately, the degree to which a cemetery reflects the organisation of the living society is unknown.

2.4.1 Working with Data from Mortuary Contexts

The mortuary contexts themselves vary widely, including mausolea, pyramids, mastaba tombs (the name derives from the Arabic word for 'bench' on account of the tomb superstructure), mortuary temples, rock-cut tombs, cemeteries reserved for the elite, mixed cemeteries, and those for burials of non-elites, simple pit graves sometimes including animal burials, as well as animal necropoleis dedicated to particular deities (see Section 3.5.5: Mummified Animals). From the Badarian period onwards (from c. 4400 BC), differentiation is observable in burials within Egypt, in terms of their size, construction, provisions, as well as treatment of the body. Burials have therefore been used extensively to inform as to the organisation of society, especially during Egyptian prehistory, when there are no written records. The first step for any potential analysis is to clarify the research questions to be asked and establish which types of data are relevant. This might simply comprise analysis of single variables (e.g. the number of ceramic vessels per grave) to create a simple frequency distribution, or might include correlation of two different variables (simple bivariate analysis, e.g. the total number of grave objects with the volume of the grave), or more complex multivariate analyses (Fletcher & Lock 2005: 139–153). Potential data types can relate to the burial itself, the grave, the grave offerings or the wider cemetery.

So, before moving on to discuss specific types of data analysis, let us look at examples of potential questions that we might want to ask of the mortuary data, and the types of variables that might be useful to examine:

1   What were the roles, positions, and importance of the deceased within the community in which they had lived? Useful indicators include the size of the grave, the complexity of the architecture, the number and type of grave goods present, as well as the existence of any textual sources. In certain situations, analysis might start by correlating the grave size with the number of grave goods or the number of different types. The type of grave goods could also indicate the roles of the deceased.

2   What are the reasons for multiple burials? In order to approach this question, each of the multiple burials needs to be compared and contrasted, in terms of the number of individuals, their positions, whether they were all buried in one contemporaneous event or whether secondary burials are included. Osteological analysis might also be undertaken to assess for evidence for the cause of death, such as resulting from an episode of violence or disease (see Section 3.3: Health and Disease).

3   Why are some groups of burials clustered together? Because clusters can occur for a wide number of reasons it is important to look at the objects as initial indicators of the date of the graves, the orientation of the interred, the type of grave, in addition to markers of sex and/ or gender and social and/or biological age, and of course the location of the clusters within the cemetery and within the natural topography of the landscape. Clusters of inhumations might result from burials occurring at a similar time, or could be family or group members returning to the same spot to bury their dead, or could be based on sex/gender age (e.g. clusters of child burials), or rank (e.g. elite burials or non-elites focused around an elite burial).

These types of basic or initial analyses consider indicators of similarity between graves in terms of sex, age, grave type, grave size, and grave goods, whilst taking spatial location into account. Other questions relating to the wider community include:

1   Was funerary ideology homogenous throughout the community (or between communities) or does it suggest the presence of groups or individuals from different backgrounds? Useful indicators might include the orientation of the burials and the treatment of the body itself, e.g. wrappings and position.
2   Was there a hierarchical structure either within the local community or regionally? Useful starting indicators can be the total numbers of goods found in each grave, or the size and types of grave.
3   Was there differential access to resources, including commodities and workforces? Is this apparent between burials of all periods or does it increase through time? Key indicators here could include the range of raw materials within each grave (e.g. stone types, metals), the size of the grave, and the complexity of the architecture (e.g. chambers, roofing, mud-bricks or stone). The objects themselves, as indicators of relative chronology (and potentially later for absolute dating), are essential within this analysis.

Most of the theoretical and methodological approaches to mortuary contexts were not conceived specifically for application to Egyptian cemeteries or, at times, even for archaeological contexts at all. From a theoretical standpoint, many approaches have been adopted from anthropological and ethnographical studies into past and present societies. These have included social evolutionary approaches as to how societies develop and become increasingly complex, such as Service (1962) critiquing the delineation of societies into bands, tribes and chiefdoms; Fried (1967) breaking down social structures into egalitarian, ranked and stratified societies; and Johnson & Earle (2000) looking at three 'critical levels of sociocultural integration' (these being the Family-Level Group, the Local Group and the Regional Polity). In the 1970s, major steps were taken to consider mortuary data in various worldwide contexts, from an archaeological point of view, notably in the edited volume by Binford (1971) examining approaches to the social dimensions of mortuary practices. This volume included a key contribution from Binford evaluating the social relations between the living and the dead. Also in the 1970s, Peebles & Kus (1977) considered how to look for social ranks in the archaeological record and Tainter (1978) used mortuary contexts to investigate prehistoric society. Mortuary analyses have been particularly informative for prehistoric Egyptian societies, although not exclusively so. Notable other work includes social status analyses of Predynastic and Early Dynastic contexts by Castillos (1982), and Ellis (1992) at the early Tarkhan cemeteries, as well as Hendrickx (1994) at Elkab. Aspects such as differential social status, effort expenditure, and the differences between male and female burials have been considered within the context of pre- and early state society in Egypt either individually or in comparison with each other, for example in multivariate analyses (e.g. Drennan 2010: 265–267). These studies all used statistical methods of varying kinds. More recent studies of mortuary archaeology have usually synthesised the archaeology more explicitly with the osteoarchaeology to form a social bioarchaeology (Agarwal & Glencross 2011), such as at Amarna (Rose & Zabecki 2010), Tell Ibrahim Awad (Phillips et al. 2010), Quesna (Rowland et al. 2010) or Dakhleh Oasis (Dupras & Tocheri 2007; Wheeler et al. 2013).

2.4.2 Types of Data and Organisation

Some funerary data are obtained from the burials themselves, whereas others are derived from the grave and its provisions. One of the most difficult types of information to extract, however, is that relating to the mortuary ritual both preceding, during and after burial, and such possibilities rely heavily upon the quality of the documentation and publication of the excavation of the cemetery. It is this integration of the above data with taphonomic studies that forms the basis for archaeothanatology (Duday 2009) and hence for the development of models of funerary treatment and ritual.

Funerary and mortuary data are usually organised in separate but linked tables in a relational database. Descriptive aspects are normally coded numerically to enable easier statistical manipulation (Fletcher & Lock 2005: 5); for example, rather than writing 'oval' for grave shape, this grave form might be coded '3', whereas a rectangular grave might be coded '1'. Furthermore grave numbers and burial numbers are not equivalent as more than one individual might be inhumed within one burial. Within the grave itself, commonly important features include shape, type or construction (both in terms of form such as pit burial, shaft burial, mastaba tomb, chambered tomb, and in terms of material such as mud-brick or stone), grave size, location within cemetery or other location, and presence (or lack thereof) of a coffin/matting/wrapping and its composition. It is important to capture the breadth of this data, in combination with other variables (Drennan 2010: 266). The burial itself usually includes, as a minimum, information as to the age and sex of the interred. These data normally include levels of confidence expressed by the bioarchaeologist, such as possible or definite female, or younger adult or distinct potential age range (such as for children). Where possible, the position of the body is also recorded, such as lying on its side, prone or supine, the orientation and positioning of the face and arms/hands, and the direction of the head relative to the rest of the body.

Additional information, such as relating to disease, diet, or cause of death, may also be added where available, allowing for more dimensions of the individual and the society to be developed. This may also comprise aspects of the natural environment in which the community lived (Section 2.3: The Environment). Information relating to familial relationships (e.g. DNA analysis, see Section 3.7.3: Population and Ethnicity) may be useful when investigating the reasons for spatial zoning of burials, or for the reasons for multiple interments within a single grave (either at a single point in time or sequentially). Multiple burials can be problematic as it can be difficult to ascertain which burial goods were originally placed with which individual. Furthermore, graves containing multiple burials cannot simply be compared with single burials as a larger grave size is obviously required for multiple interments. Multiple burials demonstrate the importance of burials and graves being numbered separately.

Information that might be compiled relating to the grave offerings includes: the main groups (and subgroups) of types of object or artefact present (including ceramics, stone vessels, jewellery, stone tools, copper objects, etc.) and their placement or orientation within the burial, the presence of any potmarks and other inscriptions on vessels or other artefacts, brief descriptions of the raw materials from which the grave offerings are made in advance of further analyses (see Chapter 4), and the presence and, where possible, identification of any plant or animal remains. Archaeologists differ greatly in their opinions as to the most appropriate ways of analysing social differentiation through burials. Castillos (1997: 251) suggests that there is not enough knowledge for the researcher to assess the relative importance of artefacts as found within burials and prefers only to consider total counts of grave goods (contra Hendrickx 1994: 217). In addition to the counting of objects, it is also possible to attempt to categorise such aspects as energy expenditure using numerical ranking of the data. It is important to note that the ascending numerical values imply only a general increase and are not 'real measurements' (e.g. some specific variables being categorised as 1 = low, 2 = medium and 3 = high). This is different from the example of giving numerical categories to shapes or types graves, as described above, where the number is purely descriptive and includes no aspect of ranking of the graves (also see Fletcher & Lock 2005: 3; Drennan 2010: 268).

To return to the subject of how archaeologists choose to approach burial data, Hendrickx (1994: 217) suggests that objects deemed to be of ritual significance should not be considered, given that these are not fully understood. One alternative, developed by Drennan (2010: 266, 268), is simply to note the presence or absence of objects considered to be more 'ceremonial' than 'practical'. Similarity or correlation coefficients can be used on presence and absence data, to detect patterns relating to, for example, the presence of objects of certain types of stone, or certain types of ceramic vessels (e.g. Drennan 2010: 277). It is subsequently possible to compare these results to assess whether, for example, alabaster commonly appears with ceramics, and whether copper objects commonly co-occur with objects of siltstone. These categories can be materials-based (alabaster, ceramic, flint, copper etc.) or object-based (jug, bowl, knife) or a combination of both, for example a siltstone palette or a copper knife. Lacking any *a priori* knowledge, it is normal to examine the widest range of object types and material types possible, in order to best investigate patterns within a full set of mortuary data. The positioning of such grave goods can be vital in developing models of the funerary ritual as these can affect local taphonomic changes.

It is necessary to have some means of dating the individual tombs and a date range for the use of the cemetery, especially if the purpose of analysis is to investigate change over time, and to interpret what this change actually means (Chapman 2005). Some cemeteries are used over long periods of time, either continuously or intermittently, whereas others only have single periods of use. The most common dating method is still relative dating, which more often than not relies upon examination of the ceramic assemblages, as seen in Box 2.1. Absolute dates, such as obtained through AMS ^{14}C, are advantageous because the presence of a certain ceramic type in the Delta and the Nile Valley might suggest that they are contemporary in absolute time, but this is not necessarily the case. Solid chronological foundations allow relationships between the spatial and temporal spread/drift of cemeteries to be analysed. If ^{14}C dates are to be used, it is important that there is sufficient coverage across the cemetery, to allow for statistical modelling (see Box 2.2). In an ideal situation, relative and absolute dates would be compared to ascertain the possible absolute date ranges for ceramic (or other artefact) types. Within funerary contexts, plenty of organic materials are available from which samples can be taken, such as basketry coffins, reed mats or the remains of fruit offerings from jars. Bundles of organic materials, such as reeds, are sometimes found between courses of mud-brick, and also permit radiocarbon dating of the brick courses (see Section 2.1.2: Absolute Dating).

As discussed above, the detail and critical nature of the questions depends upon the excavation methodology, the osteological analyses and the integrated application of archaeothanatological approaches (see Duday 2009 for details). It is not always sheer wealth or their expression in terms of large quantities of artefacts or command of manpower that indicate individual importance within a community (Stevenson 2009). The 'value' of an individual need not be directly or solely shown through socio-economic wealth within a burial (e.g. Carr 1995; Drennan 2010: 266; Stevenson 2009). The leader of a community might be distinguished through differentiation in burial. One potential example is the individual deemed to possibly be a community head at El-Omari and who was buried with a worked stick (Debono & Mortensen 1990: Pl. 28.1), whereas the other burials were most frequently supplied only with a single ceramic vessel (Childe 1952: 40–41; Hoffman 1980: 196; Midant-Reynes 2000:

122). An individual might be revered for medical/magical powers, but not be the leader of a community. One such possible example is the female buried at the site of Hierakonpolis in the non-elite cemetery of HK43, Burial 333, dating to the Predynastic period. The woman, aged between 40–50 years old, with an elaborate hairstyle, was equipped with ceramic vessels, a palette and, most notably, a basket full of objects including pendants, an amulet, minerals, bone tools and gaming pieces (Friedman 2003: 18–19). The basket also contained a wide variety of plant remains, some of which may have medicinal properties (Fahmy 2003: 20; 2005). The surrounding burials are those of children, and Friedman (2003: 19) considers that she might have been considered as suitable protection for their afterlife.

Box 2.9   Burial and Mortuary Analysis: Abydos in Practice
Sonia Zakrzewski

Funerary analysis of Egyptian cemeteries uses anthropological methods to develop hypotheses regarding the Egyptian population living at the time of use of the cemetery. But the principles upon which these analyses are based are scientific in their approach and application.

Mortuary studies in Egypt have usually focused on recognition or identification of social differentiation within or between different subsectors of society. Studies of Predynastic cemeteries have focused on grave size and depth, the amount of labour required in construction, numbers and types of grave goods included in the interment, or the position and orientation of the body. Castillos (1997: 251, 1998) has argued that there is insufficient emic knowledge for the researcher to assess the relative importance of artefacts found within burials, and suggests it is better to consider only total numbers of grave goods. By contrast, others consider the significance of the various offerings and types of goods and/or materials found in association with the age and sex or gender of the body (Anderson 1992; Bard 1989; 1994; Seidlmayer 1988; Stevenson 2009; Zegretti 2012). Identification of social hierarchy on the basis of mortuary studies is potentially easier during the Dynastic period, and certainly by the Middle Kingdom, "the mortuary environment provided an important arena for conspicuous consumption as a means of status display" (Richards 1997: 33).

Richards (1997; 2005), employing a multidimensional analysis of the Middle Kingdom Northern Cemetery from Abydos, evaluates ancient Egyptian social systems. This analysis integrates archaeological and textual materials, and then situates them within the mortuary landscape. Taking an anthropological approach, rather than focusing on the elite graves in a quasi-historical Egyptological method, Richards studies the entire range of mortuary behaviours. During the Middle Kingdom, mortuary landscapes are made up of three categories: royal, elite and non-elite burial practices. The most spatially and socially constrained were the royal burials (Richards 2005: 76), with only those highest in social rank buried near the king. Using statistical patterning techniques of analysis, the non-royal mortuary landscapes were also found to have some degree of social differentiation. By assigning a diversity score on the basis of the number of different and distinct artefact categories present within each grave assemblage (a measure of access to resources), and associating this with wealth indices derived from the amount of effort required to obtain the raw materials and the apparent values of individual grave goods based upon Middle Kingdom texts, the non-royal burials from Abydos showed that material resources may be a measure of socioeconomic differentiation and that burial assemblage diversity and levels of wealth correlate with levels of social hierarchy. Problems arise with these studies, as many tombs are reused or older burials destroyed to make space for new interments. Grajetzki (2014: 109–110) describes the jewellery and statuary found in Tomb 108 from Abydos, excavated by Garstang around 1900, and notes that the high-ranking person apparently buried within the tomb might have been male, female, or even a couple. Such difficulties in analysing old excavations demonstrate the importance of modern methods of cemetery excavation and recording.

In addition to social hierarchy, funerary archaeology enables other social clusters or groupings to be delineated. Newborns and children were commonly buried under house floors, such as at el-Lahun (Richards 2005: 66), and may not have generally been interred in the Abydos Middle Kingdom cemetery (Richards 2005: 169). At Middle Kingdom Abydos, where children were interred in the cemetery, they were buried in both shaft and surface contexts; these are hypothesised to represent both poorer and wealthier forms of burial practice (Richards 2005: 170–171). In the preceding First Intermediate Period, numerous burials of infants and small children are found in the North Abydos town, the new mortuary town at South Abydos, the pyramid town at el-Lahun (mentioned above) and the frontier town of Elephantine (Richards 2005: 174). During the

Third Intermediate period, a tightly clustered series of child burials were located in one stratum of Site 8 in northwest Abydos (Patch 2007). Such age-related patterning demonstrates that biological anthropology may permit the identification of personhood within childhood to be delineated.

More recent cemetery studies have employed complex analyses of spatial patterning, such as through the use of GIS methods. These studies focus on the internal structure of the cemetery, such as the internal spatial segregation, but the actual construction and delineation of the cemetery complex is also clearly important. Verner and Brůna (2011) analysed the formation and foundation of the 5th Dynasty cemetery at Abusir in terms of both cartographic orientation and symbolic and/or religious linkage. Using viewshed methods from GIS, both the visibility of the tombs themselves and the visibility of the surrounding region from the cemetery can be studied. Greater local (and relatively distant) land is visible from southern Levantine Chalcolithic secondary burial sites than from their associated habitations (Winter-Livneh et al. 2012); viewshed analysis has here been used for territorial study, but has been extended to include social and cognitive aspects of the funerary landscape (Ericson Lagerås 2002), but has yet to be applied to Egyptian funerary contexts, although Kemp (2007) has considered visibility and lines of sight for the South Cemetery at Amarna.

Anthropological studies (see also Chapter 3) have included archaeothanatological methods (Duday 2009). At Adaïma, movement of the labile bones, such as the finger phalanges, indicated dehydration and decomposition of the hand in a void. Furthermore, the majority of inhumations from the western cemetery at Adaïma were found with the vertebrae in perfect anatomical orientation and articulation, indicating that the sand was directly in contact with the body at burial (Crubézy et al. 2002: 445–449). Detailed archaeothanatological and taphonomic study has, thus far, rarely been undertaken during the excavation of Egyptian cemeteries, and it is hoped that, especially by combining this method with GIS, its introduction will enable greater understanding of funerary processes and social structures.

References

Anderson W. (1992) Badarian burials: evidence of social inequality in Middle Egypt during the early Predynastic era. *Journal of the American Research Center in Egypt* 29: 51–66.

Bard K. (1989) The evolution of social complexity in Predynastic Egypt. *Journal of Mediterranean Archaeology* 2: 223–248.

Bard K. (1994) *From Farmers to Pharaohs: Mortuary Evidence for the Rise of Complex Society in Egypt.* Sheffield: Sheffield Academic Press.

Castillos JJ. (1997) New Data on Egyptian Predynastic Cemeteries. *Revue d'Egyptologie* 48: 251–256.

Castillos JJ. (1998) Wealth Evaluation of Predynastic Tombs. *Göttinger Mizellen* 163: 27–33.

Crubézy E., Janin T. & Midant Reynes B. (2002) *Adaïma. 2. La nécropole prédynastique. Fouilles de l'IFAO 47.* Cairo: Institut français d'archéologie orientale.

Duday H. (2009) *The Archaeology of the Dead: Lectures in Archaeothanatology.* Oxford: Oxbow.

Ericson Lagerås K. (2002) Visible Intentions? Viewshed Analysis of Bronze Age burials mounds in western Scania, Sweden. In: Scarre C. (ed.) *Monuments and Landscape in Atlantic Europe.* London: Routledge. pp. 179–191.

Grajetzki W. (2014) *Tomb Treasures of the Late Middle Kingdom.* Philadelphia: University of Pennsylvania Press.

Kemp B. (2007) The Orientation of Burials at Tell el-Amarna. In: Hawass ZA. & Richards J. (eds) *The Archaeology and Art of Ancient Egypt. Annales du Service des Antiquités de l'Égypte, Cahier No. 36.* Cairo: SCA. pp. 21–31.

Patch DC. (2007) Third Intermediate Period Burials of Young Children at Abydos. In: Hawass ZA & Richards J. (eds) *The Archaeology and Art of Ancient Egypt. Annales du Service des Antiquités de l'Égypte, Cahier No. 36.* Cairo: SCA. pp. 237–255.

Richards JE. (1997) Ancient Egyptian Mortuary Practice and the Study of Socioeconomic Differentiation. In: Lustig J. (ed.) *Anthropology and Egyptology: A Developing Dialogue.* Sheffield: Sheffield Academic Press. pp. 33–42.

Richards JE. (2005) *Society and Death in Ancient Egypt: Mortuary Landscapes of the Middle Kingdom.* Cambridge: Cambridge University Press.

Seidlmayer S. (1988) Funerärer Aufwand und soziale Ungleichheit: eine methodische Anmerkung zum Problem der Rekonstruktion der gesellschaftlichen Gliederung aus Friedhofsfunden. *Göttinger Mizellen* 104: 25–51.

Stevenson A. (2009) *The Predynastic Cemetery of el-Gerzeh. Orientalia Lovaniensia Analecta 186.* Leuven: Peeters.

Verner M. & Brůna V. (2011) Why was the Fifth Dynasty cemetery founded at Abusir? In: Strudwick N. & Strudwick H. (eds) *Old Kingdom, New Perspectives: Egyptian Art and Archaeology 2750–2150 BC.* Oxford: Oxbow. pp. 286–294.

Winter-Livneh R., Svoray T. & Gileada I. (2012) Secondary burial cemeteries, visibility and land tenure: A view from the southern Levant

Chalcolithic period. *Journal of Anthropological Archaeology* 31: 423–438.

Zegretti C. (2012) Child Burials of the Nubian A-Group. In: Kabacínski J., Chłodnicki M. & Kobusiewicz M. (eds) *Prehistory of Northeastern Africa: New Ideas and Discoveries*. Poznán: Poznán Archaeological Museum. pp. 141–152.

3   The Biography of People

For the layman, the archaeology of Egypt is (usually) all about the people. This may be their mummies, their funerary temples and/or their tomb complexes. The scientist, however, can provide specific detail to add to the Egyptological analysis, and so may give further information on the life or death of the particular individual, their diet and lifestyle, their origins and affiliations etc. (Buzon 2012). The person, as an individual, should not be forgotten inside his/her tomb or mummy. "To understand a burial is to bear in mind, above all, that skeletons were once corpses" (Duday 2009: 7), and before that, a living a person.

3.1   Death and Burial

For much of the general public, the tombs and burial places of ancient Egyptians are *the* defining monuments and archaeological evidence of the Egyptian civilisation. Death and burial were important as, through the tomb, they provided the link to the continuation of life. Effort was expended on ensuring that deceased individuals were accorded proper burial rites and rituals in order to ensure that they would be reborn and would live happily in the afterlife, but also, one assumes, to cope with grief and facilitate social cohesion amongst the living following the person's death. Furthermore, in comparison with the modern world, where, although death is not taboo, but rather is experienced in public and in private (Sayer 2010), it is likely that death (of all people) was more socially visible to the living ancient Egyptian population.

Archaeological analyses of death are undertaken on a variety of scales, from the individual body, to the funerary facility and architecture, and the broader mortuary landscape (Chapman 2003). The richness of Egypt's burial record means that these tombs and graves provide a large proportion of our archaeological knowledge of Egyptian society and social organisation. The mortuary temples of the Royals provide information about cosmology and the role of the Pharaoh and his family within that cosmology. In contrast, tomb chapels provide details about the name, profession and family of their owner. Through the artistic record found in these funerary contexts, such as paintings, reliefs and statues, we may find hints as to diet and subsistence patterning, or as to family organisation and structure. Despite the plundering of grave goods, both in antiquity and in more recent times, the funerary equipment also provides an insight into the range and types of objects considered by the ancient Egyptians to have been important in life. Although this section will consider what we may learn of the people from the funerary context, it is important to link this to the sections that consider not only the objects themselves (see Chapter 4: Objects), but also the analysis of cemeteries (Section 2.4) and analysis of stone in terms of tomb and grave construction methods (Section 4.4.1).

A complication arises from the fact that much of what is known archaeologically (or Egyptologically) regarding death and burial derives primarily from the elite sector of society. From texts such as the *Tale of Sinuhe* (a Middle Kingdom fictional story), it is evident that burial was important for the ancient Egyptians as they could only be ensured an eternal afterlife if the correct Egyptian rituals and rites were performed. In addition, foreigners who lived an 'Egyptian' life might be accorded an 'Egyptian' burial. For example, several Middle Kingdom funerary stelae from Gebelein, such as Boston MFA 03.1848 (which specifically calls the individual depicted 'Nehesy', i.e. the ancient Egyptian name for Nubians (Kendall 1997)) and Leiden F 1938/1.6 suggest that Nubian mercenaries at Gebelein married Egyptian women (Fischer 1961) and were buried near the Egyptian community. The stelae show that they were buried in an Egyptian manner, whilst still being depicted as Nubian, thus retaining their ethnic identity.

3.1.1 Burial Rites and Rituals

The preservation of the physical body was an essential part of Egyptian funerary practice, as it was to the body that the soul, primarily the *ka*, would have to return in order to find sustenance. Although the *ka*, like the *ba* and *akh*, was a spirit, the preservation of the body in a recognisable form was essential for its survival. If the body had decayed, decomposed or was unrecognisable, the *ka* would go hungry and the afterlife would be jeopardised.

During the Predynastic period, and in poor social classes throughout Egypt's history, the interment of the deceased in the sand would generally have meant that the body was likely to be preserved through natural desiccation. It has been argued that the practice of mummification grew from the recognition of this natural desiccation process and a desire to improve upon it; however, it is more likely that the practice evolved to preserve the image of the body (Shaw & Nicholson 1995). Some Naqada II period bodies from Hierakonpolis suggest that early attempts at artificial preservation were undertaken – 'resin' was applied both to the body and to linen bandages wrapped around it (Ikram 2010; Ikram &

Dodson 1998). Furthermore, Jones et al (2014) have demonstrated the use of pine resins, plants gums/sugars and fats in Badarian linen wrappings from Mostagedda using gas chromatography and mass spectrometry (see Section 3.5: Clothing & Adornment). In that the body was necessary for the afterlife, over time, burials became more elaborate, and hence the methods employed to preserve the body became more complex. In the Old Kingdom, the body was wrapped carefully and moulded in plaster-soaked linen so that it almost resembled a statue (Ikram & Dodson 1998).

The basic principle of mummification is very simple and focuses upon artificially dehydrating the body and preserving it with natron. Natron is a salt comprising sodium carbonate, sodium bicarbonate, sodium chloride and traces of sodium sulphate. Natron also dissolves fatty tissues and protects the flesh from bacterial and fungal attack (Ikram 2010). Although the basic principles of mummification remained constant, the materials and methods used to preserve the body gradually changed. Furthermore, differences exist between the methods employed by various embalming houses. As a result, the form of mummification depended upon the personal preferences of the deceased and that individual's wealth.

The 'classic' manner of mummification involved the removal and preservation of the internal organs, the disinfection of the body with palm wine and/or a natron solution, and the desiccation of the body with powdered natron. The earliest confirmed evidence of this dates from the 4th Dynasty (Ikram 2010); the internal organs of Queen Hetep-heres were found within the canopic chest in her burial at Giza (Andrews 1984). During the 18th Dynasty, the brain was commonly removed. Initially, this was done through the foramen magnum, but later, using an iron hook, the left nostril was used by breaking through the ethmoid bone (Ikram 2010). The cranial cavity was then filled with liquid resin.

In all mummification, the internal organs were extracted, usually through an incision cut in the left flank using an obsidian blade (Shaw & Nicholson 1995). The viscera were dried, rinsed, bandaged and placed in canopic jars (although during the Third Intermediate Period, they were returned to the body cavity). The canopic jars of Queen Hetep-heres indicate that the internal organs were soaked in a solution of 3 per cent natron during mummification (Harris & Weeks 1973), although this is too weak a concentration to prevent decay, suggesting that this natron solution was merely a preservative (Andrews 1984). The body cavity was then cleaned out, rinsed with palm wine and powdered spices, and stitched up again. During the New Kingdom, the abdominal cavity was stuffed with linen soaked in resin to maintain a lifelike shape. Furthermore, the resin added to the body was heated before application to ensure that it completely coated the internal cavity, thereby killing any bacteria and forming an airtight seal to preserve the soft tissue (Harris & Weeks 1973). To speed up the dehydration process and to prevent any disfigurement of the body, the abdomen and the thorax were usually packed with temporary stuffing or packing material. Traditionally, this process of desiccation took 40 days. It is thought that natron was at first used in solution rather than as dry crystals, but the latter was preferred later as crystals enable more rapid desiccation.

After the dehydration process, the body was removed from the natron, and the temporary stuffing removed from the body cavities. This was then followed by a 30-day period of further funerary rites and preparations, including washing the body and anointing it with oils and incense, the recitation of prayers, and the wrapping of the body in linen bandaging anointed with gum prior to burial. At this stage, hair might be added in the form of wigs or extensions, missing limbs might be replaced (e.g. with rolls of cloth; Vogelsang-Eastwood 2000), and in the New Kingdom artificial eyes were even inserted (Szpakowska 2008). The body of a woman (Unknown Woman B) from Deir el-Bahari, tentatively identified as Queen Tetisheri, and dating to the 17th Dynasty, had artificial hair braided into her own white hair (Ikram & Dodson 1998). After this process, the mummified body was placed in a wooden coffin. Anthropoid coffins were first used around the middle of the 12th Dynasty, but this shape of coffin only became common within the non-elite towards the end of the 13th Dynasty. Rectangular coffins later replaced anthropoid coffins.

Less expensive methods of mummification also existed. For example, juniper or cedar oil, together with an oleo-resin akin to turpentine, was sometimes injected into the abdomen, for example through the anus, in order to dissolve the viscera (Ikram & Dodson 1998; Shaw & Nicholson 1995). The body was then mummified for the prescribed number of days, after which the oil was allowed to escape. This removes all the internal organs in a liquefied form. Furthermore, the natron eats away the flesh, so reducing the corpse to skin and bone. A third and even cheaper method of embalming comprised simply washing out the abdomen with a purge and mummifying the corpse for 70 days before it was taken away (Shaw & Nicholson 1995).

Gas chromatography-mass spectrometry (GC-MS) and thermal desorption (TD-) or pyrolysis (Py-) permit the identification of the embalming substances employed. Resins from conifers and *Pistacia*, coniferous pitch, potentially balsam, plant oils and some animal fats and beeswax have been identified in Pharaonic and Greco-Roman mummies (Buckley & Evershed 2001), with some of these same embalming agents also found in the Badarian wrappings from Mostagedda, and identified using TD/Py-GC-MS (Jones et al. 2014) Sequential TD-GC-MS and Py-GC-MS requires very small sample sizes of 'resin', bandaging or mummy tissue, and an early identification of the third component of the embalming 'resin'. The main products are usually derived from plant oils and fats, suggesting that these were the key ingredients in mummification and were probably used as a relatively cheap substrate with which to mix more exotic embalming agents to the bodies and/or their wrappings. These unsaturated oils and fats are able to polymerise, thereby stabilising the otherwise fragile tissues and/or textiles. Detection of pine wood compounds, such as methyl esters formed during the smouldering of pine wood in the presence of excessive air (using GC-MS), and/or high levels of sodium in bones (as measured by atomic emission spectrometry) can also be employed to identify bodies that have been defleshed or at least partially skeletonised prior to embalming. An example is the Idu II mummy from Giza (Pelizaeus Museum at Hildesheim no. 2639) (Koller et al. 1998).

Endoscopy may be employed to visualise the state of preservation of the body and the tissues, the presence or absence of the viscera, and the embalming methods. For example, 'fluid' from within the skull may be endoscopically removed and then biopsied in order to identify its components, be they resin or decomposing brain tissue (David 2000). Histology and transmission or analytical electron microscopy, using these endoscopically removed samples, may also be undertaken, such as when *Ascaris* and accumulated heavy metals were identified in the Philadelphia University Museum II mummy (David 2000). Near Infrared Fourier Transform (NIR-FT) Raman spectroscopy can also be used to study molecular structures of biological compounds. This analytical, non-destructive technique, based on the analysis of laser light reflected from the sample, has been employed to investigate the molecular structures of skin samples from mummified bodies, such as the Nekht-Ankh Middle Kingdom mummy (Edwards 2005: 248).

Where there has been some reason why mummification and/or intentional burial has not happened, the macroscopic visual identification of pulpal cases of flies or beetles can be used to identify delayed burial (Faulkner 1986; Nystrom et al. 2005). Having discussed the body itself, it is also important to consider what is around the body, such as the wrappings and/or coffin. These too have scientific potential for analysis, thereby informing us further about the Egyptian funerary process, and also about trade, living and burial conditions, and even religious ideology.

The materials used to wrap bodies vary both geographically and temporarily, with a woollen cloth being found around the skeleton of a man from a 1st Dynasty burial at Helwan (Vogelsang-Eastwood 2000) and a sheepskin found wrapped around an unknown male body from the royal cache of mummies at Deir el-Bahari (Andrews 1984). Textiles excavated from tombs demonstrate that as early as the Predynastic period, the Egyptians were proficient at spinning and weaving (Vogelsang-Eastwood 2000). For example, the Naqada II mummies from Hierakonpolis were wrapped in shrouds, and may have originally worn linen and leather clothing (Ikram & Dodson 1998). The majority of ancient Egyptian textiles are of linen, made from flax (Vogelsang-Eastwood 2000). Before a linen thread can be made, the flax must be spun. Spun threads may be

S-spun (anticlockwise), Z-spun (clockwise), or I-spun (no spin). When two or more spun threads are plyed together, the yarn is in the opposite direction to the original spin, so S-plyed yarn may be made up of two or more Z-spun threads. This identification can be made using a microscope. The quality of the linen cloth depended upon social rank. New, unused linen with a high thread count was used for royalty, whereas lower ranking individuals would have been wrapped in fabrics that had already been used in life (Szpakowska 2008). Computed tomography (CT-scanning) enables the number of layers of linen to be counted without destruction or unwrapping. For example, the small female mummy found in mastaba G 2220 at Giza, dating from the 5th Dynasty, had each limb and extremity individually wrapped in over 37 layers of linen bands, into which linen pads were also inserted to give the body a more lifelike appearance (Ikram & Dodson 1998). For some mummies, counting the separate layers in a reliable manner using CT is impossible due to the blending of bandages (Taconis & Maat 2005). With these mummies, however, the distribution of resin-soaked bandages can still be identified as the soaked bandages have a higher density. Some bodies are found clothed, such as the mummy of a woman buried at Giza wearing a sleeveless dress with a deep V-neckline (Vogelsang-Eastwood 2000), and scientific analysis of such clothing might permit estimation of the sewing and construction methods.

A shroud, or cloth cover, was also placed over or around the body. By the New Kingdom, some shrouds were inscribed in ink with the name of the deceased and spells or chapters from the Book of the Dead (Vogelsang-Eastwood 2000). On some occasions shrouds were more pictorial, such as the one over the 18th-Dynasty mummy of Hatnefer (Cairo JE 66218) which took the form of a linen sheet decorated with a painted life-sized figure of Osiris. At other times, actual garments were used for shrouding, such as the two tunics also found covering the outer layers of bandages around the mummy of Hatnefer.

The vessels into which the bodies were placed affected the preservation of the individual; for example, the small coffins of clay, basketry or wood used by the end of the Late Predynastic inhibited the desiccation of the corpse through direct contact with sand, thereby leading to decomposition of the body. Mummies from the Early Dynastic period were placed in a flexed position in rectangular clay or wooden coffins or stone sarcophagi (Ikram & Dodson 1998) resembling contemporary houses (Andrews 1984), although some bodies from Tarkhan, dating to the 1st Dynasty, were placed on beds.

The introduction of deliberate mummification in the Old Kingdom led to the use of rectangular coffins, as, with the body elongated, it was easier for embalmers to perform their tasks such as the removal of the internal organs. Wooden coffins from this period for the non-royal deceased were generally made from rough planks of wood fastened by dowels, with little external decoration (Andrews 1984). In contrast, higher ranking individuals were buried in sarcophagi made of granite, basalt, limestone or alabaster.

Most types of wood can only be identified microscopically to genus level, but some, such as the sycamore fig, as used for the Old Kingdom coffins BM EA46632 and BM EA46633, have diagnostic characteristics enabling their full identification (Gale et al. 2000). Coffins and funerary stelae were also made of other woods, such as Persea (used for the corner piece of the Middle Kingdom coffin BM EA24800), cyprus or turkey oak, whereas funerary fittings, such as coffin dowels or funerary garlands, were made from yet other woods such as tamarisk and willow (Gale et al. 2000). Many of these woods were imported into Egypt, potentially from as far afield as India, thereby providing evidence of trade and interaction. The woods used for the coffins can be identified, and then dated using dendrochronology. Furthermore, grain details in the wood, such as similarities in the ring patterns, allow boards to be identified as deriving from the same tree, such as the cedar boards from a 12th Dynasty sarcophagus in the Boston Museum of Fine Arts (Kuniholm 2001). Obviously, these wooden coffins require the felled timbers to be converted into wood for use. This process can leave macroscopically visible saw lines on the wood of the coffin. Early Dynastic coffins were constructed from planks cut with small handsaws, as seen depicted at the tomb of Ty at Saqqara (Gale et al. 2000). By the end of the Old Kingdom, the pullsaw had been developed which changed the cut marking on the wooden planks. In addition, by the 3rd Dynasty, lamination of layers of wood at right angles to each other had started, as in a coffin found in Gallery V at the Step pyramid of Djoser (Gale et al. 2000). Lamination and its internal complexity, such as the strengthening of the coffin above with wooden battens, gold sheet and nails, or corner joint construction, such as with tenons, tongues or dowels, may be identified non-destructively using computed tomography.

The sources of the stone for the sarcophagi may be determined using petrological analysis (using thin-section petrography), Scanning Electron Microscopy (SEM) (coupled with either energy-dispersive X-ray analysis (-EDS) or wave-length dispersive spectrometry (-WDS)), X-ray fluorescence (XRF), X-ray diffraction (XRD), electron microprobe analysis (EMPA), neutron activation analysis (NAA) and/or atomic absorption spectrophotometry (AAS) (Aston et al. 2000). Of these, the simplest to undertake in a laboratory is thin-section petrology. The stone is then matched with stone from ancient Egyptian quarries or geological maps. In the field, many types of stone can be roughly identified through various means – megascopic examination using a hand lens, carbonate content testing using hydrochloric acid, colour streak testing, testing the Mohs scratch hardness, and/or by using a portable hydrometer (to measure specific gravity or density) (Aston et al. 2000).

Basketry was used for small coffins by the end of the Late Predynastic. Microscopic study of the cross-section or epiderm patterns is required to identify plant parts, and studies of size and shape are required to identify the individual fibres. Using these methods, the shrub-like *Ceruana pratensis* has been identified as having been used in the Early Dynastic period for making basketry coffins (Wendrich 2000). Macroscopic analysis also reveals the form of basketry technique employed. For example, bound coffins, twined and bound reed matting and woven bed matting were found among the burials at Tarkhan, Fayum, Badari and Merimda (Wendrich 2000). Furthermore, the walls of a 1st Dynasty funerary chapel at Saqqara were made of flexible grass matting, with the bundles of grass made into a fabric with widely spaced rows of twining.

The decoration of coffins and other funerary items may also be analysed. Gypsum was made into thick plaster and could be applied to wood in order to disguise the grain (Gale et al. 2000). The pigments for the decoration of the coffins, such as the colour used for the hieroglyphs of the titles of the owners and the *wedjat*-eyes, may be analysed with polarised light microscopy (PLM), with scanning electron microscopy (SEM), with electron microprobe techniques (EMPA), X-ray fluorescence spectroscopy (XRF) and/or X-ray diffraction (XRD). Veneering and inlaying were used for the decoration of boxes and tomb furniture. Prior to the Late Period, dyed textiles, however, are rarely found within tombs; this is likely a result of the fact that flax, used to make linen, is difficult to dye (Vogelsang-Eastwood 2000). The earliest method of dyeing cloth was smearing, where the colour was spread over the material, possibly with the aid of a medium such as clay, mud, or honey. Later, both vat- and acid- (or non-vat-) dying were developed. Some clothes were even bleached, potentially as an indicator of social stature and/or a sign of cleanliness (Vogelsang-Eastwood 2000). There are, however, fragments of animal skins in Predynastic graves, and some of these appear to have been coloured red, black, white or yellow (Van Driel-Murray 2000). XRF analysis may identify the mineral pigments employed in this colouring, such as the red on leather shoes from Abydos which appears to be derived from iron and possibly lead compounds. The methods of curing animal skins to form leather are not fully understood, but vegetable oil, animal fats (such as cholesterol), salt, alum, ochre, calcium compounds, urine and fermenting flour appear to have been used, and can be identified using dispersive X-ray analysis or X-ray fluorescence spectroscopy (Van Driel-Murray 2000).

3.1.2   *Patterning of Burials*

The layout and format of cemeteries varies primarily across time periods, but also geographically. Some aspects of this are briefly discussed in Section 2.4: Organisation of Burial Grounds, but here the ideas are developed. As noted earlier, the patterning of burials enables us to consider social organisation, familial affiliation and religious concepts of death. Who

Box 3.1   An Inside Look at Craniotomy
Andrew Wade

Assessment of brain treatment in Egyptian mummies has often relied on circumstantial evidence, such as nasal tampon presence, or on direct internal observation following destructive drilling, sawing, or decapitation. Computed tomography (CT) and plain film radiographs provide non-destructive alternatives for detailed three-dimensional studies of craniotomy, evisceration, and other mummification features.

What Types of Brain Treatment are Present in Egyptian Mummies?

*   Transnasal craniotomy (TNC) is the best-known Egyptian craniotomy process, in which a trocar-like tool is inserted into the nose to perforate the thin table of bone to the cranium's interior. In so doing, other bones are also often affected. Following excerebration, embalmers often filled the cranial cavity with quantities of linen or resin (Figure 3.1A, B). The nasal passage and artificial foramen were typically sealed with resin-impregnated rolls (tampons) of linen (Figure 3.1A [marked #], B).
*   Where the brain is absent and the bones of the nose and face are undamaged (Figure 3.1C), mummies are often assumed to have undergone transforaminal craniotomy (TFC) – excerebration through the foramen magnum and an incision at the skull base (e.g. Merigaud 2007; Nelson 2008). This technique is neither well documented nor well understood.
*   Rarely, the brain was removed opportunistically through perimortem trauma to the skull (e.g. Marx & D'Auria 1986).
*   In many mummies, the brain was not removed (Figure 3.1D). Although early researchers questioned the possibility of brain mummification, intact brains are present in numerous Egyptian mummies (e.g. Lewin & Harwood-Nash 1977).

What Are the Difficulties in Assessing Craniotomy?

Obviously, the skull prevents easy internal examination. Gross physical examination, even when not impeded by wrappings, can elicit very little information about brain treatment. Nasal tampons are suggestive, but not conclusive, of TNC craniotomy. And

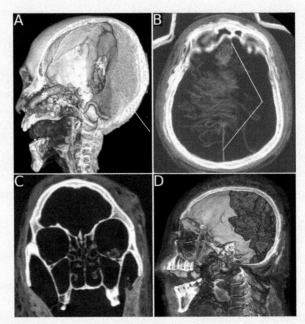

Figure 3.1 CT imaging to demonstrate craniotomy

A: lateral view, showing resin-impregnated rolls (tampons) of linen, denoted with #, with bone fracture fragment visible, denoted by *, and dural remnants marked with lines. B: superior view, with dural remnant marked with lines. C: anterior view, brain is absent and the bones of the nose and face are undamaged. D: lateral view, showing brain remaining in situ (Image courtesy of Andrew Wade).

the mere absence of a brain in a detached cranium may simply be the result of rough handling, thereby fragmenting an intact brain sufficiently for it to settle out, rather than the result of an embalming practice. Modern and ancient repairs may exacerbate this damage.

Plain film X-rays are a non-destructive means of examining the interior of objects, and have been used to examine mummified remains almost since the discovery of X-rays. By interpreting bone contours and densities, it is possible to identify brain presence or absence, packing materials and craniotomy damage. However, once the X-rays pass through three-dimensional anatomy they are recorded in a two-dimensional medium, with the X-ray shadow of one structure superimposed upon that of other structures above and below it. This superimposition can obscure contours, and superimposed structures may artificially increase the density of one another.

How Do We Get a Good Look Inside, and What Can We Expect To See?

CT scanners pass fan-shaped X-ray beams at numerous angles through an object, to rows of detectors on the other side, and measures the radiation passing through thin slices (sub-millimetre in current scanners) of the object. Reconstruction algorithms create two-dimensional cross-sectional images of the object, which are stacked together to produce detailed three-dimensional views of a skull's interior.

Craniotomy is marked by the absence of the brain from the cranial cavity. Lateral plain film projections can provide an indication of complete brain absence or of packing materials in the cranial cavity. When mummified, however, the brain is present in the posterior of the cranial cavity (Figure 3.1D), reduced to approximately one quarter of its original volume. The heterogeneous, medium density mass of the mummified brain exhibits an undulating border suggestive of gyri and sulci, and frequently has an identifiable midline fissure. This is often visible in frontal projections.

Dural membranes are often visible in excerebrated and intact crania. On plain films, these structures are not directly visible but may be inferred by their interaction with other materials in the cranial cavity, such as linen and resin. It is easy to visualise dural remnants on CT scans (marked by lines in Figure 3.1A & B).

Damaged facial bones are suggested by disrupted bone contours on plain films and can be clearly visualised in CT scans. Damage on only one side, as viewed in frontal projection, is a likely, but not universal, indicator of TNC. Subtle damage to deep or complex structures, such as the ethmoid air cells (Figure 3.1C), is difficult to visualise on plain films without multiple projections, and a non-displaced fracture may not be visible on any but a handful of plain film views. It will, more likely, only be noted on CT scans. Fracture fragments, however, are relatively easy to identify on both plain films and CT scans (marked by * in Figure 3.1A).

Evidence of TFC is often marked only by the absence of the brain and of TNC indicators (Figure 3.1C). Damage to the skin at the skull base and/or wrappings that intrude into the foramen magnum are suggestive of a TFC embalming incision, but may be the result of unrelated trauma. The absence of or a displaced fracture of superior cervical vertebrae is visible on plain films, but non-displaced and subtle damage may be more reliably noted on CT scans. Resin in the foramen magnum or vertebral canal is not necessarily indicative of TFC, as resin from the abdominal cavity may enter the vertebral canal and cranial cavity.

Resin is a high density homogeneous material (approaching that of bone) and, in the cranial cavity (Figure 3.1A), is the most likely indication of craniotomy. Its relative homogeneity distinguishes it from similar patterns of sand or mud packing. Resin may demonstrate a solidified fluid level, where the superior surface of the once liquid resin is nearly flat. Movement during solidification and secondary applications results in curved, parallel, or non-parallel surfaces of the resin. If sinuses are damaged during craniotomy, similar solidified fluid levels may be apparent in their posterior aspects.

Nasal tampons (# in Figure 3.1A) are easily identified by their placement and are made even more identifiable by resin-impregnation. Untreated linen packing may also be recognisable, especially in CT scans, as its characteristic folded and rolled structures may be visualised (Figure 3.1B).

Craniotomy is an important feature of the ancient Egyptian mummification tradition, but craniotomy reporting in the literature remains stereotyped, simplified, or simply absent. Unhindered by superimposition, computed tomography is the most effective means by which to accurately assess the details of brain treatment and thus is essential to studies of evolution in Egyptian mortuary ritual.

References

Lewin PK. & Harwood-Nash D. (1977) X-ray computed axial tomography of an ancient Egyptian brain. *IRCS Medical Science: Anatomy and Human Biology: Biomedical Technology: Nervous System* 5: 78.

Marx M. & D'Auria AH. (1986) CT examination of eleven Egyptian mummies. *RadioGraphics* 6(2): 321–330.

Merigaud S. (2007) *Étude Paleoradiologique des Deux Momies Égyptiennes du Musée des Beaux Arts et d'Archéologie de Besançon*. Unpublished MD thesis. Besançon: Université de Franche Comté.

Nelson AJ. (2008) *Preliminary Report on the Radiological Examination of Hetep-Bastet*. Unpublished internal report for Galérie de l'Université du Québec à Montréal (UQAM).

was buried in the cemetery? Was all of the society buried there? Or were certain individuals buried in other places? Were some people buried together and others buried separately? Is there patterning of individuals or graves in terms of age, sex, or in association with particular grave goods or funerary treatment? Does the patterning allow us to identify immigrants or people with specific roles or occupations? What can this tell us about the society? Can we identify any social or other hierarchy within the funerary treatment? Does the patterning both within the burial itself and of the grave within the greater cemetery context provide us with any information about the importance of the deceased as perceived by those who buried them or about the role of that person within their society?

There are two broad approaches to the interpretation of early Egyptian burials, with the first attributing funerary elaboration to the need to establish an afterlife for the deceased while the second associates investment in funerals with the social status of the deceased and thus acts as a measure of social hierarchy. Scientific methods allow each of these two approaches to be developed, but before scientific analysis may be undertaken, basic cemetery patterning needs to be evaluated. For example, Predynastic graves rarely intersect, suggesting that some form of above-ground marker existed so that graves were identified and remembered (Stevenson 2009). By the Early Dynastic, large grave structures, such as those at Hierakonpolis, are known. These prominent burials could provide a basis for further clustering over several generations, such as, for example, the circular arrangements of burials that have been observed in cemeteries such as Gerzeh. The surface superstructures (i.e. the mastabas or chapels) do not necessarily survive from poor cemeteries, and so simple assessments of geographic location must be undertaken in the light of analysis of taphonomic change. Irregularity in the alignment of grave pits and bodies across the cemetery might be evidence of shifting patterns of religious or other belief, or may be due to family linkage, or due to other social clustering (Stevenson 2009). For example, during the Middle Kingdom, most burials in proper cemeteries were in shaft tombs, with a range in the depth of the shafts, and with many having more than one horizontal chamber dug off the vertical shaft itself (Szpakowska 2008). Each of these chambers would have held a coffin, thereby potentially allowing whole families to be buried in the same tomb. Furthermore, in non-royal cemeteries, especially in Upper Egypt, the location of a tomb was an indicator of the social status of the deceased. At Haraga, in the Middle Kingdom, the higher ranked individuals were buried on higher ground (Szpakowska 2008). The quality of rock available, could, however, reverse this hierarchy, and so at Thebes, individuals who desired a carved decorated tomb had to sink their tomb below ground in the lower part of the cliff or on the valley floor (Dodson & Ikram 2008). It is through the interaction of Egyptological knowledge and scientific analysis and assessment that the potential for understanding the patterning of the cemetery can be maximised.

3.1.3 Burials and Tombs

How were burials undertaken? How much investment was required to construct a tomb? How were the tombs actually constructed or excavated?

Most of the questions that might be asked and answered within this section are covered elsewhere in this volume in more detail. This section exists primarily to highlight the potential for the integration of such analysis with the archaeological process.

Most tomb substructures are rock-cut and may include a simple shaft straight down into the rock, opening into one or more rooms, or stairways and passages leading to a series of chambers. In the Delta, where the excavation of a full substructure was impossible, the tomb itself was built from brick or stone with a full superstructure erected above. The superstructure itself consisted either of a rock-cut set of rooms or was a freestanding tomb, such as a mastaba, pyramid or temple design.

The 18th Dynasty rock-cut funerary chapels at Amarna were abandoned prior to final use and so permit an estimation of the methods employed in their excavation and decoration (Aston et al. 2000). Some of the New Kingdom tombs at Thebes, such as that of Re, TT201, had the main axis marked by stretching a red painted cord from the entrance to the end of the tomb (Dodson & Ikram 2008). Brick-built tombs obviously required pre-made moulded bricks, the sizes of which vary temporally. The clay from which these are made may be provenanced through analytical chemistry (for details, see sections in Chapter 4: Biography & Analysis of Objects). Bundles of organic materials, such as reeds, were sometimes placed between these brick courses, and hence permit radiocarbon dating of the brick courses (see Section 2.1.2: Absolute Dating). After the construction of funerary monuments in stone developed during the 1st Dynasty, provenancing of the source stone becomes simpler, requiring petrographic analysis. This is helpful as, during the Old Kingdom, limestone was sometimes painted in order to appear like granite (Dodson & Ikram 2008), such as the false door in the tomb of Seankhuiptah at Saqqara. The quarries of Egypt, such as at Tuna and El-Masara, have been much studied and provide comparative samples.

How were the tombs actually decorated? Rock walls were usually primed with mud plaster, straw and gypsum plaster prior to decoration. It is well known that grid lines, formed by dipping strings in red ink and pulling them taut across the walls, were used to structure the images. The technical materials used can be identified, such as the types of rushes used as paintbrushes (e.g. *Juncus maritimus*) through different types of scientific analysis. Identification of the composition of the paints, the adhesives and binding media, such as resin and beeswax, and especially the colours within them, requires a series of chemical analyses including PLM, SEM, XRF, XRD and/or EMPA (for further detail, see Section 4.4.6: Pigments).

How were bodies actually buried? Obviously we know that some individuals were buried in sarcophagi and then interred in burial tombs whereas others were simply buried in pits in the ground. How can we ascertain whether a coffin was originally used if no wood remains?

Bourriau (2001) has argued for a change in body position (disposition) associated with the change from rectangular to anthropoid coffins. An 'anthropologie de terrain' approach may be used to debate this by identifying the skeletal changes and movements associated with the body coverings and organisation. For example, 'verticalisation' of the clavicles occurs as a result of the body being placed in either a narrow coffin or a tight shroud (Duday 2009), and hence may occur as a result of tight mummification or burial wrapped in a shroud following Coptic norms.

In certain modern contexts, cremains may be recovered. Although intentional cremation was theologically impossible in ancient Egypt, and is proscribed by Islam and the Coptic Church, it might be possible, in the future, and especially within forensic contexts, to find cremated remains. Cremated material is difficult to study but it has potential to provide plentiful information (McKinley 2013). Burning can lead to extreme fragmentation, with osteoporotic bone being more likely to fragment than normal bone (Christensen 2002). Reconstruction of cremated remains can facilitate determination of human v. nonhuman, such as through recognition of specific skeletal elements (Ubelaker 2009). The degree of destruction of human bodies and the colour of bone fragments are related to the duration and intensity of the fire (Bohnert et al. 1998; Walker et al. 2008). Isotopic analysis, in order to determine diet or mobility, is possible. Istopic analysis of carbon and nitrogen from bone collagen after heat exposure of no more than 300°C, whereas the isotopic signal from strontium remains recognisable and unaltered even in bones exposed to very high temperatures (Harbeck et al. 2011). Furthermore, authentic, amplified DNA appears to be retrievable from experimentally cremated bones at temperatures of 600°C, but contamination may be an issue in certain contexts (Harbeck et al. 2011).

3.1.4 Age and Sex

Is it a boy or a girl? This is usually the first question asked upon hearing that someone has given birth. Indeed, the first characteristics noticed upon meeting someone new are usually their gender, their approximate age and their height. In this section, aspects of assigning biological age and sex categories are briefly considered. It is important, therefore, to remember that these are biological categories rather than socially linked equivalents. The literature on the difference between sex and gender is well known (e.g. Hollimon 2011; Sofaer 2006a; 2006b; Walker & Collins Cook 1998), but the differences between chronological age, biological age and social age are less well developed (but see Appleby 2010; 2011; Gowland 2006; Lewis 2007; Sofaer 2006a; 2011 for discussions).

Why are we interested in knowing the biological sex of an individual? Knowing the sex of a skeleton permits us to be more accurate in determining its biological age and calculating the likely stature or height.

Furthermore, the sexes differ in their responses to disease, as females usually exhibit a greater immune response (Ortner 1998), thereby leading to differences in disease frequency between the sexes (Grauer & Stuart-Macadam 1998). In addition, knowing the biological sex allows us to make assessments of gender-based differences in life course aspects such as life expectancy, activity patterning, occupation (Section 3.2) or diet (Section 3.4).

It is well known that male and female adult skeletons differ sufficiently in gross morphology for the sex of burials to be ascertained with reasonable accuracy. For mummified remains, obviously, one hopes that the external genitalia are still visible or may be ascertained from methods such as CT scanning. Methods for the assessment of sex have been and are being developed that increase the accuracy of such determinations for Egyptian and Nubian skeletons. Although the whole skeleton should be studied, and each section of the body should be analysed independently, the pelvis and the skull are the most sexually dimorphic regions of the body (Cabo et al. 2012; Garvin 2012). Although a narrow pelvis is more efficient for bipedal locomotion, female pelvic morphology is constrained by the requirement for breadth in order to facilitate childbirth (Mays 2010).

There are a series of gross morphological characteristics used to assign a sex to an adult skeleton (e.g. see recommendations in Brickley & McKinley 2004; Buikstra & Ubelaker 1994; Workshop of European Anthropologists 1980). Using these methods, there are (usually) five categories of biological sex into which an adult skeleton may be placed: male, ?male (where the features appear more male than female but sex determination is not completely certain), ?sex or unsexable (where either the features exhibit both male and female traits or there are insufficient features preserved for a sex determination to be made), ?female (where more traits are female than male) and, finally female. In most analyses, the male and ?male data are pooled (and similarly for female and ?female). Relative to other groups, Egyptians and Nubians tend to have rather inflated mastoid regions, and so the use of mastoid size as an indicator of sex can be problematic. In addition to sexing using pelvic or cranial characters, sex determination may be undertaken using metric traits, such as femoral or radial head diameter or canine crown size, or through other morphological assessment, such as constriction of the trochlea of the humerus, relative dimensions of the sternum, or evaluation of muscle robusticity and rugosity.

Using these methods and seriation to establish the sample-specific range of sexual dimorphism, it is usually possible to assign sex to an adult skeleton. Children, however, are another matter. The assessment of the biological sex of an individual who has not yet attained adulthood is extremely hard, although there are methods that have been developed which may permit probable sex to be assigned. These methods, using the morphology of the pelvis (Holcomb & Konigsberg 1995; Schutkowski

1993; Wilson et al. 2008), the skull (Loth & Henneberg 2001; Molleson et al. 1998) and of the dentition (De Vito & Saunders 1990; Hillson 1996; Saunders et al. 2007), are used by some archaeologists but are not widely and unequivocally accepted.

Ancient DNA analysis, although destructive and expensive, has proved very useful for sex estimation of non-adults, of fragmentary adults or of cremated material. The method works by detecting X and Y chromosome-specific sequences within the amelogenin gene which is present on both the X and Y chromosomes (Gibbon et al. 2009; Stone et al. 1996). Although there are issues with the non-survival of ancient DNA in Egypt, the successful use of this method to determine the sex of potential Roman infanticide victims from Israel (Faerman et al. 1998), suggests that the method may be of use in the future.

Whereas the sex of an adult can usually be determined skeletally, while the sex of a juvenile cannot, the biological age of a juvenile can usually be determined relatively accurately whereas that of an adult is much more difficult to ascertain. Biologically, people age at different rates, and it is likely that, in Egypt, social age was of greater importance than chronological age (Janssen & Janssen 2005). Given this discrepancy, why are we interested in knowing the age of an individual? It is useful to know how old people were when they died in order to interpret patterns of longevity or disease as some diseases affect age groups in different ways. For example, osteoporosis and dental caries tend to be diseases affecting older people, whereas epidemic-type diseases, such as measles, tend to affect children as they have yet to develop immunity to the specific infection.

Techniques used to age the human skeleton are based upon patterns of growth, remodelling and degeneration. Sub-adults or juveniles can be aged on the basis of their dental development and eruption, and their bone development, growth and maturation. As dental development is highly genetically canalised, methods based upon dental morphology (including crown formation, root completion, and eruption of both the deciduous and permanent dentition) are usually considered more accurate than and hence preferable to those based upon the long bones. Obviously, where teeth are still in crypts in the jaws and have yet to erupt, dental age may be more accurately assigned if radiographs of the dentition can be taken. Although no atlas of specifically Egyptian dental development currently exists, data for East African males (Chagula 1960), Native American (Ubelaker 1978), modern American (Moorrees et al. 1963a; 1963b; Smith 1991), French-Canadian (Demirjian et al. 1973) and British (AlQahtani et al. 2010) samples are generally used for comparison. More accurate but expensive and destructive methods also exist, such as counting the perikymata under SEM (Hillson 1996; 2005), and as such, it is hoped that an Egyptian-specific dental development atlas may be compiled.

Bone development may also be used to age juveniles, using the lengths and other measurements of the long bone, and the fusion states of the varying epiphyses (e.g. using charts and tables in Scheuer & Black (2000; 2004) or the recently published Egyptian-specific versions in Boccone et al. (2010)). Epiphyseal fusion starts at around 11–12 years of age with fusion of the epiphyses of the distal humerus and proximal radius (around the elbow). As a result, this method is most applicable to older juveniles and adolescents. Due to sexual maturation, the epiphyses of females tend to fuse up to 2 years earlier than those of males, thereby complicating ageing. For younger individuals, and particularly for foetal or perinatal skeletons (Scheuer et al. 1980), the length of long bone diaphyses can be used to assign age. Childhood growth, however, may be delayed due to periods of malnutrition or infection, and hence it is uncertain how accurate such methods really are for past Egyptian or Nubian samples (see also Boccone et al. 2010). Furthermore, children in the past appear to have been shorter than their modern peers (Humphrey 2000) but children who suffer stress during growth appear to 'catch up' when they have recovered or been removed from the stressor and so may end up reaching their genetic potential for adult size (Tanner 1963; 1989; Weiss 2015: 17–54).

Ageing adults, by contrast, is a more problematic issue (Garvin et al. 2012; Milner & Boldsen 2012). Whereas children can usually be aged to within 2 years (or even less for the youngest), adults are usually only assigned into one of four categories: young adult (approximately biologically 20–35 years), middle adult (approximately 36–49 years), older adult (50 years plus) or simply adult (a skeleton for which it is impossible to assign an age). Within bioarchaeology, it is normal to use several methods for age estimation, but all these methods tend to underage skeletons known to be over 70 years at death and overage those who died at less than 40 years of age (Molleson & Cox 1993).

For younger adults, such as those who died in their 20s, similar methods to the developmental methods used for children may be employed, such as eruption of the third molar or fusion of the late-fusing epiphyses such as the sternal length of the clavicles or the iliac crest on the pelvis. After this age, methods involving remodelling or bodily degeneration are required. Unlike growth-related processes, degenerative processes are strongly influenced by environmental factors and hence lead to only broad age estimation ranges. The most common methods for age estimation include analyses of the morphology of the auricular surfaces (Buckberry & Chamberlain 2002; Falys et al. 2006; Lovejoy et al. 1985; Mulhern & Jones 2005), of the pubic syntheses (Brooks & Suchey 1990; Todd 1920), of the sternal rib ends (İşcan et al. 1984; 1985; Loth & İşcan 1989; Russell et al. 1993), of the first ribs (Kunos et al. 1999), of the cranial sutures (Meindl & Lovejoy 1985; Perizonius 1984) or of the teeth (Brothwell 1989; Miles 2001). Of these methods, cranial suture closure is probably the least precise and, as such, should only be used as an addition to other methods or when no other methods are available. Furthermore, as a result of their diet and the abrasiveness of sand, older Egyptians and

Nubians tended to have more heavily worn teeth and so ageing based upon dental wear can be problematic.

There is also a series of potentially more accurate methods of assigning age to the skeleton, but currently these are not usually undertaken in Egypt due either to their cost or their destructive nature. Microscopic ageing of bone uses thin sections of bone, normally obtained from the cortex of a femur or a rib, and relies on calculation of the relative number of osteons and the proportion of lamellar bone in the sample, with an increase in osteons and fragments thereof and a decrease in lamellar bone being associated with increasing age (Robling & Stout 2000; 2003). More complex dental methods include counting incremental cementum layers and the evaluation of root translucency. The former method relies on counting the thin layers of cementum using a thin section from a single rooted tooth under a microscope (Blondiaux et al. 2006; Renz & Radlanski 2006; Wittwer-Backofen & Buba 2002; Wittwer-Backofen et al. 2004). Unfortunately, cementum preservation is poor and the periodicity of the cementum deposition is uncertain and so it is unclear whether these increments are annual (Mays 2010). Root translucency appears to occur as a result of age-related change within the dentine and progresses from the tip of the tooth root to the cemento-enamel junction. Transillumination permits root translucency to be measured without the need for teeth to be sectioned (Lamendin et al. 1992), but, although apparently accurate in forensic contexts (Lucy et al. 1995; Megyesi et al. 2006; Sengupta et al. 1998; Solheim 1989), it appears to be affected by both the burial environment and the duration of burial (Mays 2010). Future research on these or similar methods may enable the development of more accurate methods for assessing Egyptian skeletal age, and thus may permit the rougher and cheaper methods outlined above to be calibrated and hence become more accurate for Egyptian skeletal material.

3.1.5 Identification of Children and Childhood

When does a child become a person? Is there a point, in his or her life course, at which he or she is recognised as a separate social individual? And when and how does a child become an adult? The study of childhood in Egypt has usually focused upon either the artistic recognition of children (distinct from adults in that they are visually represented either naked or wearing jewellery, with sidelocks of hair and frequently with a finger towards the mouth; Szpakowska 2008) or the identification of children's burials within the funerary record, separating out of the types of grave goods that may have been associated with children. One of the most commonly perceived limitations to the study of children has been their poor preservation, leading to a common belief that children rarely survived the burial environment (Lewis 2007). Given the excellent burial preservation within Egypt and Sudan, there are sites with many nonadult skeletons, such as Adaïma (Crubézy et al. 2002), Gerzeh (Stevenson 2006), Naga-ed-Dêr (Lythgoe 1965; Podzorski 1990), Gebelein and Assiut (Masali & Chiarelli 1966) and medieval Meinarti in Sudan (Green et al. 1974; Swedlund & Armelagos 1969; 1976).

In order to undertake analyses that enable childhood to be better understood, children themselves need to be accurately aged. As summarised above, children and infant burials are assigned biological ages on the basis of their dental development, their state of epiphyseal fusion and the length of their long bones. The ages that these methods provide have a degree of uncertainty associated with them, such as plus or minus 2 years. It is, however, possible to assign a more accurate age if microscopic analysis of the teeth is undertaken, such as through counting the perikymata.

Box 3.2 Breastfeeding and Weaning Practices in Ancient Kellis: The Stable Isotope Evidence
Tosha L. Dupras

Carbon and nitrogen stable isotope analyses are accepted scientific methods of interpreting breastfeeding and weaning practices in ancient populations (see Dupras et al. (2001) for detailed explanation). Carbon stable isotopes are used to distinguish between diets based on C_3 and C_4 plants.[1] C_3 plants, such as wheat, barley, rice, grasses, trees, and most fruits and vegetables, range in $\delta^{13}C$ values from $-22‰$ to $-33‰$,[2] while C_4 plants, such as maize, sorghum, some millets, sugar cane, and tropical grasses, are enriched in ^{13}C with values that range from $-16‰$ to $-9‰$ (Smith & Epstein 1971). There is an approximately $+5‰$ difference between plant $\delta^{13}C$ values and those of bone collagen (Ambrose & Norr 1993) due to the physiological fractionation of isotopes in the body; therefore humans consuming a diet based on C_3 plants have collagen $\delta^{13}C$ values of approximately $19‰$, while those consuming only C_4 plants will have values of approximately $-8‰$, and those eating both have intermediate values. Fuller and colleagues (2005; 2006), and Richards et al. (2002; 2006) have demonstrated that breastfeeding infants in modern and archaeological contexts have $\delta^{13}C$ values that are enriched by approximately $1‰$ over adult female values, also referred to as the 'carnivore' effect (Bocherens et al. 1995), indicating that infants are most likely to be only consuming breastmilk and not complementary foods. However, it may be possible that complementary weaning foods that include C_4 plants (e.g., millet cereal, or milk from C_4 plant fed animals) may also cause enrichment in $\delta^{13}C$ values (Dupras et al. 2001).

Nitrogen stable isotopes are used to illustrate the shift between a reliance on human milk proteins to those obtained from complementary foods (e.g., Fogel et al. 1989; Katzenberg & Pfeiffer 1995). There is a stepwise increase, called the trophic level effect, of approximately 3‰ between levels of the food chain (Schoeninger & DeNiro 1984), and because a breastfeeding infant receives its protein from breastmilk, the $\delta^{15}N$ value of the infant's tissues is approximately 2–3‰ higher than that of its mother (Fogel et al. 1989). Breastfeeding infants will have the highest $\delta^{15}N$ values in the population because they are at the top of the food chain. The typical weaning process is gradual, since complementary foods are added slowly to the infant's diet as dependency on breastmilk decreases; this dietary change is detected through decreasing $\delta^{15}N$ values that eventually reach adult values once the infant is completely weaned (Schurr 1998).

Kellis, Dakhleh Oasis

The ancient village of Kellis in the Dakhleh Oasis was occupied during the Romano-Christian period.[3] Excavations of the Kellis 2 cemetery have thus far revealed 771 individuals, including 260 infants/children (Wheeler 2009), a sample large enough to make valid statements about weaning practices during this period.

Dupras (1999) and Dupras et al. (2001) conducted initial stable carbon and nitrogen isotope analyses on a cross-sectional sample of juveniles from the Kellis 2 cemetery to interpret weaning patterns at Kellis. Analyses were also conducted on botanical and non-human animal remains to reconstruct the isotopic food-web.

Figure 3.2 shows the combination of $\delta^{13}C$ and $\delta^{15}N$ isotope data (Dupras et al. 2001), with the addition of 35 individuals in the same age range. The nitrogen data shows that the $\delta^{15}N$ values increased to a peak value at six months of age, reaching one trophic level over adult females.[4] This indicates an exclusive dependence on breastmilk until approximately six months of age. Nitrogen values then begin a steady decrease until the age of approximately three years, when childhood values approach those of adult females.

The $\delta^{13}C$ values also show enrichment during the first year of life, reaching values that are approximately 1‰ enriched over adult females. Although current research suggests a 'carnivore effect' that occurs in breastfeeding infants (Fuller et al. 2006; Richards et al. 2006), Dupras et al. (2001) suggest that the enrichment in $\delta^{13}C$ values in this data may be due to the types of complementary foods that were introduced at circa six months of age. The presence of pearl millet (*Pennisetum glaucum*) in Kellis (Thanheiser 1999), and enriched C_4 signals in cow and goat remains, suggests that millet may have been

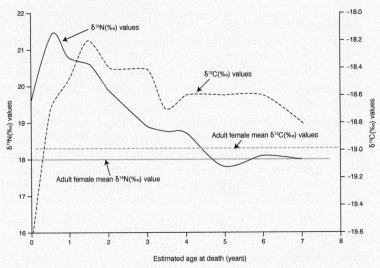

Figure 3.2 Carbon and nitrogen average isotopic values for infants from the Kellis 2 cemetery

The solid data line represents the mean nitrogen values, while the dashed line represents the carbon values. Corresponding mean values for the adult females are indicated by the straight lines with the solid line representing nitrogen (mean $\delta^{15}N$ is 18‰) and the dashed line representing carbon (mean $\delta^{13}C$ is –19‰).

present in the diet of these infants (Dupras 1999). Ancient literary sources such as Galen and Soranus suggested that mothers should introduce complementary foods at six months of age, and that gruel/pap mixtures made with cow or goat milk were favoured (Temkin 1956; Green 1951). Given this evidence, it is possible that the Kellis infants were fed complementary foods that included pearl millet, or that they drank milk from cows/goats that were fed pearl millet.

The combination of complementary foods, the possibility of physiological fractionation, i.e., the 'carnivore effect', and the trophic level effect of nitrogen, all indicate that the infants from the Kellis 2 cemetery were exclusively breastfed for the first six months of life, with a slow transitional feeding period, and weaned completely by three years of age. Further longitudinal research conducted by Dupras and Tocheri (2007) on enamel stable oxygen isotopes supports this interpretation of weaning practices at Kellis.

Notes

1 C_3 and C_4 are used to differentiate the ways that plants metabolize environmental carbon. C_3 plants use the enzyme bisphosphate decarboxylase to fix atmospheric carbon, resulting in a compound with three carbon atoms, while C_4 plants utilize the enzyme phosphoenol

pyruvate carboxylase, resulting in a compound with four carbon atoms, thus C_4 plants are more enriched in carbon.

2 Stable isotope values are expressed using the δ notation.
$\delta = [(R_{sample}/R_{standard}) - 1] \times 1000$. $R = {}^{13}C/{}^{12}C$ for $\delta^{13}C$; and
$R = {}^{15}N/{}^{14}N$ for $\delta^{15}N$. All δ values expressed in parts per mille (‰).

3 The archaeological and ^{14}C dates for Kellis are not congruent. Archaeological evidence suggests that Kellis was occupied from approximately 50 AD to 390 AD (Hope 2001; Bowen 2003), while C^{14} dates from the Kellis 2 cemetery indicates use from as early as 40 AD to as late as 540 AD (Molto 2001; Stewart et al. 2003).

4 Nitrogen values from Kellis are substantially enriched (similar to marine-based diets) in comparison to populations subsisting on terrestrial diets. This phenomenon occurs in desert environments where the base of the food chain has enriched nitrogen values (see Schwarcz et al. (1999) for further explanation).

References

Ambrose SH. & Norr L. (1993) Experimental evidence for the relationship of carbon isotope ratios of whole diet and dietary protein to those of bone collagen and carbonate. In: Lambert JB. & Grupe G. (eds) *Prehistoric Human Bone Archaeology at the Molecular Level*. Berlin: Springer-Verlag. pp. 1–37.

Bocherens H., Fogel ML., Tuross N. & Zender M. (1995) Trophic structure and climatic information from isotopic signatures in Pleistocene cave fauna of southern England. *Journal of Archaeological Science* 22: 327–340.

Bowen GE. (2003) Some observations on Christian burial practices at Kellis. In: Bowen GE. & Hope CA. (eds) *The Oasis Papers 3: The Proceedings of the Third International Conference of the Dakhleh Oasis Project*. Oxford: Oxbow Books. pp. 166–182.

Dupras TL. (1999) *Dining in the Dakhleh Oasis, Egypt: Determination of Diet using Documents and Stable Isotope Analysis*. Unpublished PhD dissertation. Hamilton: McMaster University.

Dupras TL. & Schwarcz HP. (2001) Strangers in a strange land: stable isotope evidence for human migration in the Dakhleh Oasis, Egypt. *Journal of Archaeological Science* 28: 1199–1208.

Dupras TL. & Tocheri MW. (2007) Reconstructing infant weaning histories at Roman period Kellis, Egypt using stable isotope analysis of dentition. *American Journal of Physical Anthropology* 134: 63–74.

Fogel ML., Tuross N. & Owsley D. (1989) Nitrogen isotope tracers of human lactation in modern and archaeological populations. *Annual reports of the Director, Geophysical Laboratory, 1988–1989*. Washington: Carnegie Institute of Washington. pp. 111–116.

Fuller BT., Fuller JL., Harris DA. & Hedges REM. (2006) Detection of breastfeeding and weaning in modern human infants with carbon and nitrogen stable isotope ratios. *American Journal of Physical Anthropology* 129: 279–293.

Fuller BT., Molleson TI., Harris DA., Gilmour LT. & Hedges REM. (2005) Isotopic evidence for breastfeeding and possible adult dietary differences from late/Sub-Roman Britain. *American Journal of Physical Anthropology* 129: 45–54.

Green RM. (translator). (1951) *Galen. Hygiene (De Sanitate tuenda).* Springfield: Thomas.

Hope CA. (2001) Observations on the Dating of the Occupation at Ismant el-Kharab. In: Marlow CA. & Mills AJ. (eds) *The Oasis Papers I: The Proceedings of the First Conference of the Dakhleh Oasis Project.* Oxford: Oxbow Books. pp. 43–59.

Katzenberg MA. & Pfeiffer S. (1995) Nitrogen isotope evidence for weaning. In: Grauer AL. (ed.) *Bodies of Evidence: Reconstructing History through Skeletal Analysis.* New York: Wiley. pp. 221–235.

Molto JE. (2001) The comparative skeletal biology and paleoepidemiology of the people from 'Ein Tirghi and Kellis, Dakhleh Oasis, Egypt. In: Marlow M. & AJ Mills (eds.) *The Oasis Papers 1: The Proceedings of the First International Symposium of the Dakhleh Oasis Project.* Oxford: Oxbow Books. pp. 81–100.

Richards MP., Fuller BT. & Molleson TI. (2006) Stable isotope palaeodietary study of humans and fauna from the multi-period (Iron Age, Viking and Late Medieval) site of Newark Bay, Orkney. *Journal of Archaeological Science* 33: 122–131.

Richards MP., Mays S. & Fuller BT. (2002) Stable carbon and nitrogen values of bone and teeth reflect weaning age at the Medieval Wharram Percy Site, Yorkshire, UK. *American Journal of Physical Anthropology* 119: 205–210.

Schoeninger MJ. & DeNiro MJ. (1984) Nitrogen and carbon isotopic composition of bone collagen from marine and terrestrial animals. *Geochimica e Cosmochimica Acta* 48: 625–639.

Schurr MR. (1998) Using stable nitrogen isotopes to study weaning behaviour in past populations. *World Archaeology* 30: 327–342.

Schwarcz HP., Dupras TL. & Fairgrieve SI. (1999) ^{15}N enrichment in the Sahara: in search of a global relationship. *Journal of Archaeological Science* 26: 629–636.

Stewart JD., Molto JE. & Reimer P. (2003) The chronology of Kellis 2: the interpretive significance of radiocarbon dating of human remains. In: Bowen GE. & CA Hope (eds) *The Oasis Papers 3: The Proceedings of the Third International Conference of the Dakhleh Oasis Project.* Oxford: Oxbow Books. pp. 345–364.

Smith BN. & Epstein S. (1971) Two categories of $^{13}C/^{12}C$ ratios for higher plants. *Plant Physiology* 47: 380–384.

Temkin O. (translator). (1956) *Soranus of Ephesus: Gynecology.* Baltimore: Johns Hopkins Press.

Thanheiser U. (1999) Plant remains from Ismant el-Kharab: first results. In: Hope CA. & Mills AJ. (eds) *Dakhleh Oasis Project: Preliminary Reports on the 1992–1993 and 1993–1994 Field Seasons.* Oxford: Oxbow Books. pp. 89–93.

Wheeler SM. (2009) *Bioarchaeology of infancy and childhood at the Kellis 2 cemetery, Dakhleh Oasis, Egypt.* Unpublished PhD dissertation. London: University of Western Ontario.

There are a series of stages a child passes through during his or her life. Some of these are biological transitions, whereas others are social rites of passage. The biological changes start with the period of weaning. The process of weaning, and the food used during the process of weaning, may be identified through isotopic analysis of the teeth. At Roman Kellis, in the Dakhleh Oasis, the change from complete reliance on nutrition from maternal milk to consuming either foods made with pearl millet, or cow/goat milk from animals consuming such pearl millet, occurred between six months and three years of age (see Box 3.2 for details).

It is harder to recognise the social processes that determined when infants or children become recognised as individuals or people. It is clear that in Egypt, whether the children were stillborn "or verging on adolescence, buried alone or in family vaults, there was an immediate concern for their chances of attaining an afterlife" (Meskell 1994: 37). Indeed, carefully wrapped foetal remains have been recovered from within the Roman cemetery at Dakhleh Oasis (Tocheri et al. 2005), suggesting that infants and neonates were considered part of the Kellis society. It does however appear, but only at certain periods, that there were age-determined differences within society as, for example, at Deir el-Medina, where the lowest part of the cemetery slope was reserved for infants and perinates (Meskell 1994).

Burial assemblages associated with children frequently have objects which fall within the normal range of adult grave goods (Meskell 1994). By developing tight boundaries for biological age, it is possible that narrow or minor differences in social aspects of age may be identified from the archaeology. For example, Reiter (2008) demonstrated differences in the funerary assemblage between children of different biological ages at Naga-ed-Dêr, some of which potentially could have been linked with the mid-childhood growth spurt, thereby linking the biological and social processes of childhood. Children obviously provide a link between one generation and the next, and we know women did die in childbirth or during pregnancy as female burials have been found with foetal heads in the pelvic cavities, such as a C-group woman near Khor Meris (Elliot Smith & Wood-Jones 1910). Furthermore, individual osteobiographies of women and their pregnancies may be constructed, as rapid osteoclastic resorption of bone during pregnancy and the period of lactation may be identified histologically (Denton & David 2005).

3.1.6 Kinship

How can we identify kin groups or families? Can we tell if certain people are related? Given that the Egyptian language had a set of terms for kin groups larger than simple nuclear families (Campagno 2009), it is clear that recognising affiliation and identifying its form was of emic importance. During the Predynastic, kinship probably constituted the main axis of social organisation in village communities, and the clustering of burials within cemeteries at sites such as Badari, Naqada and Hierakonpolis is considered to reflect contemporaneous descent groups (Campagno 2009).

But how might we recognise kinship? The study of nonmetric traits, such as metopism or the presence of additional ossicles, has been used to detect 'family groups' within cemeteries and also to identify approximate levels of genetic relatedness. Using both dental and cranial nonmetric traits, Lovell and colleagues discovered that, at Naqada, the apparently richer individuals in the elite cemetery, Cemetery T, appeared to be less genetically similar than those in the other Predynastic cemeteries were to each other, suggesting that Cemetery T was used by biological lineages from the local population (Johnson & Lovell 1994; Prowse & Lovell 1996). Furthermore, the apparent genetic distinctiveness of the high status Cemetery T individuals implies that, at Predynastic Naqada, social status was ascribed and potentially inherited.

Biological relatedness can also be ascertained genetically from ancient DNA (for details regarding the use of DNA analysis in the archaeology of death and burial, see Bramanti 2013). This has financial and methodological issues, and, in Egypt, the high ambient temperature also is a cause of problems with DNA survival. Even in good preservation conditions, there are difficulties in amplifying short tandem repeats (STRs) of DNA consistently from archaeological specimens (Brown & Brown 2011). To study archaeological genetic relationships, it is normal to use mitochondrial DNA as this is only inherited from the mother and has high copy numbers (Mulligan 2006). At Roman Kellis, the mitochondrial DNA studied indicated relatively high genetic diversity (Parr 2002), indicating low levels of female kinship, and as such, implying the potential in-marriage of females. To study male genetic relatedness, and hence the paternal lineage, the Y chromosome is required, but this is problematic as excellent DNA survival is required. Although not undertaken in Egypt, a good comparative example comes from Bronze Age Mycenae in Greece, where success in amplifying ancient DNA was limited, with only four of 22 skeletons providing authentic mitochondrial DNA sequences suitable for kinship analysis (Bouwman et al. 2008). As methods of amplification of ancient DNA improve, it is possible that such analyses will be possible in Egypt despite the preservation of only short DNA sequences (see also Box 3.11).

3.2   Activity and Occupation

What did the person do during his/her life? Humans manipulate and affect their environment, and their environment and the activities that they repeatedly undertake affect their bodies. It is important to separate out those bodily changes that occur as a result of normal growth and development from those changes that occur as a result of repeated activity, such as strenuous and physical work (Jurmain et al. 2012).

Activity and occupation are related to the growth and development of the individual. During childhood, the body grows and matures. In younger individuals, such as children, these very processes are used to assess the age-at-death of the body, and thus to correlate the developmental age of the skeleton or mummy with the chronological age of the person (Baker et al. 2005; Lewis 2007; Scheuer & Black 2000). Childhood growth processes, however, may be affected by activities undertaken during this period, thereby affecting the child's bodily form (Halcrow & Tayles 2011; Ruff 2005; 2007).

During childhood growth, bone remodels, leading to changes in morphology and robusticity (for full details see Scheuer & Black 2000). Similar processes occur in adult life, but, as all bony growth plates are used, such as at the articular ends of the long bones, these processes are limited by the body's ability to change its dimensions. The human body is plastic and so frequently occurring and repeated actions may lead to physical changes in the body. Highly repetitive strenuous muscle activity results in bone remodelling. This particularly occurs at sites of muscle-to-bone and tendon-to-bone attachments within the body.

The most common osteological study of activity concerns changes in bone dimensions. As noted earlier, bone tissue remodels in response to mechanical stimulation (Wolff 1892). Additional bone tissue is deposited and added to the existing bone in order to strengthen it. Increased physical activity at certain locations in the body leads to increased capillary action and hence blood flow around the bones involved, thereby stimulating bone remodelling (Hawkey & Merbs 1995). This bone remodelling can have two forms, bone formation or bone erosion and resorption. Bone formation leads to increased bone size or roughness whereas resorption can lead to pitting of the bone surface. The change in bone size can be identified through simple analysis of long bone lengths, cross-sections of long bone shafts and other external dimensions such as bone circumferences (e.g. Masali 1972; Robins 1983; Zakrzewski 2003). The larger (more robust) long bones indicate greater strength and more repeated physical activity than smaller (or less robust) long bones. This is a cheap and simple technique, but is a relatively imprecise way of measuring bone strength as it does not indicate changes in the relative distribution of bone tissue along the bone itself.

Through analysis of the cross-sectional geometric properties of a long bone, it is possible to measure the strength of the bone and hence to infer its ability to resist bending and torsion. This can be undertaken by sectioning the bone and studying the cross-sections, calculating total bone area from external dimensions or most accurately through computed tomographic (CT) scanning. This approach has been pioneered by Ruff (e.g. 1999; 2000a; 2000b) for non-Egyptian samples, but rarely has been applied to Egyptian material due to the cost of CT scanning and the destructive nature of the cheaper method of cutting the long bones. More cheaply, it is possible to combine moulds of the external dimensions of bones with X-rays, in order to obtain accuracy and consistency with bone strength (O'Neill & Ruff 2004; Stock 2002). Anterior–posterior long bone strength (particularly of the femur in males) is usually found to be greater in foraging or more mobile population samples than in more sedentary groups (Larsen 1995; 1997). Furthermore the degree of torsional or twisting strength is linked to the relative ruggedness of the terrain rather than with the subsistence strategy employed. Using the cheaper method of employing radiographs and moulds of external bone morphology, changes in the humerus and femoral midshafts have been noted, providing evidence of changes in habitual behaviour over time, with the bone structure of the Jebel Sahaba sample indicating high levels of mechanical loading during the late Pleistocene as compared with lower levels amongst Egyptian agricultural populations (Stock et al. 2011). Furthermore Stock et al. (2011) suggested that there was a change in sexual dimorphism in both humerus and femur strength, implying that the intensity of manual activity and mobility increased for men and decreased for women from the Neolithic through to the 12th Dynasty.

Changes in bone surface morphology, known as musculoskeletal stress markers (MSM) or enthesopathies can be used to identify repeated long-term activity (e.g. Zabecki 2008; 2009). These lesions occur around muscle insertion or attachment sites. The larger the bony scar or lesion, the more developed and more heavily used the muscle or group of muscles involved. This type of data needs to be carefully understood, since it is linked to the age of the individual, as the older the person being studied, the more their muscles will have been used, and hence the more their muscle markers are likely to be developed. Studies of long bone diameters and MSM will not indicate particular activities that have been undertaken, but rather will indicate which muscles have been repeatedly stressed through heavy use and reuse.

Overuse of certain joints can lead to degenerative pathological change, such as osteoarthritis (Weiss 2015: 62–70, 75–85, 89–100). Although many factors may contribute to the development of degenerative arthritis, physical stress is one of the most significant factors (Ortner & Putschar 1985). As a result, it is important to remember that activity and occupation may lead to physical impairment and what might be classified as disease. Degenerative joint disease occurs as a result of bodily ageing and the degeneration and loss of the articular cartilage. The bone changes take a long time to develop, and may only occur many years after the initial changes to cartilage within the joint. This degenerative arthritis may lead to bone destruction or bone growth. The former occurs when bones end up rubbing against other bones at joint surfaces, such as inside the knee or elbow. The latter occurs when there is a build-up of irregular bone, such as extra bony lipping forming around joints or around the vertebrae. Arthritic pathological change may be affected by the overall health of the individual concerned as aspects of health will affect their ability to undertake specific and repeated activities. For example, extensive antemortem molar tooth loss has been blamed for high frequencies of degenerative joint disease in Nubian skulls from Sayala (Strouhal & Jungwirth 1980) and Kulubnarti (Sheridan et al. 1991) due to the impact of tooth loss upon each individual's ability to chew. Such patterns can also identify sex or gender differences. The females from Kulubnarti have much higher levels of osteoarthritis at the temporomandibular joint (where the jaw hinges at the base of the skull) (Sheridan et al. 1991).

The articular surfaces of the joints of the human skeleton are adapted to normal mechanical loading. Over an individual's lifetime, the cartilage covering the surfaces of these articular joints may erode as a result of repeated joint use. Pathological skeletal changes may start to occur along the margins of the joints as a result of this use (or overuse) of the joint. The most common activity-related change to a joint surface is the development of bony spicules. More prolonged or severe joint use may lead to erosion of the joint surface itself, leading to a shiny surface developing on the cortical bone (called eburnation). These skeletal changes are symptomatic of osteoarthritis (also called degenerative joint disease), and, even early in the palaeopathological study of Egypt, was considered the most common disease found in ancient Egyptian and Nubian mummies (Elliot Smith & Dawson 1924). Certain specific patterns of osteoarthritis have been identified that are suggestive of activity (e.g. Bridges 1992).

Other pathological lesions may occur as a result of repeated activity. The most commonly recognised of these is the development of additional bony facets on the anterior surface of the tibia, known as squatting facets. These are considered to result from repeated squatting or kneeling activity. Of greatest importance to Egypt is their identification in Levantine groups associated with heavy cereal grinding activities (Molleson 1994). As noted above, certain forms of trauma are linked to certain activities. Many of these soft tissue traumas do not affect the hard tissue of the skeleton but may be identified within mummified material. Some skeletal fractures, however, are indicative of particular repeated activities. Clay-shoveller's fractures are fractures of the spinous processes of vertebrae, and are associated with heavy bending and loading work placing strain upon the lower neck and upper back (Knüsel et al. 1996). This means that work patterns, such as at Hierakonpolis, may be modelled and reconstructed skeletally (see Box 3.3).

Despite the fact that when changes in bone morphology and robusticity are considered, it is usually the long bones that are studied, the influence of behaviour and the environment on skull morphology was noted long ago. In the 5th century BC Herodotus described apparent differences in cranial robusticity between the Persians and the Egyptians:

> The skulls of Persians are so weak that if you so much as throw a pebble at one of them, you will pierce it; but the Egyptian skulls are so strong that a blow with a large stone will hardly break them.
>
> (Larsen 1997: 226)

Box 3.3   Work Patterns at Hierakonpolis, Giza, Abydos and Amarna

Melissa Zabecki

While all bioarchaeologists are well-versed in collecting age, sex, and pathological data from skeletons, there are dozens of specialized data collection techniques that allow us to know more about past peoples beyond basic demography and health characteristics. More and more, scientists have been attempting to determine what everyday life was like for the ancient masses. What were people doing all day? How hard were they working? As bone is a dynamic tissue, it reacts to various stimuli including use, overuse by strain, and trauma – all part of everyday living in ancient times. Any study focusing on these changes due to mechanical stimuli adheres to the basic assumption of 'Wolff's Law' (Ruff et al. 2006), which acknowledges that bone tissue rearranges itself due to changes in the pressure it receives as a result of a multitude of causes. With this assumption in mind, bioarchaeologists have applied studies of musculoskeletal stress markers (MSM) to learn about the amount or severity of work done by individuals in skeletal samples. Muscle, tendon, and ligament attachment scars on bone can be evaluated using a scoring system that assigns a score to each scar relative to a general pattern of use or overuse for that particular marker. Scores from 0 to 3 are usually used to describe normal use, from no use to extreme use, and scores from 4 to 6 describe varying degrees of overuse and muscle strain as evidenced by enthesopathic lesions at the attachment site. Generally, 15 or more attachments are scored per individual on the upper limb, lower limb, and sometimes the skull and pelvis. The scores are then averaged, and a mean workload score is given to that individual. The individuals' scores are then grouped into categories depending on the research question, and that whole group's mean score can be used for comparative purposes. The bioarchaeologist can develop an impression of general workload patterns within demographic units of age, sex, social status, or the entire population's general level of activity. This method assumes that muscle usage progresses on a continuum, and therefore the scores are not quantitative, but arbitrarily qualitative. The scores, however, allow for non-parametric statistical testing to be applied to the data, to make them comparative across populations. It should be noted that other, more quantitative, techniques for MSM data collection aside from the 0–6 scoring system (often referred to as the 'Hawkey Method' or the 'Hawkey–Merbs Method', from the seminal study that introduced the technique; Hawkey & Merbs 1995), are currently in the process of review. These techniques include studies of cross-sectional anatomy and attachment surface areas. Hopefully, a combination of these techniques will be used in the future to enable an accurate measure of workload to be obtained. Extreme caution must be used in hypothesizing about specific activities, due to the fact that individuals' bones will react to stimuli in inestimable ways as a result of biological/genetic, environmental, social, and personal differences (Jurmain 1999; Jurmain et al. 2012). For now, bioarchaeologists are content to agree that general degrees of workload can be gleaned from the muscle attachment data, and that this information can enable studies of gender or social patterning in workload to be undertaken.

MSM research is relatively new, so its application within ancient Egyptian contexts is just beginning. MSM data from the Predynastic cemeteries HK43 at Hierakonpolis and N7000 at Naga ed-Der, the Old Kingdom Western Mastabas at Giza, the Middle Kingdom North Cemetery at Abydos and the New Kingdom South Tombs Cemetery at Amarna have all been collected and analysed. Significant differences between or within sites have not been observed, but comparisons with other skeletal populations around the world have shown the ancient Egyptian workload to be comparatively low (Zabecki 2009). Employing the 0–6 scoring system, ancient Egyptian mean MSM scores typically fall between 0.9 and 1.2 whereas groups from many other parts of the world typically show mean scores around 2.0 and above. These comparative samples include indigenous South Americans, Alaskan Inuits and Levantine individuals from the Natufian and Neolithic periods, with the latter samples being groups that one might assume would be similar to Egyptians as a result of similarity in local environment, culture and timescale. Despite the difference in time periods and social classes, the Egyptian MSM mean scores are all low.

Significant sexual dimorphism is seen in workloads at Predynastic Hierakonpolis (cemetery HK43) (p $<$ 0.01, n_{male} = 34, n_{female} = 57), with greater activity loads seen in the males. Studying these individuals by age groupings (as MSM scores usually increase with age due to greater time for the development of such lesions), the males had higher scores than the females in all age-matched groups (Zabecki 2009). Surprisingly, significant gender dimorphism in MSM scores was not seen in the comparative samples either from Old Kingdom Giza or from Middle Kingdom Abydos. The Giza sample, in contrast, expressed significant bilateral asymmetry, with most individuals appearing right-side dominant, thereby suggesting preference for use of the right hand and arm (Zabecki 2009). The individuals from Middle Kingdom Abydos exhibited the highest mean MSM scores of the Egyptian samples studied (MSM $x^- = 1.29$, $n = 21$, $\sigma^2 = 0.28$). One might have expected individuals living during the Late Predynastic or Amarna periods to have worked harder, as major changes in social organization and situation have been known to cause stress on populations, but, surprisingly, that does not seem to be the case.

One potential interpretation of the apparent evidence for relatively light or low workloads in comparison to groups from other parts of the world is that the climate and geography of Egypt made daily life and work more predictable and hence, potentially, less intensive than for comparative prehistoric societies. While irrigation was practised, the scale, both in terms of original construction and subsequent upkeep, might not have required both repetitive and repeated heavy labour. This may be because the landscape precluded the need for huge and complex irrigation systems, and because the consistent weather (with little-to-no precipitation) and the regularity of the inundation itself might have had the effect of reducing reparative repetitive workloads. MSM study of the 'workers' who built the temples and pyramids of ancient Egypt has not yet been undertaken: those data may result in higher mean MSM scores, and it is hoped that these populations will be studied in the near future.

References

Hawkey DE. & Merbs CF. (1995) Activity-induced Musculoskeletal Stress Markers (MSM) and Subsistence Strategy Changes among Ancient Hudson Bay Eskimos. *International Journal of Osteoarchaeology* 5: 324–338.

Jurmain R. (1999) *Stories from the Skeleton: Behavioral Reconstruction in Human Osteology*. Amsterdam: Gordon and Breach.

Jurmain R., Alves Cardoso F., Henderson C. & Villotte S. (2012) Bioarchaeology's Holy Grail: The Reconstruction of Activity. In: Grauer AL. (ed.) *A Companion to Paleopathology*. Chichester: Wiley-Blackwell. pp. 531–552.

Ruff C., Holt B., & Trinkaus E. (2006) Who's Afraid of the Big Bad Wolff?: "Wolff's Law" and Bone Functional Adaptation. *American Journal of Physical Anthropology* 129: 484–498.

Zabecki M. (2009) *Late Predynastic Egyptian Workloads, Musculoskeletal Stress Markers at Hierakonpolis*. Unpublished PhD dissertation. Fayetteville: University of Arkansas.

The main influence on craniofacial morphology is masticatory behaviour. In certain populations or groups other actions have an equally important effect, e.g. artificial cranial deformation in Peruvians or the use of the teeth as a tool in Eskimo populations. Extramasticatory use of the teeth has not been recorded in Egypt, but the possibility should never be discounted. Diet, and hence mastication, has been shown to have a major impact on craniofacial morphology within Nile Valley populations. Carlson and co-workers argued a masticatory-functional hypothesis to explain the changes in craniofacial size and shape in Nubia (Armelagos et al. 1984; Carlson 1976; Carlson & van Gerven 1977; 1979; Ewing 1966; Goodman et al. 1986; van Gerven et al. 1973; 1977). They suggested that the primary factor influencing Nubian anatomy was the change in subsistence economy, from foraging to food production, with its associated shift towards the consumption of softer foodstuffs. These changes resulted in a reduction in the loading activity on the masticatory muscles, so decreasing the bone growth in the skull and leading to a reduction in facial robusticity.

Recent research has shown that lateral differences in bone morphology can be used to identify hand preference (Cashmore 2009; Cashmore et al. 2008; Cashmore & Zakrzewski 2009; 2013), although analysis of the upper limb bones from Deir an-Naqlun has suggested that the pattern of directional asymmetry found may be related to factors other than handedness (Jaskulska 2009). Asymmetry may be expressed in long bone lengths or breadths, diaphyseal (long bone shaft) circumferences, epiphyseal (articular surface) measurements, robusticity indices and internal bony architecture and structure. The last of these requires analysis using X-rays or computed tomography, whereas the other methods are simple and can be undertaken using standard osteological methods. Although not yet undertaken with Egyptian samples, structured light scanning might form a simple and affordable method to employ for this type of study in the field or laboratory (see McPherron et al. 2009). This may be important as focusing upon the bones of the hand and arm may permit the identification of individuals who undertake repeated distinctly lateralised activities, such as scribes or engravers with their distinct hand preferences.

Other repeated actions, such as extramasticatory use of the mouth and teeth, will also lead to bodily effects which can be recognised. In addition certain skeletal fractures or other soft tissue trauma lesions are indicative of particular movements or activities (Weiss 2015). This type of bioarchaeological analysis can lead to the identification and understanding of individuals with bodily modification, physical impairment and/or disability.

Although the specific activities or occupations undertaken by an individual do not leave clear markers upon the body that would permit their unambiguous identification, the human body is sufficiently plastic that it responds to repeated activity by bony or soft tissue modification.

These changes in the body can be linked to certain physical movements or stresses and, through links with other aspects of the archaeological record, may permit occupations to be hypothesised for individuals. It is likely that at least individuals who undertook repeated heavy manual labour over a long period of their lives may be identified by their greater musculoskeletal marking. This, however, is linked to the length of time that the individual undertook those repeated activities and the time between completing them and death, as if the person had a prolonged period of rest and recovery, the activity-related lesions would start to be resorbed again by the body. It should also be remembered that demanding physical activity can have other side-effects upon the local population, such as reducing female fertility and fecundity (Jasieńska & Ellison 1998), so bioarchaeological identification of high levels of activity is important for the reconstruction of the Egyptian lifecourse and social structures.

3.3   Health and Disease

When archaeologists uncover human skeletons or mummies, the first questions usually asked are 'how did he die?' and 'what was his life like?' These can be broken down into several components. This chapter addresses parts of both of these questions, but considers them in a lifecourse dimension. Other parts of the second question regarding the life of the person are covered in other sections, such as Section 3.2: Activity and Occupation or Section 3.4: Diet and Subsistence. Indeed, assessing the health or level of disease expressed by an individual or a population can form the majority of what the biological anthropologist undertakes when analysing skeletal or mummified material. This work will include identifications of infectious and non-infectious disease, and assessments of bodily trauma (such as long bone fractures). However, there are certain caveats and limitations which need to be borne in mind when studying palaeopathology in human remains (for detailed discussion, see Roberts 2013; Weiss 2015).

Skeletons and mummies do not reflect the actions of diseases in exactly the same way, hence these two data sets cannot easily be compared. A disease must be present for a prolonged period within a person in order for there to be any markers of that disease in the skeleton of that individual. This means that people (or rather skeletons of people) who exhibit palaeopathological lesions on their bones may, in fact, represent those individuals who were part of the healthier portion of their society as they lived long enough with the disease active in their bodies for it to develop markers on the bone. This is known as the osteological paradox (Wood et al. 1992). In contrast, those individuals who became very ill with a disease may have died more rapidly, and hence may have died before the illness was able to have any effect upon their bones. Mummies, however, preserve soft tissue. Disease processes occur and mark the soft tissue more rapidly than the bone, and so mummies may provide a more reliable guide to the health (or not) of a population or sample as those mummies exhibiting palaeopathologies are likely to have been the more severely ill individuals within that local group. It is also imperative to remember that the sample being studied is, by definition, all dead and hence is not necessarily representative of the indigenous living population (as it is the ill or elderly who die).

Individuals who are unhealthy and exhibit the lesions of one disease process or another, be they skeletons or mummies, are also relatively likely to be affected by another disease. This occurs because the immune response is weakened by the action of the first infection. This is known colloquially as 'ticks and fleas' as a dog may have both ticks and fleas. As a result, when studying human remains for the markers of disease, it is imperative that all possibilities are studied and that the cumulative effects of several diseases or infections are considered. A differential diagnosis must be undertaken. In so doing, all possible causes are identified, the degree of activity (or healing) at time of death is assessed, the severity is judged and the implications for the life of the person are considered.

Getting at the actual cause of death of an individual is relatively hard. It is very rare that the clear cause is preserved. More frequently, osteoarchaeologists will be able to identify a series of things that affected the individual, be they disease or traumatic events to the body. Some of these may have together led to the death of the individual, but it is rarely possible within archaeological material to assign an exact cause of death. Even when there is severe and clear trauma, such as finding projectile points associated with skeletons, as at Jebel Sahaba, this does not necessarily guarantee that the individuals died from their wounds. It is possible that the spears or arrows entered the body in the time immediately after death. In this kind of situation, the trauma is described as peri-mortem (found around the time of death) and is usually thought of as having had a major impact upon the life and death of the person.

For an infectious disease, such as tuberculosis, there is a clear but large set of factors that will affect the biological response to the infection within the person's body. These include the age of onset (i.e. the age at first infection), the portal of entry into the body, the nutritional status (and hence health) of the individual, the level of immune response (as males have a lower immune response than females), the social conditions in which that individual lived, any local cultural factors such as food taboos, the size and nature of the inoculum (e.g. the number of bacteria entering the body), etc. Within Egypt these may be exacerbated or aided by the arid and sandy conditions as these can affect the lifecycle of the infectious agent and hence affect their virulence. Furthermore, within Dynastic Egypt, some diseases were recognised by the population. If a person was recognised in life to be diseased, then the efficacy of any treatment undertaken in life may affect the manifestation of that disease in either the bones or mummified remains of that person. We should however note that conditions considered to be disease in the modern or western world may not exactly mirror those considered to be disease in the ancient Egyptian mind, but we can use items such as the Edwin Smith medical papyrus to inform us as to treatment methods.

Palaeopathology may be subdivided into a series of reasonably discrete categories: trauma, chronic stress conditions, disease (including infections and joint diseases), metabolic conditions, congenital conditions and dental disease. Some of these categories therefore link into other sections of this text and so should be read in association with them. For example, metabolic disease and dental disease will also be considered within Section 3.4: Diet and Subsistence.

3.3.1 Trauma

Trauma comprises any bodily 'injury' or wound. The most common concept or thought of bodily trauma is of either a partial or complete break in a bone (i.e. a fracture). It also includes the abnormal displacement or dislocation of a bone, a disruption in nerve and/or blood supply to a part of the body, or an artificially induced change in either contour or shape of part of the body (such as amputation or foot-binding). Gross fractures can usually be identified by macroscopic observation of the bones. Smaller-scale fractures require the use of X-ray or computed tomography (CT scanning) for identification. Some fractures may be linked with certain physical activities, such as spondylolysis (a fracture of the spine of the lower back vertebrae) or clay-shoveller's fractures (Knüsel 2000), whereas others may result from disease processes, such as age-related osteoporosis being implicated in fractures to the distal radius, the vertebrae and the femoral neck (Brickley & Ives 2008). Finally, the link between bone loss and increase risk of fracture noted above (Brickley & Ives 2008) and its impacts upon population life course have not been fully explored in Egyptian samples due to the prohibition at the time of writing on exporting samples from Egypt. Changes in bone mass occur through the life course, with peak bone mass occurring in young adulthood, followed by stasis and slow decline, with an acceleration in the rate of bone loss in the elderly (Glencross & Agarwal 2011). The greater bone loss in older females is likely the result of hormonal changes resulting from the menopause. Surprisingly, at early Neolithic Çatalhöyük in Turkey, the older women exhibited percentages of cortical bone similar to the men, which Glencross & Agarwal (2011: 520) argue may reflect the influence of gender roles and activity patterning on bone metabolism throughout the lifecourse. Furthermore, the middle-aged females from Çatalhöyük had larger medullary cavities than the young age females, which may be suggestive of endosteal bone loss associated with pregnancy and lactation (Glencross & Agarwal 2011: 518).

X-ray or computed tomography can also identify changes in the blood supply, causing changes in the internal morphology of the body. These are unlikely to be of great archaeological interest, but may be of biological interest as they may inform about aspects of bodily composition and health. Studies of fractures should include an assessment of the degree of healing as this will provide an idea of how long the individual lived after sustaining the traumatic injury. In Egypt, fractures may also be recognised through the identification of the splints used to treat breaks (such as the forearm splint from Giza dating to the 5th Dynasty), or through the use of prostheses (such as the prosthetic toe associated with a healed amputation dating to the reign of Amenhotep II).

3.3.2 Infectious Disease

When people consider studying bodies for disease, it is usually infectious disease that is in mind. Infectious disease as a category itself can be further subdivided into specific and non-specific infections. Specific infections are those to which a distinct name can be given, such as schistosomiasis or tuberculosis. In order to recognise and specify these kinds of conditions using standard osteological methods, pathognomonic features need to be identified. Pathognomonic lesions are those lesions that only occur with a particular (or specific) infection and so can be considered as definite signs of a particular and specific infection. For example, finding depression lesions with a particular characteristic shape on the frontal part of a skull (caries sicca lesions) shows that that person suffered from syphilis. By contrast, non-specific infections are those which cannot be so definitively linked to a causative infection and therefore have no specific aetiology. This means that the skeleton shows markers that are characteristic of infection, such as new surface bone growth (periostitis), but these are not pathognomonic for any particular infection. This is the more common situation both in life and, for Egyptian skeletons, in death.

In the field, the most practical methods of assessing skeletal material for the evidence of disease are all macroscopic in nature. This requires the identification and description of all lesions found upon the body, and thus, most frequently, is of skeletal lesions noted upon the bones. Pathognomonic lesions may be used to identify specific infections, such as the presence of all the components of the rhinomaxillary syndrome for lepromatous leprosy (Aufderheide & Rodríguez-Martin 1998) but for some diseases, such as TB, there are no pathognomonic skeletal lesions, but instead there are lesions consistent with a diagnosis of TB (such as Pott's disease of the spine). Differential diagnosis may be carried out using the patterning and macroscopic morphologies of the gross skeletal lesions, or through radiological, microscopic or histological analysis of the lesions. These methods obviously require laboratory facilities, but are not very costly. More expensive but more precise methods of disease identification are also available. For example, biomolecular methods may aid in the identification of diseases such as tuberculosis. This disease can be identified through ancient DNA analysis (e.g. of Granville's mummy; Donoghue et al. 2010) and the ancient DNA of the *Mycobacterium* may even be recovered from skeletons that lack any skeletal indicators of tuberculosis. Mycolic acids, which are types of fatty acids found on the *Mycobacterium* cell wall, can also be identified, and are particularly useful in the investigation of co-infection with both TB and leprosy (Minnikin et al. 2011; Taylor et al. 2009).

The infection that leaves marks on the bone may or may not have contributed towards the death of that individual. It is more likely that it was involved in the death of the person if the disease lesions appear to have been active at the time of death. If, however, the lesions show signs of healing, it suggests that the person is much less likely to have died as a result of the infection. Healing of pathological lesions (i.e. the bone starts to remodel into its normal state) occurs when a person survives the initial period of infection. This means that finding skeletal markers of disease may actually indicate that that person was strong enough or healthy enough to survive the disease process long enough for it to imprint upon the bone.

Infectious diseases require certain population sizes in order for them to have a large enough pool of susceptible individuals. This is why people often consider cities to harbour more diseases than rural areas. This is something of a simplification. Large urban areas, such as developed at Hierakonpolis or Abydos, would have provided large potential communities in which infections might be transmitted. Amarna, however, might not have had such a large reservoir of potential infections as the city was occupied for such a short period. Rural areas, or rather areas where people are living in close proximity to their livestock, provide other mechanisms for diseases to spread using animals as proxies or required stages for disease transmission. For example, schistosomiasis, found in the Nile's waters, requires a transmission stage for the parasites through snails (see Box 3.4). Many infections originate with livestock, and it is possible that measles and tuberculosis became human infections from increased human–animal contact following animal domestication.

Within mummified remains, it is possible to identify diseases that affect only the soft tissues and so cannot be so easily recognised from skeletal remains. Destructive sampling may improve the detail in the identification. For example, specific strains of malaria have been identified through ancient DNA analysis of Egyptian mummy soft tissues (Lalremruata et al. 2013; Nerlich et al. 2008).

Box 3.4 Schistosomiasis – The Immunocytochemistry and ELISA Evidence

Patricia Rutherford

Schistosomiasis is endemic in the world today, infecting more than 300 million people, mainly in the developing world. The schistosoma parasites continually breed within their primary host, producing thousands of spiny ova. Half of these ova are released back into the water via faeces or urine, while the other half remain in the body, causing continuous damage, resulting in pathological changes such as inflammation, fibrosis, cirrhosis of the liver, diarrhoea and haemorrhaging (Cheever 1969). Although the majority of the species only infect animals, a few do successfully infect humans, the most prevalent being *Schistosoma haematobium, S. mansoni* and *S. japonicum*.

Although *S. haematobium* and *S. mansoni* are prevalent in Egypt today, it is not a new disease as the study of ancient Egyptian remains and literature indicates that schistosomes were present in the past. Ancient Egyptian remains have shown the presence of calcified *S. haematobium* ova (Ruffer 1910; Millet et al. 1980), calcification of the bladder (a typical symptom of *S. haematobiuml* Isherwood et al. 1979) and the presence of the schistosomes circulating anodic antigens (CAA) in Predynastic tissue (Deelder et al. 1990; Miller et al. 1992; 1993). Ancient art and literature also suggests its presence, as stela show the use of penile sheaths in water, sculptures show distended stomachs and the ancient Egyptians wrote of boys becoming men when blood was seen in their urine, and likened this to the young females' first menstruation (Despommier et al. 1995). The medical papyri also describe this classic symptom of schistosomiasis over fifty times (Farooq 1973; Contis & David 1996). The translation of such literature is however open to interpretation, and some scholars dispute the connection to schistosomiasis.

A multidisciplinary approach to mummy research has been adopted at the University of Manchester for many years, thus inspiring the creation of the international Ancient Egyptian Mummy Tissue Bank in 1996 (Lambert-Zazulak 2000). One main aim of the bank was to provide a resource of small samples for research into ancient schistosomiasis. Robust, cost effective, reproducible tests that could be applied to a large-scale study of ancient samples were therefore required (Lambert-Zazulak et al. 2003).

Detection of schistosomiasis in modern patients is usually carried out by microscopically observing ova in faeces, urine or rectal mucosa. The ova shapes are species specific. As mummified samples do not lend themselves to such tests, immunocytochemistry was used as a cost-effective, reproducible and sensitive method. In contrast to previous histological research (e.g. Fulcheri et al. 1992; Krypczyk & Tapp 1986; Nerlich et al. 1993), blocking small samples in hot paraffin wax was replaced by a cold medium called 2-hydroxyethyl methacrylate, which polymerises at 4°C (Taab, UK). The cold preparation was chosen as hot wax can have deleterious effects upon tissue, causing diffusion, loss and even chemical alterations to the antigens of interest. Also, as the ancient schistosoma antigens are already degraded, damagingly high temperatures should be avoided. The hardened resin also enables thinner tissue sections to be cut (2 μm), thereby enhancing the sensitivity of the test (Heryet & Gatter 1992). Using antisera to both *S. mansoni* and *S. haematobium* worm and egg antigens, visualisation of schistosome antigens in positive controls and ancient Egyptian tissues was achieved (Rutherford 1997; 1999; 2000; 2002; 2005; 2008a).

The acquisition of ancient samples is challenging, as only visceral samples harbour the parasitic worms and ova. Samples were taken from whole mummies, from which small sections of bladder, liver, intestines and viscera tissue were excised. As access was sometimes inhibited by cartonnage, a combination of endoscopy and radiology was used to overcome such issues. By contrast, some samples were easily accessible from canopic jars and viscera packages, and some Early Dynastic samples were free from contaminating mummification resins and sand. These form ideal samples but are rarely available.

To reinforce the positive immunocytochemistry results, other diagnostic tests were employed, specifically enzyme-linked immuno assay (ELISA), as previously successfully used (Deelder et al. 1990; Miller et al. 1992; 1993) to show active infections; histology to show tissue quality and DNA content; and DNA analysis to show which species had caused the infection. As the ELISA test and DNA analysis entail the destruction of finite samples, this was approached with caution. Furthermore, ancient DNA analysis is problematic, due to considerable problems arising from post-mortem diagenetic changes (Pääbo 1989; Pääbo et al. 2004; Rutherford 2002; 2008b).

Analysis of provenanced samples has shown the presence of schistosomiasis in both ancient Upper and Middle Egypt (Rutherford 2002; 2008a). Examples include the positive immunostaining of ova and the amplification of a small fragment of the *S. haematobium* Cytochrome Oxidase gene (CO1) found in pelvic tissue excised from a priest who lived at Akhmim 2700 years ago. *S. haematobium* has also been seen in bladder tissue from mummy 7700/1766 buried in the Fayum Oasis 1800 years ago and from mummy 7700/1777 buried in Luxor 2700 years ago. Recent tests have been carried out on burial groups found in Sudanese Nubia and the Dakhleh Oasis, and several samples have been positively immunostained.

Immunocytochemistry has been the mainstay of the schistosomiasis research, its results dictating what other tests are carried out upon each sample. It can be easily adapted for the investigation of other diseases simply by raising the appropriate antiserum towards the disease of interest. It is sensitive and specific, and thus is an invaluable tool for research into ancient schistosomiasis. Ultimately, as the number of provenanced results increases, a distribution pattern should emerge, thus contributing to an array of research fields.

References

Cheever AW. (1969) Quantitative comparison of intensity of *Schistosoma mansoni* infections in man and experimental animals. *Transactions of the Royal Society of Tropical Medicine and Hygiene* 63: 781–795.

Contis G. & David AR. (1996) The epidemiology of *Bilharzia* in ancient Egypt, 5000 years of schistosomiasis. *Parasitology Today* 12: 253–255.

Deelder AM., Miller RL., Dejonge N. & Krijger FW. (1990) Detection of schistosome antigens in mummies. *The Lancet* 335: 724–725.

Despommier DD., Gwadz RW. & Hotez PJ. (1995) *Parasitic diseases* (3rd edn). New York: Springer-Verlag.

Farooq N. (1973) Historical development. In: Ansari N. (ed.) *Epidemiology and control of schistosomiasis (Bilharziasis)*. Basel: S. Karger. pp. 1–16.

Fulcheri E., Baracchini P. & Rabino Massa E. (1992) Immunocytochemistry in histopaleopathology. *Abstract in Proceedings of First World Congress of Mummy Studies* 2. Tenerife. pp. 559.

Heryet AR. & Gatter KC. (1992) Immunocytochemistry for light microscopy. In: Herrington CS. & McGee JO'D. (eds) *Diagnostic molecular pathology*, vol 1. Oxford: IRL Press. pp. 7–46.

Isherwood I., Jarvis H. & Fawcett RA. (1979) Radiology of the Manchester mummies. In: David AR. (ed.) *Manchester Museum Mummy Project*. Manchester: Manchester University Press. pp. 25–64.

Krypczyk A. & Tapp E. (1986) Immunocytochemistry and electron microscopy of Egyptian mummies. In: David AR. (ed.) *Science in Egyptology*. Manchester: Manchester University Press. pp. 361–365.

Lambert-Zazulak PI. (2000) The International Ancient Egyptian Mummy Tissue Bank at the Manchester Museum. *Antiquity* 74: 44–48.

Lambert-Zazulak PI., Rutherford P. & David AR. (2003) The International Ancient Egyptian Mummy Tissue Bank at the Manchester Museum as a resource for the palaeoepidemiological study of schistosomiasis. *World Archaeology* 35: 223–240.

Miller RL., Armelagos J., Ikram S., De Jonge N., Krijer FW. & Deelder AM. (1992) Palaeoepidemiology of *Schistosoma* infection in mummies. *British Medical Journal* 304: 555–556.

Miller RL., Dejonge N., Krijger FW. & Deelder AM. (1993) Predynastic Schistosomiasis. In: Davies WV. & Walker R. (eds) *Biological anthropology and the study of ancient Egypt*. London: British Museum Press. pp. 55–60.

Millet NB., Hart GD., Reyman TA., Zimmerman MR. & Lewin PK. (1980) ROM1: mummification for the common people. In: Cockburn A. & Cockburn E. (eds) *Mummies, disease and ancient cultures*. Cambridge: Cambridge University Press. pp. 71–84.

Nerlich AG., Parsche F., Kirsch T., Wiest I. & von der Mark K. (1993) Immunohistochemical detection of intestinal collagens in bones and cartilage tissue remnants in an infant Peruvian mummy. *American Journal of Physical Anthropology* 91: 269–285.

Pääbo S. (1989) Ancient DNA: Extraction, characterization, molecular cloning and enzymatic amplification. *Proceedings of the National Academy of Sciences of the USA* 86: 1939–1943.

Pääbo S., Pioner H., Serre D., Jaenicke-Despres V., Hebler J., Rohland N., Kuch M., Krause J., Vigilant L. & Hofreiter M. (2004) Genetic analysis from ancient DNA. *Annual Review of Genetics* 38: 645–679.

Ruffer MA. (1910) Notes on the presence of *Bilharzia haematobium* in Egyptian mummies of the XXth Dynasty. *British Medical Journal* 1: 16.

Rutherford P. (1997) *The diagnosis of schistosomiasis by means of immunocytochemistry upon appropriately prepared modern and ancient mummified tissues*. Unpublished MSc thesis. Manchester: University of Manchester.

Rutherford P. (1999) Immunocytochemistry and the diagnosis of schistosomiasis; ancient and modern. *Parasitology Today* 15: 390–391.

Rutherford P. (2000) The diagnosis of schistosomiasis in modern and ancient tissues by means of immunocytochemistry. *Chungara Revista de antropologia Chilena* 32: 127–131.

Rutherford P. (2002) *Schistosomiasis: The dynamics of diagnosing a parasitic disease in ancient Egyptian tissue*. Unpublished PhD thesis. Manchester: University of Manchester.

Rutherford P. (2005) Schistosomiasis in modern and ancient tissues. *Journal of Biological Research* 80: 80–83.

Rutherford P. (2008a) The use of immunocytochemistry to diagnose disease in mummies. In: David AR. (ed.) *Egyptian mummies and modern science*. New York: Cambridge University Press. pp. 99–115.

Rutherford P. (2008b) DNA identification in mummies and associated material. In: David AR. (ed.) *Egyptian mummies and modern science*. New York: Cambridge University Press. pp. 116–132.

Box 3.5   Ancient DNA Identification of Infectious Diseases in Egyptian Mummies
Albert Zink

The molecular detection of microbial infections in ancient human remains presents a unique way to study diseases in Egyptian mummies. This scientific approach has developed, within recent years, from reports on the retrieval of pathogen DNA in single cases, to more extended studies on disease frequencies in ancient populations and evolutionary aspects of host–pathogen interaction. The first successful ancient DNA (aDNA) study of Egyptian mummies was reported in the 1990s using a mummy from the New Kingdom (1550–1080 BC) (Nerlich et al. 1997), where *Mycobacterium tuberculosis* DNA was detected. This confirmed supposed tuberculosis in this mummy. In subsequent years, more major infectious diseases were identified using ancient DNA techniques, including leishmaniasis (*Leishmania donovani*) (Zink et al. 2006) and malaria (*Plasmodium falciparum*) (Nerlich et al. 2008). Furthermore, studies focusing on the molecular analyis of *M. tuberculosis* complex DNA increasingly investigated larger series, including samples obtained from mummies or skeletons with non-specific lesions or without disease-related morphological alterations (Zink et al. 2001; 2003). This work has significantly improved the diagnostic value of such approaches with ancient material. It is now possible to investigate both the frequency and the distribution of infectious diseases in ancient populations. This has opened a completely new field of research in which modern data on tuberculosis epidemiology can be compared with the situation in earlier time periods. Information on the evolution of pathogens and the interaction with their hosts may be obtained. These are crucial to understand the transmission and spread of specific infectious diseases.

In this box, the presence and occurrence of tuberculosis, leishmaniasis and malaria in ancient Egypt are described, based on ancient DNA investigations of Egyptian mummies and skeletons from different time periods, with special emphasis on the relevance of these results for the study of the frequency and genetic evolution of infectious diseases in both ancient and modern times.

Tuberculosis

The first evidence that human tuberculosis was present in ancient Egypt came from typical macroscopic osseous changes in Egyptian mummies. One of the first cases was described by Sir Marc Armand Ruffer (Ruffer 1910). The mummy of the Amun's priest, Nesperhan, showed typical ventral destruction of the lower thoracic spine, leading to the typical gibbus formation of spinal tuberculosis. More recently, aDNA studies have revealed high tuberculosis frequencies in Predynastic to Early Dynastic, Middle Kingdom and New Kingdom to Late Period material, suggesting comparable infection rates in the various populations of ancient Egypt over a time period of almost 2500 years. In order to evaluate evolutionary aspects of the *Mycobacterium tuberculosis* complex and to understand the spread and means of transmission of tuberculosis in various human populations, the genetic differentiation of tuberculosis strains is of major importance. Such data on mycobacterial strain distribution have been reported from ancient Egypt (Zink et al. 2003), Medieval Britain (Taylor et al. 1999) and Hungary (Fletcher et al. 2003). In all these studies, spoligotyping was used to characterise the detected cases of tuberculosis. This technique provides a typical signature of spacers which are unique to different mycobacteria strains. For ancient Egyptian mummies, only a few successful spoligotyping results have been obtained from Pre- to Early Dynastic material (Abydos, c. 3500–2800 BC). More results are available for the Middle Kingdom (primarily c. 2050–1650 BC) and the New Kingdom to Late Period (c. 1500–500 BC). Interestingly, in none of the investigated samples has any evidence for bovine tuberculosis been detected. The spoligotyping patterns revealed ancestral strains of *M. tuberculosis* in the Pre- to Early Dynastic material, *M. africanum* strains in the Middle Kingdom material and modern strains of *M. tuberculosis* in the New Kingdom to Late Period material. These results suggest that a transition from 'old' to 'modern' *M. tuberculosis* could have occurred between 2000 and 1500 BC. The first appearance of *M. africanum* in Egypt can also be determined to have occurred at least 4000 years ago.

Leishmaniasis

Evidence for the presence of leishmaniasis in ancient Egypt was solely found in samples from the Middle Kingdom, but not in any specimens from the Pre- and Early Dynastic periods or the New Kingdom (Zink et al. 2006). This infectious disease most probably never became endemic in the Egyptian Nile Valley: leishmaniasis distribution is closely linked to that of its vector, the phlebotomine sandfly, and the distribution of Acacia-Balanites woodland (Thomson et al. 1999). On the other hand, Sudan is one of the highly endemic countries for visceral leishmaniasis, which is thought to have originated in East Africa and later spread to the Indian subcontinent and the New World (Pratlong et al. 2001). During the Middle Kingdom, the Egyptians extended their trade relationships and military expeditions into Nubia, with particular interest in the gold resources and slaves for use as servants or soldiers in the Pharaoh's army. This could indicate that those Egyptians who became infected due to the close trade contacts and associated travel with Nubia introduced leishmaniasis to Egypt at this time. The concomitant retrieval of Leishmania DNA in Nubian mummies from the Early Christian period suggests that leishmaniasis was endemic in Nubia for several thousand years, thereby supporting the theory that Sudan could have been the original focus of visceral leishmaniasis.

Malaria

Until relatively recently, there was little conclusive evidence for the presence of malaria in ancient Egyptian populations. The use of PCR amplification of the pfcrt gene in Egyptian mummies enabled the clear identification of *Plasmodium falciparum* DNA, thereby confirming the presence of malaria in ancient Egypt (Nerlich et al. 2008). This was further substantiated by the detection of *P. falciparum* DNA in king Tutankhamen and three further royal mummies of the New Kingdom (Hawass et al. 2010). On the base of these data it is still impossible to determine the frequency of the disease. It does, however, provide additional insight into the spectrum of infectious diseases that plagued the ancient Egyptian populations and could have contributed to their low life expectancy (at birth).

References

Fletcher HA., Donoghue HD., Taylor GM., van der Zanden AGM. & Spigelman M. (2003) Molecular analysis of Mycobacterium tuberculosis DNA from a family of 18th century Hungarians. *Microbiology* 149: 143–151.

Hawass Z., Gad YZ., Ismail S., Khairat R., Fathalla D., Hasan N., Ahmed A., Elleithy H., Ball M., Gaballah F., Wasef S., Fateen M., Amer H., Gostner P., Selim A., Zink A. & Pusch CM. (2010) Ancestry and pathology in King Tutankhamun's family. *Journal of the American Medical Association* 303: 638–647.

Nerlich AG., Haas CJ., Zink A., Szeimies U. & Hagedorn HG. (1997) Molecular evidence for tuberculosis in an ancient Egyptian mummy. *The Lancet* 350: 1404.

Nerlich AG., Schraut B., Dittrich S., Jelinek T. & Zink AR. (2008) Plasmodium falciparum in ancient Egypt. *Emerging and Infectious Diseases* 14: 1317–1319.

Pratlong F., Dereure J. & Bucheton B. (2001) Sudan: the possible original focus of visceral leishmaniasis. *Parasitology* 122: 599–605.

Ruffer MA. (1910) Potts'che Krankheit an Einer Ägyptischer Mumie aus der Zeit der 21 Dynastie. *Zur Historischen Biologie der Krankheiserreger* 2/3: 9–16.

Taylor GM., Goyal M., Legge AJ., Shaw RJ. & Young D. (1999) Genotypic analysis of Mycobacterium tuberculosis from medieval human remains. *Microbiology* 145: 899–904.

Thomson MC., Elnaiem DA., Ashford RW. & Connor SJ. (1999) Towards a kala azar risk map for Sudan: mapping the potential distribution of Phlebotomus orientalis using digital data of environmental variables. *Tropical Medicine & International Health* 4: 105–111.

Zink A., Haas CJ., Reischl U., Szeimies U. & Nerlich A. (2001) Molecular analysis of skeletal tuberculosis in an ancient Egyptian population. *Journal of Medical Microbiology* 50: 355–366.

Zink AR., Sola C., Reischl U., Grabner W., Rastogi N., Wolf H. & Nerlich AG. (2003) Characterization of Mycobacterium tuberculosis complex DNAs from Egyptian Mummies by Spoligotyping. *Journal of Clinical Microbiology* 41: 359–367.

Zink AR., Spigelman M., Schraut B., Greenblatt CL., Nerlich AG. & Donoghue HD. (2006) Leishmaniasis in ancient Egypt and Upper Nubia. *Emerging and Infectious Diseases* 12: 1616–1617.

3.3.3　Chronic Conditions and Joint Disease

These are conditions which affect the individual for a prolonged period during their life and may have some impact upon their ability to undertake certain actions or activities. This section should therefore be read in association with Section 3.2: Activity and Occupation. As bone remodels throughout life, most infectious diseases that leave a mark on the skeleton are chronic conditions, as it requires long-term disease activity for pathological lesions to form and remain on bone. Other chronic conditions are those which affect the joints and occur either as a result of a definite disease process or as a result of ageing or repeated activity.

The ageing process affects bone turnover in the body. As a person ages, their bone remodels more slowly and so it takes them longer to recover from traumatic events, such as bone fractures. The body may also lay down extra bone in a 'supportive' manner in order to stabilise areas of weakness. This can be good in principle, but can cause problems for the person. The most common form of extra bone laid down in life is in bony spicules or osteophytes around bone surfaces. These osteophytes, such as those that cause lipping of the vertebrae in the back, may inhibit some movements.

Degenerative joint disease (DJD) is the term given to a catalogue of associated osseous changes including osteophyte development affecting joints, and particularly the spine. As a result of repeated contact between the bones, this type of joint disease is usually accompanied by eburnation at the joint surface. DJD and osteoarthritis (OA) frequently occur in combination and are often considered to be synonymous.

Some of the most common pathological lesions seen are Schmorl's nodes (Weiss 2015: 75–77). These are depressions on the vertebrae of the spine. They are thought to be associated with herniation of the intervertebral disks into the vertebral bodies, and potentially with either bearing heavy loads or with the actions of heavy forces upon the spine, such as through jumping great distances. Rose (2006) has suggested from a study of Schmorl's nodes and DJD in the vertebrae from Amarna that back stress would have affected between 2 and 8 per cent of the adult population.

Osteoporosis frequently affects older individuals, particularly post-menopausal women due to the effects of lowered levels of oestrogen in the body leading to lowered calcium absorption in these individuals (Agarwal & Stout 2003; Brickley & Ives 2008). The associated reduction in bone density can lead to increased chance of skeletal fracture. Older individuals, especially females, with multiple healed or healing bone fractures should not necessarily be considered accident-prone or injury recidivists, but this may be the result of such age-related osteoporotic bone changes.

Chronic conditions also include stress-related disorders. These can be linked to diet, discussed below, to chronic inflammation, or to chronic parasitic infection etc. Chronic intestinal infection with parasites, such as *Trichuris* or *Ascaris*, may lead to chronic blood loss through the gut and hence to anaemia within the individual. This may be expressed through the body as an increase in porosity and the development of pathological lesions such as cribra orbitalia and porotic hyperostosis (discussed further below).

3.3.4 Metabolic Conditions

These disorders are primarily associated with diet and hence with malnutrition. Malnutrition does not necessarily mean a lack of overall foodstuffs, but rather a deficiency in certain aspects of the diet. These conditions also tend to be inter-related as a diet deficient in one nutrient may be associated with a deficiency of another nutrient or may lead to a reduction in absorption of some other nutrient. In addition, it is important to realise that malnutrition can incorporate over-consumption of certain foodstuffs or specific nutrients.

The most common disorders assumed to be the result of diet are cribra orbitalia and porotic hyperostosis. These pathological lesions, seen as an increase in bone porosity in the orbit region and upon the cranial vault resulting from hypertrophy of the bone marrow (Aufderheide & Rodríguez-Martín 1998; Ortner & Putschar 1985), are symptoms of anaemia. This anaemia, however, need not necessarily be the result of iron deficiency, although this is the most common cause (Larsen 1997; Stuart-Macadam 1992a), but may have other causes: genetic anaemias, such as thalassaemia or sickle cell anaemia; other nutrient deficiencies, such as folic acid or vitamin C; or high parasite load, such as *Ascaris* and *Giardia* (Stuart-Macadam 1992a; 1992b). Furthermore, increased bone porosity, especially when located in other regions, can be the result of other dietary factors. For example, the classic expression of scurvy, due to vitamin C deficiency, is increased porosity around the temporomandibular joint and around the palate (Ortner & Ericksen 1997; Ortner et al. 1999; 2001; Ortner & Putschar 1985). Along the Nile Valley, skeletal changes associated with anaemia have usually been considered to result primarily from iron deficiency (e.g. Mittler & van Gerven 1994), although intestinal infections and malaria have also been identified as potential causative agents (e.g. Buzon 2006). The presence of porotic hyperostosis in Roman perinatal individuals from Dakhleh Oasis, which occurred younger than would have been possible due to insufficient iron absorption, suggests that anaemia may have developed in these young children as a result of weaning with goats' milk (Fairgrieve & Molto 2000).

As noted above, vitamin C deficiency leads to scurvy. This has not been commonly recorded in Egypt, most likely because of the relative ease of access to fruits such as dates. Childhood vitamin D deficiency leads to rickets, whereas adult deficiency leads to osteomalacia. Within ancient Egypt, vitamin D deficiency is considered highly unlikely due to the body's ability to synthesis vitamin D from sunlight through the skin. As such, long bone deformities usually associated with rickets and osteomalacia generally require other causative explanations.

Overconsumption of certain nutrients can also have a detrimental effect upon bodily health. For example, too large an intake of vitamin A may lead, amongst other effects, to birth defects and a reduction in bone density. Gout is the result of a build-up of uric acid in the bloodstream and may result from the repeated consumption of large amounts of protein. Although not described within ancient Egypt, it has been recognised in European medieval samples, particularly among wealthy clergy and monastic groups (Aufderheide & Rodríguez-Martin 1998: 108–111; Patrick 2007).

It should also be remembered that there is a significant link between malnutrition and increased vulnerability to infectious disease. Prolonged malnutrition is likely to lead to a depressed immune response and hence to an increase in susceptibility to infection (which may or may not leave lesions upon the skeleton). In addition, gender or age differences in access to food resources may also impact upon dietary intake and hence upon disease processes.

3.3.5   Congenital Conditions

These are relatively rare conditions and are usually only found within skeletally 'unusual' individuals, such as dwarfing. Within Egypt it is known that, at least in certain time periods, dwarfs had high status (Dasen 1993); however little research has been undertaken to assess whether this was true for all forms of congenital malformation. A variety of congenital conditions affect the skeleton and tend to be systematic in their effect upon the skeleton (Steele & Bramblett 1988). The most common are osteogenesis imperfecta, in which the bone and/or teeth are thin and poorly formed, and achondroplastic and pituitary dwarfing.

There are also inherited diseases which are passed from parent to offspring. The most common include inherited breast cancer or cystic fibrosis which cannot be recognised within the skeletal material. It is, however, possible that the mutations that cause inherited diseases can be identified within ancient human DNA extracted from the skeleton. Although anaemia may be caused by iron deficiency and hence be an indicator of malnutrition or poor diet, it may also result from the inherited (or genetic) disease thalassaemia (Filon et al. 1995). Thalassaemia is important for epidemiologists and archaeologists since a high incidence of the disorder can indicate that the local population is or has been exposed to malaria. Sickle cell anaemia is a form of thalassaemia in which heterozygous individuals exhibit a mild form of the disease and are more resistant to malaria than those unaffected by sickling of red blood cells. As a result, being able to identify the form or type of thalassaemia affecting an individual not only tells us about that person, but may also tell us something about the health of the rest of the population cohort.

3.3.6   Tumours

Tumours are both extremely rare and not always malignant. Most tumours and tumour-like lesions are benign and of relatively unknown aetiology. Most tumours found upon bone are either primary tumours or metastases from cancers of internal organs (Aufderheide & Rodríguez-Martin 1998; Brothwell 2012; Ortner & Putschar 1985). Within ancient Egypt, it has been argued that nasopharyngeal cancers were relatively common due to the frequent consumption of salt fish and exposure to household smoke, bladder cancers due to schistosomiasis infection, and liver cancer due to the consumption of infected grain (Capasso 2005; David & Zimmerman 2010), but these hypotheses cannot be tested skeletally.

Skeletal or mummified evidence of disease does not necessarily inform as to the health of the individual. It is imperative to remember from the osteological paradox that those individuals exhibiting pathological lesions may actually have been the healthier individuals within the sample as they were those who survived to develop the disease lesions. In addition, the cumulative effects of disease interactions or the interactions between health and diet and/or activities undertaken during life should also be remembered since these will affect the expression of disease lesions upon the body (Buzon 2012; Katzenberg 2012).

Assessing a body for the gross lesions of disease (i.e. standard skeletal analysis) requires an on-site biological anthropologist or osteoarchaeologist, so that most pathological conditions within a body, such as non-specific infection or brucellosis, can be identified. Use of in-situ 3D scanning, such as laser or structured light scanning (McPherron et al. 2009), might be a suitable mechanism by which complex recording might be undertaken and (collaborative) differential diagnosis simplified. In certain situations, the precise nature of the pathology affecting the individual will not be possible to ascertain. Sampling the remains, either for thin-section analysis with high-resolution microscopy or ancient DNA analysis, may permit full identification of the disease process.

3.4   Diet and Subsistence

What did people actually eat and drink? Information as to what people ate and drank in Egypt provides us with a great source of potential knowledge of subsistence practices. The most direct way of identifying the foods consumed by past Egyptian populations is by analysing the archaeological remains of foods in the forms of animal bones or plant remnants such as grains or seeds. Data obtained from skeletal or mummified material, however, can be linked with other more classical Egyptian archaeological or Egyptological knowledge, such as that derived from texts or artistic representations. For example, the first recorded dentist was Hesy-Ra in Dynasty III *c.* 2650 BC (Nunn 1996). Weeks (1980) lists the only six dentists known, and all save one are from the Old Kingdom. This type of knowledge provides us with some indication of the way in which ancient Egyptians actually viewed their health and thus their diet. There is a range of approaches for characterising Egyptian diets, including analysis of human skeletal and dental remains, analysis of plant and animal remains, analysis of coprolites, and studies of residues found on ceramics and storage vessels, or indeed on the tools used for processing items into foodstuffs. These approaches can be used to characterise diet and subsistence strategy, and then be further developed to infer nutritional status across a group.

It is obvious that an outline of the subsistence strategy employed by a group can be derived from analysis of the plant and animal remains. The minimum number of individual animals present at the site can be estimated from the zooarchaeological bone assemblage, and this can then be used to calculate the amount of meat that was available to the people, and thus to approximate the energy or other nutrients that could have been obtained by its consumption. By linking these results with monumental and funerary art (e.g. the relief from the 5th Dynasty tomb at Saqqara belonging to Ti), architecture, figurines or statuettes, models of cooking activities (such as of breadmaking at the Middle Kingdom tomb of Meketra), models of diet and subsistence may be developed. Studying the actual human remains themselves, or items associated with the human remains, can improve these models of subsistence and individual diet. For example, clay models of cattle exist in Predynastic Nile sites, but there is little definitive faunal evidence of domestication until the very late Predynastic period in the Delta and the start of the Dynastic period in the south (Darby et al. 1977; Linseele et al. 2014). Depictions of animal husbandry and cows being milked are frequent throughout the Dynastic period (Darby et al. 1977). It is also important to consider the age structure of the zooarchaeological assemblage, as the presence of young adult animals usually implies meat production whereas a high frequency of adult females usually suggests that the animals were being kept as a source of milk rather than of meat. Animals that might have been primarily used for milk production or for traction may also sometimes have actually been consumed. It is, however, likely that the meat more commonly eaten by peasants was of smaller species; lamb, goat, pork, fish, and fowl (Darby et al. 1977). At Hierakonpolis, Nile perch dominate the piscine fauna, but large quantities of turtle and crocodile were also found, suggesting that the community was heavily dependent on riverine resources (Brewer & Friedman 1989). But how much was actually consumed by the inhabitants of Hierakonpolis?

Traditionally, skeletal analysis has used features like dental markers to get at aspects of the actual diet eaten by a particular individual. For example, dental wear provides a measure of the coarseness of the diet and hence provides guidance as to the kinds of foods being consumed by those actual individuals. Aspects of skeletal health, such as metabolic disease markers, can also aid in the reconstruction of diet and hence subsistence. Furthermore, when mummies are recovered, it is possible to study the actual intestinal contents themselves. These obviously indicate what the individual actually consumed in life. We can also look at the coprolites left by the people or the archaeoparasites found either in the body or at the archaeological site. More recently, chemical analyses of their bones, teeth and hair have provided chemical signatures of the diets consumed by the individual people themselves (Katzenberg 2012). Isotopic study of the collagen within human bone has shown that people ranging in date from the Predynastic through to the 30th Dynasty had diets based on C_3 plants (i.e. temperate plants and grasses), associated with high nitrogen intakes (Thompson et al. 2005). These results suggest that wheat and barley, in association with freshwater fish or arid-based animals, formed a significant proportion of the diet. Chemical analyses of other objects found at the site can also tell us about diet. For example, lipid analysis of the inside of ceramic vessels can indicate the sorts of foods or foodstuffs being stored in those vessels and that thus were available to the population as a whole. This, of course, does not mean they were actually consumed by the people at the site, but does show that they were available to Egyptians overall. Lastly, it is important not to forget other forms of evidence available to Egyptian archaeologists, including documentary or textual evidence, such as written lists of foods, letters referring to foods, or artistic representations of subsistence practices such as those depicted in funerary contexts.

3.4.1 Dental and Skeletal Markers

The human skeleton itself can provide basic raw data about diet and subsistence. This can obviously be complemented by analyses of the zooarchaeological and archaeobotanical assemblage. The clearest and simplest data from the human skeleton come from the teeth. For example, dental wear provides information as to the level of abrasion of the foods in diet. Dental caries indicates a high level of carbohydrate consumption; this is usually assumed to be linked to sugar and other complex starches in the diet. Severe caries or severe dental wear may lead to infection and hence the formation of dental abscesses. The factors involved in tooth wear include:

1   the nature of the diet,
2   the relative toughness of the diet,
3   the abrasiveness of the food,
4   the fibrousness of the food, and
5   the way in which the food is prepared (Ibrahim 1987).

Most adult teeth show some evidence of dental wear, and most of this due to the abrasive nature of the food consumed. Gross analyses of dental wear can provide a good estimate of the types of food that have been consumed during the individual's lifetime. This is usually a part of basic osteological analysis, and thus a rough estimate of certain aspects of the Egyptian diet may be found in most excavation reports. Dental microwear analysis, using a scanning electron microscope or texture analysis (Ungar et al. 2008), can improve the resolution of these estimates of diet through the identification of scratches and other types of damage to the tooth surface that have built up over an individual's lifetime. The form of damage reflects the types of food that have been eaten, so a diet consisting primarily of hard seeds tends to leave the surfaces of the molar teeth pitted with small indentations as a result of the cracking of these seeds within the mouth (Brown & Brown 2011).

In Egyptian populations, most abrasive matter results from grinding corn with stone, but there is also contamination of the grain with wind-blown sand (Leek 1972a; 1972b; Ruffer 1920). If we assume a staple Egyptian diet of wheat and barley, usually consumed in the form of bread and beer (Samuel 1993), and that much of the grain processing was undertaken using stone quern and mortar implements (Ruffer 1920; Samuel 1993) and was associated with the storage of grain in granaries in the sand (Ibrahim 1987), this could have led to contamination of the grain with grit, thus accelerating dental attrition. Furthermore, the quality of the food determines the amount of bite force required to perform mastication. Tough fibrous food requires a greater amount of masticatory stress to be exerted on and by the teeth, resulting in heavier wear on the biting surfaces than for soft, refined foodstuffs. Softer diets require less masticatory effort, and thus may enable a reduction in the size of muscular attachments, a reduction in the size of the posterior dentition and a reduction in facial prognathism (projection).

Cavities in teeth, due to caries, are rarely found in Egyptian archaeological samples (Grilleto 1977; 1978; 1979; Koritzer 1968; Leek 1972a; 1972b; Miller 2008; Ruffer 1920; Zakrzewski 2012), despite being the most commonly found dental disease in archaeological populations (Roberts & Manchester 1995). The prevalence of caries not only depends on diet, but also on several other factors including oral hygiene, the fluoride level in the drinking water, and the general immune status of the individual (Burt & Ismail 1986; Lukacs 2012). In Egypt, the low caries rate can probably be explained by a diet low in cariogenic foods, and implies a relatively low level of carbohydrate consumption (Hillson 1996). It is also possible that the accidental intake of the antibiotic tetracycline in either food or beer may also have had some effect on caries frequencies in Egyptians, as its presence has been linked to very low levels of bone diseases (Armelagos et al. 1981); this has been suggested as a possible cause of the low level of caries infection in Egyptian samples (Rose et al. 1993). The location of the caries within the dentition may also be important, as lesions on the occlusal (biting) and interproximal (between teeth) surfaces suggest the consumption of soft, non-abrasive food that sticks to the teeth and remains adherent long enough to cause dental decay (Hillson 1996). This shows that although caries frequencies provide evidence of the type of diet consumed by the individual, the relationship between caries presence and diet is not completely clear.

Getting at diet and subsistence from dental evidence is further complicated by antemortem tooth loss. This may occur due to:

1   caries,
2   alveolar destruction, or
3   abscessing.

Antemortem loss is related to a lack of vitamins C and D, resulting in swelling and bleeding of the gums, thereby promoting periodontal disorders and increasing the likelihood of antemortem tooth loss. It follows that periodontal disease is a reasonable marker of poor dental care and hygiene. However, tooth loss is also likely to be affected by level of wear as severe dental wear or attrition may make gingival infection more likely. Furthermore, dental caries may precipitate the formation of periapical abscesses, and hence potentially lead to antemortem tooth loss.

Teeth may also develop plaque in life that hardens into calculus in archaeological material. This is frequently accidentally removed in long-excavated collections, but sampling immediately after excavation permits some identification of aspects of the diet. Dental calculus may contain inorganic micro-remains called phytoliths. These are rigid plant microstructures composed of silica or calcium oxalate and provide a guide to long-term diet (Fox et al. 1996; Piperno 2006).

Masticatory behaviour may also modify craniofacial morphology. For example, specific actions may lead to a plastic response in the bone. This includes activities such as using the teeth as a tool, such as for stripping or treating hides. This obviously will modify the teeth and their dental wear, but will also affect the development of the muscular attachments in the skull associated with mastication. Eating a relatively coarse diet will require a great deal of chewing, and hence will result in large masticatory apparatus in the skull. This can be seen as a highly prognathic face (with a great deal of forward projection, particularly under the nose area) and/ or large muscle attachments in the parietal regions of the skull.

Metabolic disorders, such as scurvy and rickets due to insufficient intake of vitamins C and D respectively, can be diagnosed from a variety of skeletal lesions. These skeletal lesions themselves, most commonly areas of increased bone porosity, therefore provide evidence of diet and thus indirectly of subsistence practices. Of potentially greatest importance is cribra orbitalia. This lesion, indicated by porosity in the roof of the eye sockets, is an indicator of blood-related disorders such as anaemia, thalassaemia, schistosomiasis and intestinal parasites. In the context of Egyptian diet, anaemia resulting from iron deficiency is relevant. Again, it is possible that the accidental intake of tetracycline may have mitigated against cribra orbitalia within Egyptian populations.

Metabolic disorders may also affect animals being kept by humans. Animals being kept in captivity may be provided with foodstuffs that do not match with their optimal diet. This might then result in malnutrition due to either the lack of certain nutrients or the consumption of excessive quantities of particular nutrients. If these animals were then consumed by humans themselves, this might have indications for the health of those people. Although not kept for consumption, the mummified baboons kept in captivity at Tuna-el-Gebel show evidence of scurvy (Nerlich et al. 1993).

As a final note, it is important to remember that the diet of an individual will affect their susceptibility to disease and infection, as a malnourished individual is more liable to infection and hence to a further reduction in food consumption. Where a nutritional balance is not achieved, bone turnover may be altered (Brickley & Ives 2008), thereby having implications upon isotopic analyses. Furthermore, prolonged nutritional inadequacy or poor dietary quality, such as low dietary protein, can result in metabolic bone diseases such as osteopenia and osteoporosis. The actual diet that an individual consumes may therefore have a clear effect upon their overall health and wellbeing.

3.4.2 Isotopic Markers

You are what you eat, or so they say. In this sense, the actual chemicals that are consumed within the long-term diet can be found, and traced, within the human body. Most typically this involves analysis of the carbon and nitrogen concentrations within the collagen in the bones, but can also include analyses of these or other elements within the hair or the teeth (Eriksson 2013; Katzenberg 2012). The nitrogen concentration provides an indication of the amount of protein in the diet, and hence is a marker of the trophic level of the individual from which the bone sample derives. The higher the nitrogen value, the greater the amount of protein consumed in the diet of that particular person. Individuals with high intakes of meat or fish can therefore be recognised by their high nitrogen (^{15}N) intake. The carbon value provides an indication of the type of plant material being consumed by either that particular person, or by the animals upon which that person subsisted. Scientifically, plants are divided into two photosynthetic pathways, C_3 and C_4, with the former generally covering temperate plants and the latter more arid-loving tropical plants. Within Egypt, the carbon signal from the isotope analysis provides an indication of the types of grain being consumed either by the humans or their animals. An individual whose bones have relatively high values of carbon isotope (^{13}C) will have consumed mainly tropical grasses (such as sorghum or millet) or animals that ate such tropical grasses. Most Egyptians appear to have had a diet based upon more temperate plants, such as emmer wheat and barley (Thompson et al. 2005). The important thing to remember when undertaking stable isotope analysis is that comparative faunal material must also be analysed in order to provide a baseline for comparison. It is also important to note that isotopic analysis cannot distinguish between actual consumption of meat and that of secondary products such as milk and cheese.

Due to the slow turn-over of bone in the body, stable isotope reconstruction of diet from bone collagen only permits long-term diet to be assessed (usually averaging out the diet of the last 10 years of an individual's life), and will not enable recent or rapid changes in diet to be identified. By contrast, teeth develop in childhood and their enamel is not turned over during life. This means that the isotope ratios within dental enamel indicate the person's dietary intake during the period of dental development (i.e. during their childhood). As different teeth form at different times, isotope measurements taken from different teeth within the same jaw can provide a series of differing isotope values for the different periods of that person's childhood life. Furthermore, micro-sampling through the layers of the enamel can indicate seasonal variations in isotope intake (Brown & Brown 2011). Hair, which is made of keratin, also contains a record of the carbon and nitrogen isotopes in the diet in the months during which that hair grew. As hair grows at approximately 1 cm in length per month, a time series of isotope measurements of diet can be obtained by studying sections along the length of the hair (Wilson & Gilbert 2007). Analysing hair from a mummy's head can therefore provide a month-by-month record, or even a week-by-week record, of an individual's diet in the months preceding their death. In the same manner, finger or toe nails also contain this same isotopic sequencing potential, albeit with some potential nitrogen enrichment and carbon depletion (Wilson & Gilbert 2007). Other short-term changes might also be noted in such ways, such as the effects upon diet as a result of pregnancy in terms of both morning sickness and appetite changes. $\delta^{15}N$ values increase late in pregnancy due to bodily nitrogen being shifted towards tissue production rather than being excreted (Fuller et al. 2006). Lactation also may increase $\delta^{15}N$ values as a result of potential short-term water stress (Ambrose & DeNiro 1986). Similarly White & Armelagos (1997) found higher $\delta^{15}N$ values in osteoporotic females than in non-osteoporotic females, which they attributed to water stress arising from the arid Nubian setting.

Although carbon and nitrogen are the most commonly studied isotopes, it is also possible to study concentrations of other elements such as sulphur or lead. In addition, when studying bone, there is a difference between studying the bulk collagen within the bone and studying the inorganic bioapatite fraction. The collagen, as a protein, contains both carbon and nitrogen isotopes, and experimental feeding studies have shown that the collagen is formed mainly from amino acids contained in dietary proteins (Brown & Brown 2011). In contrast, bioapatite contains carbon isotopes but not nitrogen isotopes, and this derives from ingested protein, carbohydrates and fats. As a result, the results derived from the collagen and the bioapatite provide information on differing parts of the individual's diet. Similarly, the dentine within a tooth contains proteins, thereby providing both carbon and nitrogen values, whereas the enamel on the surface is composed of bioapatite and so only contains carbon. Hair, as noted earlier, contains both carbon and nitrogen isotopes, and thus informs about the amino acids contained in the dietary protein consumed.

3.4.3   *Intestinal Contents, Archaeoparasites and Coprolites*

When bodies naturally desiccate to form mummies, the material left in the stomach and colon can be examined to provide an indication of the last few meals consumed (Fahmy 2001). This provides details of the actual contents of the last few meals, but the representation and completeness will depend upon the relative preservation and decomposition of the foodstuffs (both by the digestive processes in life and by taphonomic processes after death). Intestinal contents may be studied by palaeobotanists to enable identification of organic macro-remains such as fruits, seeds, and grains. Some skeletons also contain well-preserved remains of ingested foods, such as from the 'working-class' cemetery at Hierakonpolis. While bodies preserved through intentional mummification are less likely to contain these remnants, and hence cannot so easily be studied in this manner, naturally desiccated mummies may have their gut contents assessed through endoscopy.

Box 3.6   Gut Contents: A Case Study

Ahmed G. Fahmy
† late of Helwan University

Understanding of the human interaction with plants in ancient Egypt, resulting in diet and nutrition, and reflected in food preparation and farming practices, is based on inferences derived from a variety of sources: e.g., tooth wear, bone growth, food remains. These are found as offerings in graves, botanical remains retrieved from archaeological sediments and, in the Dynastic period, from depictions on tomb walls. The recovery of a sizable and significant sample of well-preserved ingested food remains in the abdominal cavities of 'working-class' human burials at Predynastic Hierakonpolis (*c.* 3750–3300 BC) (Friedman et al. 1999) allows a rare opportunity to test and refine assumptions on subsistence, food preparation and land use in this formative period, with direct and non-circumstantial evidence. Extracting the full range of information from this material required an integrated approach to analysis, including the identification of both preserved organic macro-remains (e.g., fruits, seeds, drupes and chaff) and plant micro-remains, i.e., phytoliths and starches.

Phytoliths are distinct from macro-remains in preservation since they are inorganic cell inclusions, are unaffected by micro-organisms and do not decay (Piperno 2006: 5). In other words, phytolith analysis can generate data on plant occurrence in contexts where organic botanical material (such as tubers and leaves) is poorly preserved. While the analysis of ancient starch has a long history, it is only in the last two decades that ancient starch studies have become a significant sub-discipline within palaeobotanical approaches to archaeology (Pearsall 1989; Piperno 2006; Samuel 2000). Starch proved to be particularly enlightening with regard to food preparation at Hierakonpolis, augmenting the results of macro and phytolith analysis with otherwise undetectable indicators of the ways in which the plant material was modified into edible food.

Previous archaeological and archaeobotanical studies of Egyptian Predynastic cemeteries mainly focused on the analysis of grave goods, such as pottery contents (Brunton 1948; Brunton & Caton-Thompson 1928; Fahmy 1995; 1997). Netolitzky was one of the first to undertake a good macro-botanical study of gut contents from human bodies excavated from the Predynastic cemetery N7000 at Naga-ed-Dêr (Netolitzky 1943). The analysis of human gut contents from Predynastic Hierakonpolis can now present clearer insights into the dietary components and subsistence activities of the workers during the Middle Predynastic Period (3750–3300 BC) (Fahmy 2001; 2003; 2008). Samples of desiccated gut contents gathered from the thoracic and pelvic areas of 19 well-preserved human burials were studied. All samples were weighed and photographed before any laboratory procedures were undertaken. Each sample was divided into two halves; half of each sample has been retained untreated for future reference or study. The sample portions being studied were re-hydrated by addition of a 0.5 per cent tri-sodium phosphate aqueous solution (Holden 1994: 71). Bryant (1974: 409) recommends that the strength of the tri-sodium phosphate solution must be exactly 0.5 per cent, as a solution strength that exceeds this may destroy the middle lamella of plant cells and cause the destruction of the delicate plant tissue, thereby hampering the identification process. After a period of 4–10 days, a red/brown solution fluid is obtained, decanted into a separate container and replaced with an alcohol/ water mixture. This wet condition prevents the deformation of the morphological and anatomical features of the plant remains.

Identification of the plant macro remains is based on their gross morphological features. For this study, these were compared with modern reference collections kept at the University of Helwan. A variety of taxonomic keys for seed and fruit identification and useful drawings and illustrations are also used in this process (e.g. Boulos 1999–2005). Phytolith morphotypes can be identified following Pearsall (1989: 311–438), Piperno (2006: 23–44) and Rosen (1992; 1996). Identification of starch grains follows Gassner (1973). The genera of *Triticum*, *Hordeum* and *Aegilops*, amongst others, produce starch grains that can be distinguished in archaeological samples (Henry & Piperno 2009; Piperno et al. 2004).

Analysis of the re-hydrated gut contents revealed the presence of botanical materials that have been classified into two major groups: (I) plant macro-remains and (II) plant micro-remains (phytoliths and starch grains).

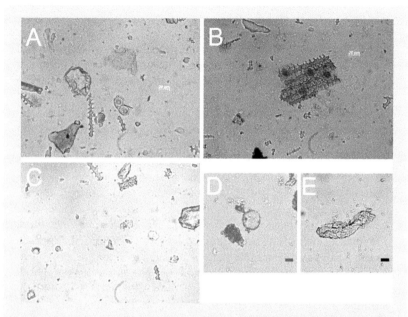

Figure 3.3 (A) Phytoliths, dendritic forms, spiny in shape from human gut contents HK43. (B) Epidermal cells from cereals from human gut contents HK43. (C) Phytoliths, rondel forms from human gut contents HK43. (D) Cereal starch grain from human gut contents HK43 (scale bar 20 μ). (E) Gelatinized starch from human gut contents HK43 (scale bar 20 μ).

I   Four seeds of melon (*Cucumis melo* L.) were found among the gut contents of burials 110, 387 and 456, while six seeds of *Cucumis* sp. were recovered from burial 394. It seems that these seeds were swallowed accidentally while eating the fleshy parts (mesocarp) of the melon fruits. This popular fruit has been recorded from archaeological sites in Egypt dating from Predynastic (4500–3200 BC) to Graeco-Roman times (332 BC – AD 395) (Fahmy 1995; Vartavan & Asensi Amoros 1997; Zohary & Hopf 1993).

II   Phytolith analysis of the gut contents from HK43 showed the presence of some morphotypes including dendritic forms (Figure 3.3A), epidermal cells (Figure 3.3B), rondels (Figure 3.3C), long cells and bulliform. These morphotypes characterised husks of cereals (Piperno 1988; 2006), suggesting that some hulled grains of emmer wheat and barley were subjected whole for grinding (without dehusking); they may have been germinated or malted, a process important in the production of beer (Samuel 2000: 540).

Cereal starches (Figure 3.3D) and phytoliths from emmer wheat and barley were also recovered from dental calculus, stomach and intestinal contents, and from a coprolite sample from the burials at HK43, thereby indicating the use of grain-based food products in the inhabitants' diets. Starch grains from these contexts had been degraded by gelatinisation (Figure 3.3E) indicating that the food was cooked prior to ingestion. The presence of starch grains from oats (*Avena* sp.) and melon (*Cucumis* sp.) and saddle phytoliths from chloridoid grasses represent variation in the diet and are in line with previous studies of the macro-remains from these contexts (Fahmy 1995; 2003; 2005). The oats (*Avena* sp.), and perhaps the chloridoid grasses, were probably present as field weeds and were collected and processed accidentally with the cereal grains into food.

Notable success has been attained from this integrated approach to the analysis of botanical macro- and micro-remains (phytoliths and starch) retrieved from human gut contents from the Predynastic cemetery HK43 at Hierakonpolis (Fahmy 2001; 2003; 2008). In addition to the macro-remains present in the identifiable food offerings in pottery and baskets, i.e., fruits of Christ's thorn (*Ziziphus spina-christi* (L.) Desf.), Egyptian plum (*Balanites aegyptiaca* Boiss.) and tubers of *Cyperus esculentus* L., starch grains and phytoliths from gut contents were also analysed. This study confirmed the existence of a mixed subsistence strategy, which included the cultivation of cereals (mainly emmer wheat) and vegetables (melons), as well as the gathering of wild fruits and tubers. Much like in Pharaonic Egypt, the inhabitants of Predynastic Hierakonpolis subsisted on a grain-based diet with variety derived from cultivated vegetables in association with the gathering of wild edible fruits.

References

Boulos L. (1999–2005) *Flora of Egypt, 4 volumes*. Cairo: El Hadara.

Brunton G. (1948) *Matma, British Museum Expedition to Middle Egypt, 1929–1931*. London: Bernard Quaritch.

Brunton G. & Caton-Thompson G. (1928) *The Badarian Civilization and Prehistoric Remains Near Badari*. London: Bernard Quaritch.

Bryant VM. (1974) Diet in southwest Texas: the coprolite evidence. *American Antiquity* 39: 407–420.

Fahmy AG. (1995) *A Historical Flora of Egypt: Preliminary Survey*. Unpublished thesis. Cairo: University of Cairo.

Fahmy AG. (1997) Evaluation of the weed flora of Egypt from Predynastic to Graeco-Roman times. *Vegetation History and Archaeobotany* 6: 241–247.

Fahmy AG. (2001) Plant remains in gut contents of ancient Egyptian Predynastic Mummies. *Online Journal of Biological Sciences* 1: 772.

Fahmy AG. (2003) Palaeoethnobotanical studies of Egyptian Predynastic cemeteries: new dimensions and contributions. In: Neumann K., Butler A. & Kahlheber S. (eds) *Food, Fuel and Fields.* Africa Praehistorica 15. Köln: Heinrich Barth Institute.

Fahmy AG. (2005) Missing plant macro remains as indicators of plant exploitation in Predynastic Egypt. *Vegetation History and Archaeobotany* 14: 287–294.

Fahmy AG. (2008) Analysis of mummies' gut contents from Predynastic Hierakonpolis, Egypt (3750–3300 BC). In: Midant-Reynes B. & Tristant Y. (eds) *Egypt at its Origins 2.* Orientalia Lovaniensia Analecta 172. Leuven: Peeters. pp. 419–426.

Friedman R., Maish A., Fahmy AG., Darnell D. & Johnson ED. (1999) Preliminary report on field work at Hierakonpolis: 1996–1998. *Journal of the American Research Center in Egypt* 36: 1–35.

Gassner G. (1973) *Mikroskopische Untersuchung Pflanzlicher lebensmittel,* 4th edn. Stuttgart: Gustav Fischer Verlag.

Henry A. & Piperno D. (2009) Changes in starch grains from cooking. *Journal of Archaeological Science* 36: 915–922.

Holden T. (1994) Dietary evidence from the intestinal contents of ancient humans with particular reference to desiccated remains from northern Chile. In: Hather J. (ed.) *Tropical Archaeobotany: Applications and New Developments.* London: Routledge. pp. 65–85.

Netolitzky F. (1943) *Nachweise* von nahrungs – und heilmitteln in den trockenleichen von Naga-Ed-Der (Ägypten). *Mitteilungen des Deutschen Instituts für Ägyptische Altertumskunde in Kairo* 11: 1–33.

Pearsall D. (1989) *Paleoethnobotany: A Handbook of Procedures.* London: Academic Press.

Piperno D. (1988) *Phytolith Analysis: An Archaeological and Geological Perspective.* San Diego: Academic Press.

Piperno D. (2006) *Phytoliths: A Comprhensive Guide for Archaeologists and Palaeoecologists.* Lanham: AltaMira.

Piperno D., Weiss, E. Holst I. & Nadel I. (2004) Processing of wild cereal grains in the Upper Palaeolithic revealed by starch grain analysis. *Nature* 430: 670–673.

Rosen A. (1992) Preliminary identification of silica skeletons from Near Eastern Archaeological sites: an anatomical approach. In: Rapp G. & Mulholland S. (eds) *Phytoliths Systematics Emerging Issues. Advances in Archaeological Science, vol. 1.* London: Plenum Press. pp. 129–147.

Rosen A. (1996) Phytoliths in Predynastic: A microbotanical analysis of plant use at HG, in the Hu Semaineh region, Egypt. *Archeo-Nil* 6: 77–82.

Samuel D. (2000) Brewing and Baking. In: Nicholson PT. & Shaw I. (eds) *Ancient Egyptian Materials and Technology.* Cambridge: Cambridge University Press. pp. 537–576.

Vartavan C. de & Asensi Amoros V. (1997) *Codex of Ancient Egyptian Plant Remains.* London: Triade Exploration.

Zohary M. & Hopf M. (1993) *Domestication of Plants in the Old World.* Oxford: Oxford University Press.

Occasionally human faecal material is preserved in a desiccated form as coprolites. The composition of these too can be studied in order to ascertain the components of foodstuffs consumed previously. They, obviously, cannot be linked to one individual person, but provide evidence of the diet of the local group. Traces of bone or hair are usually understood as evidence of meat being consumed, whereas the identification of seeds, pollen and phytoliths indicate which types of plants were eaten (Brown & Brown 2011). The faeces of animals, faunal coprolites, not only indicate the animals being kept by the human population, such as pigs at the workmen's village at Amarna (Panagiotakopulu 1999), but also demonstrate what was consumed by those animals themselves. In this way a food web model of diet may be developed. The presence of human myoglobin in a human coprolite found in the prehistoric Southwest of the USA demonstrates that human remains had been consumed in the area (Marlar et al. 2000)! Immunological detection assay methods (enzyme-linked immunosorbent assay or ELISA) can be used to identify meat residues, and ELISA successfully detected human myoglobin in the human coprolite mentioned above (Marlar et al. 2000). As myoglobin is only found in skeletal and cardiac muscle cells, the only way for it to be present in a coprolite is if flesh has been consumed. Using this approach the actual meats consumed by humans or animals can be determined very precisely. Due to the current prohibition on the export of archaeological samples from Egypt, this type of research has not yet developed within Egyptian archaeology.

It is important not to forget other locations where archaeobotanical remains may be preserved and hence employed to ascertain diet. The contents of a basket buried beside a woman at Hierakonpolis included tubers and cereal grains, showing that the subsistence pattern probably comprised cultivation of cereals associated with gathering of wild fruits and tubers and herding of livestock (Fahmy 2005). Helminth parasite eggs were found in embalming materials from Saqqara (Harter et al. 2003), suggesting close proximity with livestock such as pigs or cattle, and these authors suggest that the shrouds of mummies, clothes and/ or other wrapping tissues should be searched for archaeoparasites. Furthermore, helminth eggs are very durable and can be easily retrieved from archaeological sediments and coprolites and so can aid in the reconstruction of the living environment and hence the diet (Reinhard & Araújo 2008).

3.4.4   Other Evidence

Indirect evidence for diet and subsistence may be obtained from other objects recovered, for example ceramic vessels and cooking pots. Beer vats, such as those at Hierakonpolis and Tell el-Farkha, or bread moulds may be recovered and identified during actual excavation. Other items may only provide dietary evidence through scientific study. The internal surfaces of vessels may provide lipid or other organic residues which permit the identification of the types of foodstuffs stored within them, such as palm fruits in vessels from Qasr Ibrim (Copley et al. 2001). These residues will be of fats or other lipids, or occasionally proteins, which became absorbed into the wall of the original vessel during the cooking process. Some residues, such as those of wine or cooking oils, can also become absorbed into the internal walls of the vessel in which they are stored. These absorbed residues are not visible to the naked eye, and the compounds present within them need to be identified using techniques such as gas chromatography–mass spectrometry. By contrast, visible deposits on the surfaces of potsherds or vessels tend to be sooty. Deposits found on the external surface may derive from the wood or other fuels used to cook the food, whereas those on the internal surface generally result from foodstuffs becoming burnt during cooking. Of course, those food remnant deposits found on the internal surface are of foods which, by definition, were not actually eaten.

What constitutes food? And what is diet? Eating and drinking involves making specific choices as to what is consumed. The choice of foodstuffs will differ in times of hardship from those selected in times of plenty. It is important to remember that diet is more than simply food. This is because not only does it involve an understanding of not only the edible nature of the item, but also potentially some understanding of its nutritional use, such as for protein. The concept of what constitutes food is strongly influenced by cultural preferences and normative practices. Items that are recognised as foodstuffs today may not tally with those identified as such in the past. We are able to match these up through artistic depiction, and hence obtain some impression of the relative social importance of food types within Egypt. Furthermore the processes of transforming food items into foodstuffs also need to be considered. The very activities of eating and drinking, and the preceding activities of producing the comestible items, are intensely social processes involving repeated activities and a certain degree of skill and learning.

How representative was the general Egyptian diet of bread and beer? Were all individuals equally able to access these food resources? Questions of food distribution within groups are linked to the very social nature of eating and drinking, and the potential social rules and taboos associated with consumption. Without studying the human remains themselves, however, we cannot obtain a reliable and accurate measure of the foods and drinks actually consumed by each individual. Without a written record of the social mores surrounding eating and drinking, we can only assess the nature of these from comparisons of access to food items and the nature of the actual items consumed by each individual.

The study of diet provides some guidance as to the subsistence strategy employed and trades undertaken within the local community. Furthermore, diet can interact with health. As noted earlier, a diet deficient in iron may lead to anaemia and a diet deficient in vitamin C may lead to scurvy. But the local environmental conditions may also affect health through diet. For example, chronic parasitic intestinal infection, as a result of eating contaminated food, may also lead to disease, such as occurred to the mummy Asru at Manchester Museum, who had a hydatid cyst in her lung caused by infection with the dog tapeworm *Echinococcus granulosus* (Tapp 1984). Artistic representation of domestication and cultivation can provide guidance as to the subsistence processes. These may be in idealised form, rather than in the actual format that was practised in the area at the time. Studying the biological signals of diet therefore ties in to studies of economy and exchange, and hence into the overall understanding of social organisation. Understanding variation within dietary intake within a community, such as by age, sex or social rank, allows us to understand differences in access to such resources as food, and therefore improve our understanding of the social roles and organisation within that grouping of people.

Ideally we would hope always to undertake an integrated analysis of the diet of each Egyptian group excavated. In an ideal world this would be done by linking stable isotope results with other markers, such as those of dental health. The stable isotope signature would provide the actual concentrations of the types of food consumed, in terms of the relative proportions of protein and carbohydrates in the diet. Phytolith analysis from the dental calculus would add to our knowledge of the starches consumed. The dental wear analysis would provide an indication of the robustness of the actual foods eaten, and hence of the food processing techniques used. These would be compared with the archaeological remains, e.g. bread moulds. In addition, any gut contents recovered would be analysed as a clear remnant of the last meals consumed (in contrast to the long-term dietary reconstruction from stable isotopes). Finally, these would be linked with any local palaeobotanical and zooarchaeological remains to assess both the local economic feasibility of the modelled reconstructed diet and the local environmental living conditions.

3.5   Clothing and Adornment

What can we learn about a person from the way they were dressed or decorated in death? How were their clothes or mummy wrappings made? What fabrics were used? Considering aspects of clothing and bodily adornment or decoration involves analysis that links through to Chapter 4 of this volume: Biography & Analysis of Objects. Here, the biographies of the amulets or clothes are considered in relation to the biographies of the humans with which they are buried.

3.5.1 Fabric Remains and Wrappings

As noted earlier, analysis of the fabrics found in burial contexts can inform as to date of burial, to the quality and type of cloth used, and to the method of production of that fabric. Most bandages were reused clothes or linen sheets, often bearing laundry marks giving the names or titles of the deceased (Ikram & Dodson 1998). The earliest form of mummy wrapping was an animal skin, usually that of a goat or occasionally a gazelle, into which the flexed body was placed prior to burial. Animal skin wrappings were followed by mat wrappings, although mat wrappings continued to be used for poor burials throughout the Pharaonic period. For wealthier burials, strips of linen soon started to be used, with varying thicknesses and qualities being used. The format of the wrapping provides some evidence of the date. In the Middle Kingdom, large folded sheets were included in the wrappings to give the mummy bulk and protect it from damage during placement in the coffin. In the New Kingdom, close overlapping spiral-bound wrapping of individual limbs became the norm. After this, polychrome textiles became used as shrouds. In the Late Period, bodies were wrapped in narrow bandages which were wound to form complex and elaborate patterns, with geometric patterning of this bandaging, such as to form lozenge shapes, reaching its peak in the Greco-Roman period (Ikram & Dodson 1998). Cloth was also placed into the burial in the form of covers for amulets and statues and was wrapped around meats and other foodstuffs. As discussed earlier (see Section 3.1.1: Burial Rites and Ritual), details of the fibres used to make the textiles, such as their thicknesses, the degree of spinning of the threads, and the sources of the yarns, may be revealed through high magnification brightfield and polarisation microscopy. Dating is possible using radiocarbon analysis.

3.5.2 Ornamentation and Amulets

The placing of amulets, jewellery or other ornaments on the corpse, interspersed with the linen wrappings, was an important part of the mummification and burial process. They can basically be considered as charms and were believed to protect the wearer or provide magical benefits (Ikram & Dodson 1998). The power of an amulet was transmitted by its bodily location, material, colour and shape. Each of these aspects must therefore be considered.

In the past, amulets were recovered during the unwrapping of mummies, and robbers dug holes in the chests of many mummies to obtain the heart scarab (Ikram & Dodson 1998). Using CT, ceramic or metal amulets placed within mummy wrappings can be easily identified and their exact position located. As noted earlier, during the burial ritual, the body was adorned with jewellery. This was either specifically funerary or burial jewellery or was jewellery used during the lifetime of the deceased

Box 3.7   Textile Bindings at Hierakonpolis
Jana Jones

The research potential of microscopic analysis for the study of archaeological textiles and fibres is well documented (e.g. Jakes 2000). It is essential in the analysis of Predynastic textiles, which are generally fragmentary, fragile and in various stages of degradation. Using a microscope, technical information can be obtained from samples as small as 1 cm², and thread counters and hand lenses do not directly touch the textile during observation. Inclusions invisible to the naked eye (e.g. resins or minute granules of minerals, such as malachite or iron oxide) are easily identified and can provide information on the function or context of the textile. Fibre identification, often a hit-or-miss process based on observationby eye, is conducted with a greater degree of certainty, despite limitations imposed by the condition of the material. Furthermore, the physical or chemical tests used to distinguish modern fibres cannot be applied successfully to aged material (Goodway 1987).

In the absence of textual evidence and limited explicit evidence for manufacture during the Predynastic period, the detailed examination of textile remains provides significant information on the technical features of production, changing technologies, use and re-use, and by inference, on craft specialisation and economic organisation.

At Hierakonpolis textile evidence was studied from three localities:

- HK43, the non-elite cemetery with good textile preservation (214 samples)
- HK6, the elite cemetery with poor preservation (21 samples)
- HK11 (Operation G), the settlement area with good organic preservation in the debris (7 samples).

The morphological examination and analysis of textile structure was carried out using a Leica MZ6 stereomicroscope, with a magnification range between 6.3× and 64×. A Leica CLS100 fibre optics unit provided illumination because a cold-light source is essential when observing aged, desiccated textiles. The recording and cataloguing were based on guidelines by Walton and Eastwood (1988), CIETA (1964; 1997) and Emery (1995).

The fibre was examined with a Zeiss Universal microscope system equipped with transmitted light and epi-illumination optics.

Figure 3.4 Hand (shown palm up) wrapped in 'resin'-soaked layers of textile. HK43, Burial 16 (Photograph courtesy of Ron Oldfield)

Polarisation light microscopy is the most useful of the various transmitted light techniques as it shows internal fibre cell detail in very high contrast and marked colour relief. Flax has characteristic 'X' shaped dislocations and a narrow central lumen (or canal), and diagnostic surface striations slanting from left to right (\). The latter are visible with Differential Interference Contrast (DIC), another transmitted light technique (Jones 2008: Figure 7). Micrographs can be compared with modern fibre atlases to confirm the identification of the fibre as *Linum usitatissum*, the domesticated species of flax grown to the present day (Catling & Grayson 1982: 12–17, pl. 11; The Textile Institute 1985: 16, Figure 55; for methodology see Jones 2008: 102–5). Photographic recording was carried out using a photomacrographic system and photomicrographic attachments connected to the stereo and light microscopes.

In cemetery HK43, textiles in various states of preservation were present in almost every burial. During the cemetery's use in Naqada IIB, generally only two qualities of textile were used as wrappings – fine and medium (Jones 2001). They were plain (tabby) weaves of flax yarns spun in the 'S' direction (from left to right) (Jones 2008: Figure 8; figs 3a, 4a). The remains of shrouds that had covered the bodies were all the same medium quality. Both qualities were combined in the thick layers that encased the base of the skull, jaw, chin and hands of three intact female burials. The internal organs were also wrapped and repositioned inside a number of bodies (Friedman et al. 2002: 65–6; Jones 2002a). Macro and micro examination of the wrappings showed that they were not applied as bandages or strips. They comprised square or rectangular pieces of textile folded to form pads of 8–10 layers, soaked in an unidentified melted resinous substance, and methodically pressed and moulded to the body. On drying, the layers consolidated into a solid mass (Figure 3.4; also Jones 2002a; 2007).

Recent biochemical analysis by gas chromatography–mass spectrometry (GC-MS) and thermal desorption/pyrolysis (TD/Py)-GC-MS has revealed the presence of complex processed 'recipes' consisting of resins, oils and fats in wrappings from Badarian and Predynastic period graves at Mostagedda (Jones et al. 2014). These results validate the premise that experimentation that would evolve into full artificial mummification was already taking place up to 500 years earlier at Mostagedda than at Hierakonpolis (Jones 2002b; 2007: 979–84).

There was no evidence for the re-use of discarded material in the HK43 wrappings, although the small size of the specimens may contribute to the lack of diagnostic features such as hems and seams. The uniform quality of the burial cloths appears to suggest specialised funerary production and a recognised standard for the grading of linen into qualities. These factors imply that there was some form of centralised control during Naqada IIB.

In the elite cemetery HK6 (Naqada IC-IIA), despite the smaller sample studied, greater variability in decoration and weaving techniques is evident. Samples include the earliest recorded example of linen dyed with red iron oxide, and fragments of an elaborately woven textile hat may have been a pouch for malachite in the tomb of an aurochs (Jones 2002a). The quality of the funerary wrappings is identical to the shrouds covering most of the bodies at HK43, suggesting that the weavers who supplied the elite used the same material in their own burials.

The trash deposits in the domestic area HK11 (Operation C) (Naqada IC-IIA) contained evidence of textile production in woven pieces, hanks of unspun fibre, spun yarn prepared for weaving, and pottery spindle whorls (Friedman et al. 2002: 55–60; Jones 2001). One of the textile specimens was identical in structure to the finer quality found in HK43 burials.

A radical technological innovation took place during the period Naqada IC-IIA – a change in spinning yarn from the 'Z' (/) to 'S' (\) direction, with the latter becoming the standard in Egyptian production from *c.* Naqada IIB (Jones 2008: 107–8; 111–16). A number of textiles from the settlement debris and the early phase at HK6 incorporated both 'Z' and 'S' spun yarns in the one weave

(Jones 2008: 109), attesting to a transitional phase in the process that would result in the manufacture of finer, stronger, evenly woven linen. There is insufficient evidence to deduce whether this collection of discarded material was associated with a specialist workshop or with domestic-based production.

References

Catling D. & Grayson J. (1982) *Identification of Vegetable Fibres*. London: Chapman and Hall.

CIETA (1964) *Vocabulary of Technical Terms English–French–Italian–Spanish*. Lyon: Centre International d'Étude des Textiles Anciens (CIETA).

CIETA (1997) *Vocabulaire Français–Allemand–Anglais–Espagnol–Italien–Portugais–Suedois*. Lyon: Centre International d'Étude des Textiles Anciens (CIETA).

Emery I. (1995) *The Primary Structures of Fabrics: an Illustrated Classification*, 4th edn. New York: Watson-Guptill.

Friedman R., Wattrall E., Jones J., Fahmy A., Van Neer W. & Linseele V. (2002) Excavations at Hierakonpolis. *Archéo-Nil* 12: 55–68.

Goodway M. (1987) Fibre identification in practice. *Journal of the American Institute for Conservation* 26: 27–44.

Jakes K. (2000) Microanalytical methods for studying prehistoric textile fibers. In: Drooker PB. & Webster LD. (eds) *Beyond Cloth and Cordage. Archaeological Textile Research in the Americas*. Salt Lake City: University of Utah Press. pp. 51–68.

Jones J. (2001) Bound for eternity: examination of the textiles from HK43. *Nekhen News* 13: 13–14.

Jones J. (2002a) Funerary textiles of the rich and poor. *Nekhen News* 14: 13.

Jones J. (2002b) Towards Mummification: New evidence for early developments. *Egyptian Archaeology* 21: 5–7.

Jones J. (2007) New perspectives on the development of mummification and funerary practices during the Pre- and Early Dynastic periods. In: Goyon J-C. & Cardin C. (eds) *Proceedings of the Ninth International Congress of Egyptologists*. Orientalia Lovaniensia Analecta 150. Leuven: Peeters. pp. 979–989.

Jones J. (2008) Pre- and Early Dynastic textiles. Technology, specialisation and administration during the process of State formation. In: Midant-Reynes B. & Tristant Y. (eds) *Egypt at its Origins 2*. Orientalia Lovaniensia Analecta 172. Leuven: Peeters. pp. 99–131.

Jones J., Higham TF., Oldfield R., O'Connor TP. & Buckley SA. (2014). Evidence for prehistoric origins of Egyptian mummification in late Neolithic burials. *PloS one* 9(8), e103608.

The Textile Institute. (1985) *Identification of Textile Materials*, 7th edn. Manchester: The Textile Institute.

Walton P. & Eastwood G. (1988) *A Brief Guide to the Cataloguing of Archaeological Textiles*. London: Institute of Archaeology.

individual. When compared with real 'lifetime' jewellery, burial jewellery was relatively flimsily constructed, for example, made from thin sheet-gold or gold leaf upon wood, as in the tomb of the Lady Sitwerut at Dahshur (Ikram & Dodson 1998).

Green and blue stones or faience signified resurrection and rebirth, haematite referred to strength and support, whereas carnelian, jasper, red glass or faience symbolised blood, energy, strength and power (Ikram & Dodson 1998). During the Old Kingdom and the First Intermediate Period, *djed* pillar amulets were usually made of faience or lapis lazuli (Ikram & Dodson 1998). Although faience is discussed in detail elsewhere (see Chapter 4: Biography & Analysis of Objects), its production varied temporally and in quality, with differing proportions of silica, calcium oxide (lime), soda (natron or other alkali), cobalt and/or other colorants etc., and differing methods of glazing, such as efflorescence or cementation (Nicholson & Peltenburg 2000). These varying aspects may be teased out using X-ray fluorescence (XRF) and scanning electron microscopy (SEM). The constituents of other amulets and their provenances may be obtained through thin-section petrography, X-ray diffraction (XRD), XRF, SEM, neutron activation analysis (NAA), atomic absorption spectroscopy (AAS), or inductively coupled plasma mass spectrometry (ICP-MS).

Masks and other external ornamentation range from crude plastered images to solid gold death masks. As noted earlier, the earliest body wrappings were animal skins, followed by mats, and later linen. By the Old Kingdom, plaster was added to the outer covering of the bandages to enable external physical features to be modelled. By the 8th Dynasty, helmet masks made of cartonnage (plaster-soaked linen) covering the entire head were common. Rigid cartonnage bands were added to the outer ornamentation during the New Kingdom (Ikram & Dodson 1998). The cartonnage was decorated with pigments and/or gold, and thus the pigment compositions, such as Egyptian blue, green earth or green chrysocolla, may be analysed using XRF and XRD. The binding material, such as egg or plant gum, may be identified through enzyme-linked immunosorbent assay (ELISA) and gas chromatography–mass spectrometry (GC-MS) (Scott et al. 2009). Fourier transform infrared spectroscopy (FTIR) can also identify the composition of the top layers of cartonnage, such as beeswax, in which it is known that Petrie used to steep artefacts prior to transport to European collections (Scott et al. 2009).

Box 3.8   A Chrysocolla Amulet in a Child Mummy from the Early Dynastic Period

Raffaella Bianucci & Grazia Mattutino

Some 4700 years ago, an Early Dynastic child from Gebelein died from an acute cerebral malaria attack (Bianucci et al. 2008). The corpse of this 15- to 18-month-old infant was loosely wrapped in a bundle of linen bandages and buried in a shallow pit grave in the northern ridges of Gebelein's second hill. His naturally mummified body was unearthed during an excavation campaign carried out by Schiapparelli and co-workers in 1914, and is currently stored at the Museum of Anthropology and Ethnography of the University of Turin (Donadoni-Roveri 1990).

When the mummy was CT-scanned and a virtual 3D reconstruction was performed, two pyriform images were observed (12.9 × 10.2 mm – the upper left item in Figure 3.5a; 5 × 4 mm for the lower right item in Figure 3.5a). They were placed dorsally within the bandages at the level of the left thigh. The upper one was removed from the bandages and examined.

Radiological and stereomicroscopic observation showed that the pyriform formations corresponded to a small leather bag closed by a knot. The rope used to close the bag was made out of vegetal fibres, most likely linen. The bag, which showed a marked indentation in its surface, appeared to contain a small ellipsoidal, radiopaque object (4–5 mm × 2.5–3 mm) with bone-like density. Through a crevice in the bag's surface, the object was observed to be an emerald-green mineral of unknown composition.

Scanning electron microscopy (SEM) and energy dispersive X-ray detection (EDX), equipped with a Cameo™ program for X-ray colour imaging, were used to study the morphology of this green coloured area and to analyse the mineral's chemical composition directly in variable pressure (Reed 2005; Torre & Mattutino 2000). For reference identification purposes, with the aim of identifying the chemical composition of the mineral kept inside the bag, several different green minerals known to have been used for adornment and in funerary contexts in ancient Egypt (from the Predynastic to the Ptolemaic period) were examined. Apart from turquoise, the principal green stones employed by Egyptian lapidaries were azurite, chrysocolla, chrysoprase, green feldspar, green jasper, malachite, prase, serpentine and, in the Graeco-Roman Period, beryl and olivine (Lucas & Harris 1962; Nicholson & Shaw 2000).

Scanning electron microscopy showed the mineral inside the leather bag to be an amorphous cryptocrystalline material with a

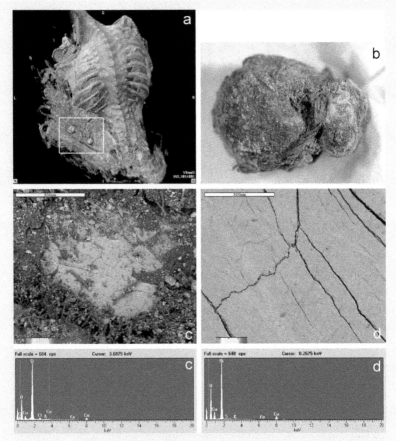

Figure 3.5 (a) 3D virtual reconstruction result, dorsal image; the two pyriform formations are indicated by a square. (b) The leather bag closed by a knot as it appears to the stereomicroscopy observation. (c, d) SEM-EDX results (LEO 1430VP scanning microscope, LEO Electron Microscopy Ltd, Cambridge, UK with a Link ISIS 300 dispersive X-ray analyser, Oxford Instruments, High Wycombe, UK). c: image of the green area appearing from the bag's crevice and its micro-analytical spectrum. d: image of a chrysocolla mineral along with its micro-analytical spectrum.

surface traversed by fissures. The microanalysis spectrum showed that it was composed of silicon, oxygen and copper, with traces of iron. By comparison with microanalysis spectra of the above green minerals, chrysocolla was found to be the only green mineral whose composition is superimposable with the spectrum obtained from the item inside the leather bag of the child mummy (Bianucci et al. 2009).

Chrysocolla is a blue-green natural cryptocrystalline hydro-silicate of copper found in secondary copper ore deposits together with malachite and azurite (Roberts et al. 1990; Van Oosterwyck-Gastuche 1970; Van Oosterwyck-Gastuche & Grégoire 1971). Chrysocolla and malachite have been mined from copper ores both in Sinai and in the Eastern Egyptian Desert since 4000 BC (Crane et al. 2001; Nicholson & Shaw 2000). In its natural state, chrysocolla appears very similar to malachite, except that its colour is more bluish.

From the Badarian Period (4500–3800 BC) onwards, beads made from different minerals were used as elements of funerary paraphernalia. Crudely shaped beads made from green and black minerals (such as malachite and galena) were kept in small lumps in little bags of leather. These were placed somewhere between the throat and the pelvis of the deceased, deposited in their hands, or, especially in the case of children, worn around the neck as 'apotropaic' amulets.

Most of the materials used as amulets in Ancient Egypt were chosen for the magical properties of their colours. Green was the colour of growing crops, new vegetation and fertility, hence of new life and resurrection. Green colour was also associated with Osiris, the god of rebirth and resurrection, who played a key role in Ancient Egyptian funerary beliefs. Emerald-green malachite, used as an eye paint cosmetic, was frequently associated with Pre- and Early Dynastic burials. Until the 4th Dynasty, the green cosmetic was used for both adults and children of both sexes. At the end of the Old Kingdom, this colour assumed a further religious meaning in funerary practices, where it became the colour symbol of Horus' 'good-eye' (Bianucci et al. 2009).

While the presence of malachite amulets in the form of scarab-beetles placed at the heart of the mummy was reasonably common, the use of chrysocolla was rare; this is most likely due to the lesser extent of chrysocolla ores. Until now, only one other example of chrysocolla from ancient Egypt can be traced, and is represented by a small figure of a seated infant. This minute figure was found, in 1898, by Quibell and Petrie in a grave near Hierakonpolis (Quibell & Green 1900, 1902). Since both the archaeological findings indicate the presence of chrysocolla amulets in infant burials, it is possible that the exploitation of this specific mineral was limited, in the Early Dynastic Period at least, to use as a protective amulet to ward off unwanted influences.

References

Bianucci R., Mattutino G., Lallo R., Charlier P., Jouin-Spriet H., Peluso A., Higham T., Torre C. & Rabino Massa E. (2008). Immunological

Evidence of *Plasmodium falciparum* infection in an Egyptian child
mummy from the Early Dynastic period. *Journal of Archaeological
Science* 35: 1880–1885.

Bianucci R., Mattutino G., Lallo R. & Torre C. (2009). Identification of
a chrysocolla amulet in an Early Dynastic child mummy. *Journal of
Archaeological Science* 36: 592–595.

Crane MJ., Sharpe JL. & Williams PA. (2001) Formation of chrysocolla
and secondary copper phosphates in the highly weathered supergene
zones of some Australian deposits. *Records of the Australian Museum*
53: 49–56.

Donadoni-Roveri AM. (1990) Gebelein. In: Robins G. (ed.) *Beyond the
Pyramids: Egyptian Regional Art from the Museo Egizio*. Atlanta: Emory
University Museum of Art and Archaeology Press. pp. 23–29.

Lucas A. & Harris JR. (1962) *Ancient Egyptian Materials and Industries*.
London: Edward Arnold.

Nicholson PT. & Shaw I. (2000) (eds) *Ancient Egyptian Materials and
Techniques*. New York: Cambridge University Press.

Quibell JE. & Green WF. (1900) *Hierakonpolis, Vol. I*. London: British
School of Archaeology in Egypt.

Quibell JE. & Green WF. (1902) *Hierakonpolis, Vol. II*. London: British
School of Archaeology in Egypt.

Reed SJB. (2005) *Electron microprobe analysis and scanning electron
microscopy in geology*. New York: Cambridge University Press, second
edition.

Roberts WL., Campbell TJ. & Rapp GR Jr. (1990) *Encyclopedia of Minerals*.
New York: Van Nostrand Reinhold.

Torre C. & Mattutino G. (2000) Application of true color x-ray vision for
electron microscopy in fired bullets and gunshot residue investigation.
Journal of Forensic Science 45: 865–871.

Van Oosterwyck-Gastuche MC. (1970) La structure de chrysocolle.
Comptes Rendus de l'Académie des Sciences de Paris (Séries D) 271:
1837–1840.

Van Oosterwyck-Gastuche MC. & Grégoire C. (1971) Electron microscopy
and diffraction identification of some copper silicate. *Mineralogical
Society of Japan* Special Paper 1: 196–205.

3.5.3   Bodily Adornment and Modification

Bodies are plastic and hence may be modified and manipulated during life.
Although body modification may include head or foot binding, piercing,
tattooing and scarification are more relevant to Egyptian contexts.
Elaborate tattoos are visible on the arms and abdomens of some dancing
girls from the 11th Dynasty buried at Deir el-Bahari (Ikram & Dodson
1998). Bodily markers, such as tattooing and scarification, may be used as
markers of identity, such as ethnicity, social group or ownership. Although

both iconographic evidence and tattooed human remains exist, there has been no positive identification of actual tattooing needles (Tassie 2003). Given that many artefacts can only be inferred from their context and/or association, it is likely that tattooing needles have simply been identified as ordinary needles or awls. It may be hoped that if recognised and found well preserved, scientific analysis of the tip might be possible to identify the pigment used to create the tattoo, such as charcoal, or obtain blood residues (Tassie 2003). Even when a tattoo upon the mummy skin is not visible under natural light, tattoos may be identified and imaged using infra-red reflectography, with black or green colours being particularly easily identified (Starkie et al. 2011). Reflectance Transformation Imaging (RTI) might also be possible (see Box 4.9).

Unlike some Nubians, who had nose and lip studs, the ancient Egyptians seem to have had few body piercings, and those that they did have were mostly limited to their ear lobes. The practice seems to have been most popular during the late Middle Kingdom and early New Kingdom, such as the round ivory ear-stud found at Lahun (Szpakowska 2008). Other items of jewellery, such as beads, rings, pendants and pectorals, are more common. These may be of faience or semi-precious stones, such as amethyst, garnet or quartz, and may have had amuletic properties or may have been markers of ethnicity (Szpakowska 2008).

Other forms of bodily modification are hard to identify archaeologically. Circumcision is rarely discussed in the Egyptian literature (exceptions are Montserrat 1996; Szpakowska 2008), although Herodotus noted it among Egypt's priestly caste (Hodges 2001). Identifying circumcision, or absence thereof, relies on sufficient flesh remaining upon the mummified body to make a definitive determination. Body modification also includes deliberate skeletal trauma, such as head binding, foot binding or trephination. Head binding has been argued as occurring, at least to the royal lineage, during the Amarna period (Gerszten & Gerszten 1995). Although also not common in Egypt, trephined skulls have been noted at Saqqara and Aswan (Arensburg & Hershkovitz 1988). Such modifications may be identified in wrapped mummies using X-ray or CT imaging. Artificially induced changes in shape might be identified using geometric morphometric methods, when comparison of variation in anatomical landmarks and/or semilandmarks is undertaken (Weber & Bookstein 2011: 169–230).

3.5.4 Hair

Hair is commonly the last soft tissue to disappear after death (Aufderheide 2003) and hence is frequently preserved in Egyptian contexts. Furthermore, hair played an important role in rituals, and it was used as a marker of age, gender, ethnicity, class, social rank etc. Hairstyles were changed through the addition of wigs and extensions and thus it is important to be able to identify and distinguish different forms of hair. Furthermore, hair was treated, such as with oils, and these too have chemical signatures that may be recognised.

Using radioimmunoassay and GC-MS of the hair substrate, the chemical constituents may be identified. Studying time-sequenced sections of hair, such as for their carbon and nitrogen values, enables the individual's dietary history to be sequenced and the season in which they died may be identified (White 1993). For example, X-group and Christian Nubians from the Wadi Halfa area appear to have had seasonal variation in their diet and most died during the summer period (White 1993). Other, potentially less nutritional, items may have been consumed. For example, in Chilean mummies, coca-chewing practices were exposed from the cocaine presence in mummy hair (Cartmell et al. 1991).

The effects of sun exposure, modern fungal growth or local environmental conditions upon hair colour must be remembered when recording mummy hair colour. It is most common that dark brown or black hair may be bleached to a red or, less frequently, blonde hue as the eumelanin is lost and the red or yellow pheomelanin granules become more evident (Aufderheide 2003). Dried adult lice (*Pediculus capitis* or *P. pubis*) or their egg casings ('nits') may be found within mummy hair. These may be identified using simple magnification. Identifying different types of hair, such as animal hair being used as a wig or a hair extension, requires light and comparison microscopy or scanning electron microscopy for species identification (Wilson & Gilbert 2007). Hair may also be an indicator of disease, such as corkscrew hair form that may be found in adult scurvy (Aufderheide & Rodríguez-Martín 1998).

At least by the late Middle Kingdom, short or closely cropped hair seems to have been the norm, potentially as a mechanism to deal with head lice (Szpakowska 2008). As a result, wigs and hairpieces may have been of primary importance and were probably decorated with beads and held in place with hairpins, such as ivory ones shaped with animal heads or a human hand at one end.

3.5.5 *Mummified Animals*

What animals were mummified? Is the animal inside the mummy wrapping actually what it purports to be? There are at least four different types of animal mummies: pets, sacred animals, votive offerings, and food offerings (Ikram 2005; 2010). The animals were generally mummified using the same methods as humans, although birds tended to be prepared in simpler ways. The mixture of balms used appear to have been of comparable complexity to those used for mummified humans, with fats, oils, beeswax, sugar gum, petroleum bitumen, coniferous resins, *Pistacia* and possibly cedar resins being identified through GC-MS, TD-GC-MS and Py-GC-MS (Buckley et al. 2004). Embalming materials chemically identified in animal mummies, using these methods of analysis, were either sugar gum (polysaccharide-based plant exudate, such as gum arabic) and/or lipid based. The fatty acid compositions are indicative of both animal and plant lipid origins, but the ratios in each vary by animal mummified. An ibis from Saqqara (Liverpool museum no 1969.112.42) was mummified using a treatment that was largely sugar-based, with only trace amounts of fatty acids of plant origin. A mummified cat from Beni Hassan (Liverpool museum no. 56.22.224), by contrast, had wrappings impregnated with animal fatty acids. Furthermore, a red material had been packed into the ears of this cat, with a fatty acid composition suggestive of a plant origin (Buckley et al. 2004). Gas chromatography may also identify the geographical source of the materials. Bitumen was identified in a 'resin'-coated bandage taken from the cat mummy, with its fractional composition suggesting that it derived from the Gulf of Suez, albeit potentially mixed with bitumen from the Dead Sea. Analysis of the composition therefore may indicate the use of certain sources and thus trading routes, and indeed demonstrate, the absence of use of other sources, such as bitumen from Gebel Zeit. Gross macroscopic analysis of the mummies can also indicate their treatment; some bird and cat mummies, such as from the Bubasteion at Saqqara, have sand and gravel on their bandaging, suggesting that they were placed into the ground while they dried (Ikram 2005; Zivie & Lichtenberg 2005). It also appears that some birds from Saqqara were mummified by placing them in vats of melted resin, pitch or bitumen (Nicholson 2005). Fourier Transform Infra-Red (FT-IR) Raman spectroscopy has been employed to identify the composition of objects placed within animal mummies, such as the composition of artificial eyes found in mummified cats from Middle Kingdom Beni Hassan. Prior to Raman spectroscopy, two such mummified cats were thought to have had amber placed in their eye sockets as artificial eyes, but analysis indicated that the artificial eyes were organic in nature but did not derive from amber (Edwards 2005: 277).

Pets appear to have been mummified and buried with their owners, or outside their owner's tomb, so that they could accompany their owner in the afterlife. Votive animal mummies, deriving from the Late Period and particularly the Greco-Roman period, are offerings consisting of a specific mummified animal that was dedicated to its corresponding deity, such as the ibis and Thoth, although these animals may be the image of the god, and the animal may have been buried in a sacred cemetery as it had died on sacred soil (Ikram 2010). Some animals, especially cats, show evidence of having been deliberately killed as X-ray analysis has demonstrated broken necks, suggesting strangulation, and/or cranial vault damage (Ikram 2005; Ikram & Dodson 1998). Animals buried as food were generally not buried whole. In the New Kingdom, this food, such as joints of meat or poultry, was prepared for consumption, and then buried in elite tombs in order to sustain the individual's *ka*. Although resin has been found on these food-offering mummies, it may come from pitch applied to the interior of the small coffins shaped to the form and dimensions of the meat or joint rather than from the mummies themselves (Ikram & Dodson 1998).

3.5.6 Other Tomb Items and Equipment

In addition to the obvious canopic equipment, many other items were placed in tombs. Grave goods require contextualisation and detailed consideration (Ekengren 2013). These items and objects may reflect social ranking and identity, or may reflect occupation and diet. However, other items may have other purposes, for example charcoal, the burnt by-product of timber, has been found in Badarian graves (Brunton & Caton-Thompson 1928). These other items are considered in Chapter 4: Biography & Analysis of Objects.

3.6 Migration and Mobility

Where did the people who were buried in the graves actually come from? Were they indigenous Egyptians or did they migrate into the Nile Valley from other areas, such as from Egypt's neighbours? Was there internal migration within Egypt? Did people move from the desert margins into the Nile Valley and the oases? If so, who were these people? Did they all belong to the same 'ethnic' or other grouping? Did people move during their lives? Was there seasonal movement of people and/or animals? Migration and mobility potentially can be recognised in a variety of ways archaeologically, but can also be pinpointed more accurately through scientific methods.

Identifying migration and mobility requires two potentially distinct aspects to be archaeologically recognisable; human groups and migration or movement. Although migration is a well-studied aspect of human behaviour (Anthony 1990: 895), the unpredictability of migrations and the difficulty of recognising them archaeologically have meant that they have been avoided as an explanatory construct. Within Egypt, this is of even greater importance as, historically, migration has been linked to racist or colonial explanations for the Egyptian civilisation.

How can populations or groups be recognised? This may seem a relatively simple question, but in order to identify movement, specific subgroups or clusters within an overarching population must be identified. Archaeologists have traditionally relied on typologies to define artefact 'cultures', such as ceramic assemblages. In the past, these have been associated with their manufacturers to form culturally-defined groups and populations which then may move around the landscape (Johnson 1999; Trigger 1989). Some of this is considered in more detail in Chapter 4 in this volume: Biography & Analysis of Objects. Can populations or other groups (and thus their movements) be recognised by other mechanisms?

It is clear that grave goods such as jewellery and palettes may be used to recognise human-derived groups within funerary contexts, but people are not necessarily buried with material culture that demarcates their ethnic or other grouping. Furthermore, this material culture may have been looted and thus be unavailable for analysis. People can manipulate or modify their bodies in order to imprint upon themselves (or their children) a marker of their group membership. This might be done through tattooing or, more commonly in Egypt, hair styling. These are artificial changes made to the body that enable group membership to be identified by those who understand them or are able to 'read' them. Some of this was discussed earlier, in Section 3.5: Clothing and Adornment. For example, Nubians can usually be recognised visually in representations, such as on funerary stelae, by their distinctive bushy hair, darker skin colour, and typical clothing pattern. Using these methods, the Nubian mercenaries buried at Gebelein can be distinguished from the indigenous Egyptian population buried at the same site. There are also other traits and characteristics that can, in the right circumstances, identify and delineate group membership from a scientific standpoint.

3.6.1 Isotopic Methods

Stable isotopes have been mentioned earlier in relation to dietary analysis. But there are also radiogenic isotopes. These are isotopes which, like radiocarbon, either decay over time or are produced by the decay of the isotopes of other elements. Most commonly used for analyses of mobility and migration are strontium isotopes. Of the four naturally occurring isotopes of strontium, one, ^{87}Sr, is produced by the radioactive decay of rubidium-87 (^{87}Rb). The quantity of ^{84}Sr, ^{86}Sr and ^{88}Sr within rock formations is relatively constant, whereas the amount of ^{87}Sr depends upon the ^{87}Rb content of the rock at the time of its geological formation. As a result, the relative proportions of the strontium isotopes differ depending upon local geology, and hence vary geographically.

Whereas the stable isotope ratios in the body discussed earlier, in relation to diet, are a reflection (usually in the bone) of the food eaten, the isotopes used to assess migration are a reflection primarily of the local water. This is because the composition of the local water is usually dependent upon the local geology. In contrast to the analysis of dietary stable isotopes, where the collagen in the bone is usually studied for assessment of migration and mobility, it is the isotopic ratios of the dental enamel that are analysed. As the teeth develop in childhood, if there is a difference between the isotopic composition of the dental enamel and that of the local geology, the individual is assumed to have migrated into that particular locality after their childhood growth period. Furthermore, if there are differences between the isotope ratios of the dental enamel and the ratios of the bone of the individual, the individual must have

Box 3.9   **Investigating Ancient Egyptian Migration in Nubia at Tombos using Strontium ($^{87}Sr/^{86}Sr$) and Oxygen ($\delta^{18}O$) Isotope Analysis**
Michele R. Buzon

Strontium ($^{87}Sr/^{86}Sr$) and oxygen ($\delta^{18}O$) isotope analyses of human dental and skeletal tissues are biogeochemical methods that can be used to investigate past human mobility. First-generation immigrants can be identified (individuals who have moved during their own lifetimes), given that the isotopic signatures of the areas differ and local foods and water were consumed. Dental enamel samples reflect one's childhood location during tooth formation (each tooth crown develops at various times during the first 12 years of life). Alternatively, bone samples reflect approximately the last 10 years of life. However, enamel is less prone to contamination and is thus often the preferred tissue (Steele and Bramblett 1988; Budd et al. 2000). These methods have been successfully used to study past mobility in many areas and recently have been applied to questions addressing the movement of people in the Egyptian empire (Buzon et al. 2007; Buzon and Bowen 2010; Buzon and Simonetti 2013; Dupras and Schwarcz 2001; Iacumin et al. 1996; Prowse et al. 2007; White et al. 2004).

Strontium concentrations and ratios differ according to the composition of rocks and the time elapsed since formation or deposition. Older rocks, such as granite, have higher $^{87}Sr/^{86}Sr$ ratios than areas with geologically younger rocks, such as basalt. Strontium in rock, groundwater and soil is incorporated into the plants and animals consumed by humans (Faure 1986). As such, the strontium isotopic composition of an individual's diet will be reflected in his or her dental and skeletal tissues (Ericson 1985). In the Nile Valley, it is expected that strontium isotope values will vary between cataract areas (with outcrops of igneous rock, such as granite) and non-cataract areas (composed of sedimentary rock, such as chalk, shale and limestone) (Burke et al. 1982; Said 1962).

Oxygen isotope ratios (expressed as $\delta^{18}O$) are strongly related to the composition of consumed water sources, which vary due to hydrological, geographical and climatological factors (Dansgaard 1964; Gat 1996; Longinelli 1984; Bowen et al. 2007). For humans, cultural factors such as differential use of various water sources, as well as water storage and handling practices, could also affect the $\delta^{18}O$ values. In Egypt and Nubia, Nile water differs in oxygen isotope ratios from aquifer water (groundwater) (Thorweihe 1990; Sultan et al. 2007). For areas using primarily Nile water, sites farther from the source in central Africa may show lower $\delta^{18}O$ values due to evaporation as the water flows north (Iacumin et al. 1996; White et al. 2004).

These techniques have been applied to questions regarding human mobility during the Egyptian New Kingdom Empire in Nubia using samples from Tombos (Buzon et al. 2007; Buzon and Bowen 2010; Buzon and Simonetti 2013). This site is located at the Third Cataract of the Nile in modern-day Sudan (ancient Upper Nubia); samples used in the isotopic studies date from the mid-18th Dynasty until the Third Intermediate Period. These individuals were buried in Egyptian-style underground mudbrick chambers and appear to be middle class (Smith 2003). Although originally thought to be exclusively a cemetery for Egyptian colonists, archaeological features, artifacts, burial ritual, as well as biological identity (based on cranial morphology) suggest that the Tombos population was composed of immigrant Egyptians, Nubians, and their offspring (Buzon 2006; Smith 2003). Strontium and oxygen isotope analyses, while still preliminary in this region, provide additional complementary means for investigating the movement of individuals in the Egyptian Empire during this period.

The first strontium isotope study addressing Egyptian mobility via Tombos individuals established the feasibility for using this method in the region (Buzon et al. 2007). Continuing studies (Buzon and Simonetti 2013) have upheld this idea and expanded our understanding of the local strontium ranges in the region. Based on archaeological and modern faunal samples, the local strontium signature of Tombos was determined to be $^{87}Sr/^{86}Sr = 0.70710$ to 0.70783. Analyses of human samples from various archaeological sites in Egypt and Nubia demonstrate that there are differences in $^{87}Sr/^{86}Sr$ values between sites located in these areas (Buzon and Simonetti 2013); Egyptian sites have a statistically higher mean value (0.70777) than Nubian sites (0.70753). The individuals from Tombos range from $^{87}Sr/^{86}Sr = 0.70712$ to 0.70912, with 20 of 53 individuals falling outside of this local range, confirming the mixed composition of the Tombos sample. It is suggested that colonists from Egypt settled at Tombos and appear to have intermarried with local Nubians. Three individuals in the sample were buried in a Nubian, rather than Egyptian, burial position; all three fall within the local range. The cranial morphology of the local individuals varies, suggesting that individuals born locally were both biologically Egyptian and Nubian (Buzon 2006). This result is not surprising given the cemetery would have been composed of more than one generation, including the offspring of colonial immigrants and local Nubians, due to its length of usage (Buzon et al. 2007; Buzon and Simonetti 2013). Additional studies using strontium isotope analysis on samples from other sites in the Nile Valley region will expand our understanding of $^{87}Sr/^{86}Sr$ variability and our ability to identify immigrants.

Oxygen isotope analysis of a subset of the Tombos sample (n = 30) reveals a wide range of values, $\delta^{18}O$ = 29.2‰ to 35.3‰ (values reported relative to V-SMOW; Buzon and Bowen 2010). In comparison with other studies (e.g., White et al. 2004), this large range of values may be indicative of the presence of non-local individuals. The average $\delta^{18}O$ value at Tombos, 31.4‰, is similar to values found at Egyptian sites near Thebes (Iacumin et al. 1996), while three samples with particularly high $\delta^{18}O$ values (>33‰) are more similar to values found in Nubia (Turner et al. 2007). Preliminarily, these data could imply that the majority of individuals at Tombos are non-local. However, when considering oxygen isotope data from other Nile Valley sites it is suggested that cultural differences in the usage of water (e.g., irrigation, boiled beverages, storage) and hydrological factors such as the seepage of isotopically different aquifer water into the Nile may play important roles in the patterns seen in these data (Buzon and Bowen 2010).

While still in development, strontium and oxygen isotope analyses in the Nile Valley offer optimistic results regarding the utilization of these methods for understanding residential mobility in the region. Using these techniques, the activities of the New Kingdom Egyptian Empire including the movement of peoples from Egypt to Nubia and the interaction with local communities, have been elucidated. The promise of these methods presents many opportunities to explore socioeconomic and political processes in Ancient Egypt.

References

Bowen GJ., Ehleringer JR., Chesson LA., Stange E. & Cerling TE. (2007) Stable isotope ratios of tap water in the contiguous USA. *Water Resources Research* 43: W03419.

Budd P., Montgomery J., Barreiro B. & RG Thomas. (2000) Differential diagenesis of strontium in archaeological human tissues. *Applied Geochemistry* 15: 687–694.

Burke WH., Denison RE., Hetherington EA., Koepnick RB., Nelson NF. & Otto JB. (1982) Variation of seawater $^{87}Sr/^{86}Sr$ throughout Phanerozoic time. *Geology* 10: 516–519.

Buzon MR. (2006) Biological and ethnic identity in New Kingdom Nubia: a case study from Tombos *Current Anthropology* 47: 683–695.

Buzon MR. & Bowen GJ. (2010) Oxygen and carbon isotope analysis of human tooth enamel from the New Kingdom site of Tombos in Nubia. *Archaeometry* 55: 855–868.

Buzon M.R., and Simonetti A. (2013) 'Strontium Isotope ($^{87}Sr/^{86}Sr$) Variability in the Nile Valley: identifying residential mobility during Ancient Egyptian and Nubian sociopolitical changes in the New Kingdom and Napatan Periods', *American Journal of Physical Anthropology*, 151: 1–9.

Buzon M R., Simonetti A. & Creaser RA. (2007) Migration in the Nile Valley during the New Kingdom period: a preliminary strontium isotope study. *Journal of Archaeological Science* 34: 1391–401.

Dansgaard W. (1964) Stable isotopes in precipitation. *Tellus* 16: 436–68.

Dupras T L. & Schwarcz HP. (2001) Strangers in a strange land: stable isotope evidence for human migration in the Dakhleh Oasis, Egypt. *Journal of Archaeological Science* 28: 1199–208.

Ericson JE. (1985) Strontium isotope characterization in the study of prehistoric human ecology. *Journal of Human Evolution* 14: 503–14.

Faure G. (1986) *Principles of Isotope Geology*. New York: Wiley-Liss.

Gat JR. (1996) Oxygen and hydrogen isotopes in the hydrologic cycle. *Annual Review of Earth and Planetary Sciences* 24: 225–62.

Iacumin P., Bocherens H., Mariotti A. & Longinelli A. (1996) An isotopic palaeoenvironmental study of human skeletal remains from the Nile Valley', *Palaeogeography, Palaeoclimatology, Palaeoecology* 126: 15–30.

Longinelli A. (1984) Oxygen isotopes in mammal bone phosphate: a new tool for paleohydrological and paleoclimatological research? *Geochimica et Cosmochimica Acta* 48: 385–90.

Prowse TL., Schwarcz HP., Garnsey P., Knyf M., Macchiarelli R. & Bondioli L. (2007) Isotopic evidence for age-related immigration to imperial Rome. *American Journal of Physical Anthropology* 132: 510–19.

Said R. (1962) *The Geology of Egypt*. Amsterdam: Elsevier.

Smith ST. (2003) *Wretched Kush: Ethnic Identities and Boundaries in Egypt's Nubian Empire*. London: Routledge.

Steele DG. & Bramblett CA. (1988) *The Anatomy and Biology of the Human Skeleton*. College Station, Texas: Texas A&M University.

Sultan M., Yan E., Sturchio N., Wagdy A., Abdel Gelil K., Becker R., Manocha N. & Milewski A. (2007) Natural discharge: a key to sustainable utilization of fossil groundwater. *Journal of Hydrology* 335: 25–36.

Thorweihe U. (1990) Nubian aquifer system. In: Said R. (ed.) *Geology of Egypt*. Rotterdam: Balkema. pp. 601–614.

Turner BL., Edwards JL., Quinn EA., Kingston JD. & Van Gerven DP. (2007) Age-related variation in isotopic indicators of diet at medieval Kulubnarti, Sudanese Nubia. *International Journal of Osteoarchaeology* 17: 1–25.

White C., Longstaffe F J. & Law KR. (2004) Exploring the effects of environment, physiology and diet on oxygen isotope ratios in ancient Nubian bones and teeth. *Journal of Archaeological Science* 31: 233–250.

moved, after infancy, from an area with one isotopic signature to an area with a different signature. This is because, in contrast to teeth, bone continuously remodels throughout life. The second method of identifying migration, through intra-individual differences in isotope ratios, is problematic due to issues with diagenetic change as bone is susceptible to taphonomic change. As a result, it is preferable to compare human dental enamel signatures with those derived from either local geology or from the enamel of animals known to have lived in that locale.

Can animal movements and migration also be identified? Exactly the same methods can be used to investigate the movement of animals. Although isotopic analyses have most commonly been used to study human mobility, it is also possible to study domestic animals in order to discover whether transhumance was practised. If a series of enamel signatures is obtained along the length of a tooth, particularly for an animal with a tooth that grows over a long time period, such as a horse, it is possible to identify differences in isotopic signature which may reflect seasonal patterns of movement. This might include moving from local grazing to grazing in more distant areas with other geological signatures. Seasonal movements of humans can also be identified in the same way, by undertaking isotopic sampling either along one tooth, or from different teeth that are known to develop at different times in childhood. Laser ablation (coupled with isotope ratio mass spectrometry) is usually used to obtain an enamel sample, especially when sampling repeatedly along the length of a tooth.

Although strontium ratios ($^{87}Sr/^{86}Sr$) are most commonly studied, other isotope ratios have also been analysed. The most frequent of these is oxygen, as the isotopic ratio of $^{18}O/^{16}O$ in rainwater varies depending upon climate and other factors including altitude and distance from the sea. In Egypt, oxygen isotopes have been used to identify potential migrants to Roman Kellis (Dupras & Schwarz 2001), one of whom was identified as a lepromatous male and was therefore hypothesised to have been exiled from the Nile Valley. Although used for humans, the method is better for fauna due to differences between animals which require water to drink, obligate water drinkers, and occasional water drinkers. Drought tolerant animals are occasional water drinkers and obtain all the water that they require from the plants that they consume. As a result their $^{18}O/^{16}O$ ratios reflect the relative humidity or aridity of the environment. Obligate drinkers, by contrast, depend on locally available water, and so their isotope signatures reflect this water (Malainey 2011). In addition, animals which browse (such as cows) rather than graze (such as goats) tend to have higher ^{18}O signatures than comparative grazers (such as sheep). This method, therefore, has the potential to distinguish between sheep and goats, and has been used (although not yet in Egypt) to look at herd management (Bocherens et al. 2001) and seasonality in faunal birth patterning (Balasse et al. 2003). Some other isotopes, such as lead

(^{206}Pb/^{207}Pb), also have been analysed, but these are usually considered as environmental contaminants within bones or teeth (e.g. Farmer et al. 1994), and so have not been widely studied for mobility analyses.

3.6.2 Other Compositional Signals

There are a series of other elements and compounds which might also be studied given that their recognition and identification has the potential to identify migrants. Although not yet frequently studied in archaeology, the most common of these are the rare earth elements (REE) – the 14 naturally-occurring elements of the lanthanide series (La to Lu) (Linsalata et al. 1986; 1991; Tütken et al. 2008). The REE are also useful for characterising the effects of taphonomy and diagenesis, and for describing any postmortem treatment of the body (Trueman 1999; Trueman & Tuross 2002; Trueman et al. 2006). They can therefore be used as controls for migration studies.

3.6.3 Skeletal Metric Methods

There are bodily traits that may provide an indication of biological affinity or ethnicity and hence can be used to identify migration and mobility. These traits, such as measurements of specific portions of the body or the presence of specific minor skeletal or dental anomalies, cannot (generally) be manipulated by the individual or by their parent through bodily or cultural modification. This section considers the first of these, metric variation, and the next section discusses nonmetric variation.

It is worth noting that biological and physical differences exist between humans on a global geographic scale (Howells 1973; 1989; 1995; Lahr 1996), and between groups within non-human species (Ridley 1993). Although some attention has been paid to physical and biological definitions of groups or populations for humans, this has been laden with racist overtones, particularly with regard to Egypt. This book is not a venue for a discussion either of race or the relative merits of Afrocentrism. It is, however, rather a discussion of aspects of human variation and diversity and their uses to inform as to population history and hence migration.

Although body shape and proportion are linked to genetics, in association with climatic patterning, most studies using metric variation to ascertain biodistance have focused on the skull. Modern studies use multivariate statistical analyses (rather than historical and outdated studies using either fixed typologies or indices, such as the cranial index). Most commonly, within Egypt, these modern studies use traditional craniometric landmark data (e.g. Keita 1992; 2004; Rösing 1990; Zakrzewski 2007), although three-dimensional landmark data are starting to be employed. Traditional craniometric data have been more commonly employed as these simply require the use of sliding and spreading calipers, and hence can easily be undertaken in the field, whereas three-dimensional data can only be obtained using a microscribe, laser, structured light or CT scanner.

Craniometric analyses commonly employ statistical methods including principal components and discriminant function analyses and a variety of clustering methods. Principal components analysis (PCA) acts by considering groups of variables, and attempts to explain the variance seen in those variables. PCA aims to identify the underlying factors (variables) explaining the pattern of correlations within a set of observed variables. It can therefore be employed to ascertain which variables are of greatest importance in explaining the variance seen within the ellipse of data points in multidimensional space. By contrast, the purpose of discriminant function analysis (DFA) is to assign group membership from a number of predictor variables. The main aim of DFA is to find the dimension or dimensions by which the groups differ, and then derive mathematical classification functions from this to predict group membership. DFA forms a string of these functions and judges whether the groups it predicts from these functions match those imposed upon the data. Thus, in craniometric DFA, the raw measurements for each individual are converted into functions relating to cranial dimensions. A coefficient (weighting) is given to each measurement (variable) and the individual's actual measurement is multiplied by this coefficient. The sum of these weighted measurements comprises the individual's 'discriminant score'. For craniometric studies, PCA screens the cranial measurements in order to understand how great a contribution each variable (measurement) makes to the overall variance in cranial morphology, and DFA assesses whether cranial variables can be used to predict the particular group membership of the cranial sample. These groups might be temporal or geographic.

In addition, craniometric studies may also employ morphologically-derived distance measures (such as the Mahalanobis D^2) as proxies for the genetic or biological distance between cranial samples. Based upon these distance measures, clustering algorithms may then be employed to develop phenetic taxonomies of morphology (for explanation, see Pietrusewsky 2008). Furthermore, these computed 'biological' distances can also be compared with actual geographic distances between samples, such as when using isolation-by-distance models for cemeteries (Zakrzewski 2007) in order to identify potential migration.

3.6.4   *Nonmetric Skeletal and Dental Traits*

Nonmetric traits are morphological variants of anatomy, and are of archaeological interest when they are found in the hard tissues such as tooth or bone. They are usually considered to be minor skeletal variants as they are unlikely to affect normal function and hence are not considered pathological (for detail see Saunders & Rainey (2008) for skeletal nonmetric traits and Scott (2008) for dental nonmetric traits).

In the literature, they may also be termed discrete, quasi-continuous or epigenetic traits. Cranial, post-cranial and dental nonmetric traits are known to have a genetic basis, and hence varying frequencies of such traits in samples may be used to identify biological relationships. Groups who share similar frequencies of such traits, i.e. have similar phenotypes, are considered more closely related genetically than groups with very different frequencies of such traits. It is important that the correct traits are studied for the questions being asked, as certain traits are affected by activity or age, but even some of these traits are useful when studying migration as different groups may practise different activities.

Familial relationships may also be assessed using an argument that assumes that related individuals are more likely to possess similar patterns of such anatomical variants than individuals who are not related. When this has been done within a cemetery, such as at Roman Kellis (Haddow personal communication), these concepts are then married to ideas that geographic distance within the cemetery is also linked to and inversely correlated with genetic kinship (close kin are buried close together). Furthermore, greater diversity of such traits within one sex or the other would imply exogamy, with the more varying sex 'marrying in' to the indigenous community. In Egypt, this method has been used to demonstrate both similarities and differences between cemeteries. For the former, on the basis of both metric and nonmetric methods, the late Roman–early Byzantine cemeteries at Sayala have been recognised as morpho-genetically similar (Strouhal & Jungwirth 1979). In contrast, the elite cemetery at Predynastic Naqada (cemetery T) has been shown to differ from two local contemporaneous non-elite cemeteries (Johnson & Lovell 1994; Prowse & Lovell 1996) and, furthermore, the nonmetric trait frequencies suggest that the individuals buried in the elite cemetery practised endogamy (Prowse & Lovell 1996).

Recently, dental nonmetric methods have been preferred due to the genetic canalisation of dental development (Hillson 2005). This means that dental development is usually less affected by environmental factors than skeletal development. Furthermore, teeth are more commonly preserved archaeologically than bone. Root traits, such as the number of tooth roots present for specific teeth, may be obtainable from the alveoli in the mandible or maxilla, even when the tooth is not itself preserved. Within Egypt, however, the main problem with dental nonmetric studies is the high degree of dental wear found in many Egyptian samples. North Africans, including ancient Egyptians, tend towards a relatively simple dental pattern (Irish 1997; 1998a; 1998b; 2006), and this morphology enables Nubians to be distinguished from Egyptians, such as at Hierakonpolis (Irish & Friedman 2010).

Box 3.10 Additional Insight into Post-Pleistocene Nubian Population History
Joel D. Irish

As a geographic region, Nubia extends from the Nile River's First Cataract at Aswan, Egypt, to the confluence of the Blue and White Niles near Omdurman, Sudan (Adams 1977; Greene 1967). Like ancient Egypt, the region is then bisected into two sub-regions: Lower and Upper Nubia (Nielsen 1970).

Also like Egypt, Nubia is among the best documented of regions in Africa in terms of its ancient inhabitants; this knowledge is based on an abundance of archaeological and bioanthropological studies of samples recovered during the UNESCO/High Aswan Dam project and, recently, dam construction at the Fourth Cataract. This work revealed that Nubia's population history parallels that of Egypt in terms of antiquity as well as, in some cases, affinity, complexity, and levels of achievement. Many of these ties are related to the two regions' shared and often intimate history in terms of trade, conflict and, on several occasions, conquest and occupation of the other's lands (Davies 2003; Newman 1995; Sherif 1981; Smith 2003; Trigger 1976; Williams 1997).

However, unlike Egypt, more is known and written about early Nubian origins and ancestry, i.e., during the late Pleistocene/early Holocene. This inequity in study is due to the recovery of two important Late Palaeolithic (ca. 14,000–12,000 BP) Lower Nubian skeletal samples from Jebel Sahaba (Wendorf 1968) and Wadi Halfa (Greene & Armelagos 1972). Adequate samples of contemporary Egyptian remains have yet to be found.

Some argue for Nubian population continuity from the late Pleistocene onward (Calcagno 1989; Carlson & Van Gerven 1977; 1979; Greene 1972; Greene et al. 1967; Small 1981; Smith & Shegev 1988); however, subsequent dental, cranial, and post-cranial studies demonstrate that the Wadi Halfa and, particularly, Jebel Sahaba remains differ significantly from later groups. If both samples are representative of the population at that time, then regional genetic discontinuity is implied after the Pleistocene (e.g., Franciscus 1995 & personal communication 1995; Groves & Thorne 1999; Hillson 1978; Holliday 1995; Irish 1993; 1998a; 1998b; 1998c; 2000; 2005; Irish & Turner 1990, Turner & Markowitz 1990); recent work suggests that this break probably occurred in the early Holocene – before the Final Neolithic (Irish 2005).

Who then were the ancestors of Neolithic and later (e.g. A-Group, Kerma, C-Group, Meroitic etc.) Nubians? This box addresses that question. The mean measure of divergence (MMD) statistic (Berry & Berry 1967; Harris & Sjøvold 2004; Irish 2010; Sjøvold 1977) was used to compare 36 Arizona State University Dental System traits (ASUDAS) (Scott & Turner 1997; Turner et al. 1991) in 12 late Pleistocene through historic Nubian samples with a newly discovered "early" sample from Upper Nubia. This approach yields estimates of inter-sample biological affinities – assuming that dental phenetic similarity provides an estimate of genetic relatedness (Scott et al. 1983). The comparative data were recorded in dentitions from: 1) Jebel Sahaba and seven Neolithic through Christian period Lower Nubian samples (ca. 5700–600 BP), and 2) four Kerma period through post-Meroitic Upper Nubian samples (ca. 4400–400 BP). See Irish (2005) for a full description of the methods, rationale, and comparative samples.

The "new" early Nubian sample comes from the site of al Khiday, south of Omdurman. It *may* be the first sample of a Late Paleolithic population recovered in the region in over 40 years. The word "may" is used because dating of the heavily fossilized and deflated remains is problematic. As such, until additional work can be done, the excavators, Drs Sandro Salvatori and Donatella Usai, of the Archaeological Mission at El Salha, Istituto Italiano per l'Africa e l'Oriente, prefer the more conservative designation of "pre-Mesolithic". At present the remains can only be said to be >9000 years old, based on dating of organics from a pit cut into one early burial (Di Matteo et al. n.d.; Usai personal communication 2010; Usai & Salvatori n.d.). Detailed archaeological and bioanthropological data will be published by the excavators and project physical anthropologist, Dr Tina Jakob, at a later date.

The al Khiday sample consists of 40 individuals of both sexes that retain permanent teeth. A smaller ($n = 25$) more heterogeneous sample from the site, consisting of Mesolithic and Neolithic individuals (ca. <9000 BP), was also included in the analysis. After the 36 traits were recorded in both, frequencies were calculated and compared with those in the comparative samples using the MMD.

Given this report's brevity, the 14 × 14 matrix of inter-sample MMD distances is not included (though available from the author). Instead, the most effective and least biased way to illustrate succinctly the patterning of affinities is via multi-dimensional scaling (Kruskal & Wish 1978). The Alscal procedure in SPSS 16 was used to create the two-dimensional spatial representation of sample relatedness in Figure 3.6.

As can be seen, Jebel Sahaba (JSA) is widely divergent from the remaining 13 samples – the reasons for which are detailed elsewhere

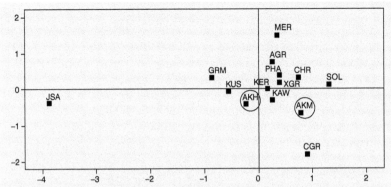

Figure 3.6 Multi-dimensional scaling 2D plot of the 36-trait MMD distances among the 14 Nubian samples

The al Khiday pre-Mesolithic sample (AKH) and the al Khiday Mesolithic–Neolithic comparative sample (AKM) are circled. The other three-letter sample abbreviations are either identified in the text or in Irish (2005).

(Irish 2005). The other early sample in this study, pre-Mesolithic al Khiday (AKH), is positioned, however, within the cluster of Neolithic (GRM) and later Nubians; the younger al Khiday sample (AKM) is nearby. As might be expected, AKH is closest to three geographically proximate Kerman through post-Meroitic samples from Upper Nubia (KER, KAW, KUS). Yet the overall patterning indicates that, unlike Jebel Sahaba, the individuals of al Khiday are dentally akin to all later Nubians. Based on individual MMD values in the aforementioned matrix, there are no significant differences (p = 0.025) between AKH and seven of the 11 Holocene comparative samples.

Therefore, in answer to the question posed above it appears that, based on this preliminary analysis of the al Khiday sample, ancestors of post-Pleistocene Nubians were likely present in the region – although clearly not at Jebel Sahaba or Wadi Halfa. It is not necessary to posit an immigration of outsiders during the early Holocene. Of course, it cannot be conclusively stated that the people of al Khiday were directly related, but assuming the dental affinities are indicators of genetic variation, then they are a good representative of what the common ancestor to later Nubians might have been.

References

Adams WY. (1977) *Nubia: Corridor to Africa*. Princeton: Princeton University Press.
Berry AC. & Berry RJ. (1967) Epigenetic variation in the human cranium. *Journal of Anatomy* 101: 361–379.

Calcagno JM. (1989) *Mechanisms of Human Dental Reduction*. Lawrence KS: University of Kansas Publications in Anthropology 18.

Carlson DS. & Van Gerven DP. (1977) Masticatory function and post-Pleistocene evolution in Nubia. *American Journal of Physical Anthropology* 46: 495–506.

Carlson DS. & Van Gerven DP. (1979) Diffusion, biological determinism, and biocultural adaptation in the Nubian corridor. *American Anthropologist* 81: 561–580.

Davies V. (2003) Kush in Egypt: A new historical inscription. *Sudan & Nubia* 7: 52–54.

Di Matteo A., Iacumin P., Salvatori S. & Usai D. (n.d.) Emerging complexity: a view from the Late Palaeolithic, Mesolithic, Neolithic and Post Meroitic cemetery of Al-Khiday 2, Central Sudan. Unpublished manuscript in possession of the author.

Franciscus RG. (1995) *Later Pleistocene nasofacial variation in western Eurasia and Africa and modern human origins*. Unpublished PhD dissertation. Albuquerque: University of New Mexico.

Greene DL. (1967) *Dentition of Meroitic, X-Group, and Christian Populations from Wadi Halfa, Sudan*. Anthropological Papers 85, Nubian Series 1. Salt Lake City: University of Utah Press.

Greene DL. (1972) Dental anthropology of early Egypt and Nubia. *Journal of Human Evolution* 1: 315–324.

Greene DL. & Armelagos G. (1972) *The Wadi Halfa Mesolithic Population*. Research Report No. 11. Department of Anthropology. Amherst: University of Massachusetts.

Greene DL., Ewing GH. & Armelagos GJ. (1967) Dentition of a Mesolithic population from Wadi Halfa, Sudan. *American Journal of Physical Anthropology* 27: 41–56.

Groves CP. & Thorne A. (1999) The terminal Pleistocene and early Holocene populations of northern Africa. *Homo* 50: 249–262.

Harris EF. & Sjøvold T. (2004) Calculation of Smith's mean measure of divergence for intergroup comparisons using nonmetric data. *Dental Anthropology* 17: 83–93.

Holliday TW. (1995) *Body size and proportions in the Late Pleistocene western Old World and the origins of modern humans*. Unpublished PhD dissertation. Albuquerque: University of New Mexico.

Hillson SW. (1978) *Human biological variation in the Nile Valley in relation to environmental factors*. Unpublished PhD thesis. Institute of Archaeology. London: University of London.

Irish JD. (1993) *Biological affinities of late Pleistocene through modern African aboriginal populations: the dental evidence*. Unpublished PhD dissertation. Tempe: Arizona State University.

Irish JD. (1998a) Dental morphological affinities of late Pleistocene through recent sub-Saharan and North African peoples. *Bulletins and Memoirs of the Society of Anthropology of Paris* 10: 237–272.

Irish JD. (1998b) Diachronic and synchronic dental trait affinities of Late and post-Pleistocene peoples from North Africa. *Homo* 49: 138–155.

Irish JD. (1998c) Dental morphological indications of population discontinuity and Egyptian gene flow in post-Paleolithic Nubia. In: Lukacs JR. (ed.) *Human Dental Development, Morphology, and Pathology: A Tribute to Albert A. Dahlberg.* University of Oregon Anthropological Papers 54. Eugene: University of Oregon Press. pp. 155–172.

Irish JD. (2000) The Iberomaurusian enigma: North African progenitor or dead end? *Journal of Human Evolution* 39: 393–410.

Irish JD. (2005) Population continuity versus discontinuity revisited: Dental affinities among Late Paleolithic through Christian era Nubians. *American Journal of Physical Anthropology* 128: 520–535.

Irish JD. (2010) The mean measure of divergence (MMD): Its utility in model-free and model-bound analyses relative to the Mahalanobis D2 distance for nonmetric traits. *American Journal of Human Biology* 22: 378–395.

Irish JD. & Turner CG II. (1990) West African dental affinity of late Pleistocene Nubians: Peopling of the Eurafrican–South Asian triangle. *Homo* 41: 42–53.

Kruskal JB. & Wish M. (1978) *Multidimensional Scaling.* Beverly Hills, CA: Sage Publications.

Newman JL. (1995) *The Peopling of Africa: A Geographic Interpretation.* New Haven: Yale University Press.

Nielsen OV. (1970) *Human Remains: Metrical and Non-Metrical Anatomical Variations.* Odense, Denmark: Scandinavian Joint Expedition to Sudanese Nubia.

Scott GR. & Turner CG II. (1997) *The Anthropology of Modern Human Teeth: Dental Morphology and its Variation in Recent Human Populations.* Cambridge: Cambridge University Press.

Scott GR., Yap Potter RH., Noss JF., Dahlberg AA. & Dahlberg T. (1983) The dental morphology of Pima Indians. *American Journal of Physical Anthropology* 61: 13–31.

Sherif NM. (1981) Nubia before Napata. In: Mokhtar G. (ed.) *General History of Africa II: Ancient Civilizations of Africa.* Berkeley: University of California Press. pp. 245–277.

Sjøvold T. (1977) Non-metrical divergence between skeletal populations: The theoretical foundation and biological importance of C.A.B. Smith's mean measure of divergence. *Ossa* 4 (Suppl.1): 1–133.

Small MF. (1981) The Nubian Mesolithic: A consideration of the Wadi Halfa remains. *Journal of Human Evolution* 10: 159–162.

Smith P. & Shegev M. (1988) The dentition of Nubians from Wadi Halfa, Sudan: an evolutionary perspective. *Journal of the Dental Association of South Africa* 43: 539–541.

Smith ST. (2003) *Wretched Kush: Ethnic Identities and Boundaries in Egypt's Nubian Empire.* New York: Routledge.

Trigger BG. (1976) *Nubia Under the Pharaohs.* Boulder: Westview Press.

Turner CG. II & Markowitz M. (1990) Dental discontinuity between late Pleistocene and recent Nubians. I. Peopling of the Eurafrican-South Asian triangle. *Homo* 41: 42–53.

Turner CG. II, Nichol CR. & Scott GR. (1991) Scoring procedures for key morphological traits of the permanent dentition: The Arizona State University dental anthropology system. In: Kelley MA. & Larsen CS. (eds) *Advances in Dental Anthropology*. New York: Wiley-Liss. pp. 13–32.

Usai D. & Salvatori S. (n.d.) *The late Palaeolithic–Mesolithic–Neolithic and post-Meroitic cemetery of al Khiday 2, the associated Mesolithic settlement of al Khiday 1, and the post-Meroitic cemetery 16-C-2, central Sudan*. Unpublished site report to the Istituto Italiano per l'Africa e l'Oriente, 14/02/2008.

Wendorf F. (1968) A Nubian final Paleolithic graveyard near Jebel Sahaba, Sudan. In: Wendorf F. (ed.) *The Prehistory of Nubia, vol 2*. Dallas, TX: Fort Burgwin Research Center. pp. 954–995.

Williams B. (1997) Egypt and sub-Saharan Africa: their interaction. In: Vogel JO. (ed.) *Encyclopedia of Precolonial Africa: Archaeology, History, Languages, Cultures, and Environments*. Walnut Creek: AltaMira. pp. 465–472.

3.7 Social Organisation

How were people organised? What was the scale of the social unit? Did social roles change as someone aged and passed through different life stages? Here we are looking at social systems as comprised of groups, with social relations and the people's interlocking social positions and roles. These social relationships vary through an individual's life cycle, and are delineated by his/her social category or social group. As a result, these may reflect factors important in their lives, such as social role, that do not represent biological descent systems or affines. Furthermore, it is important to consider the difference between etic and emic views of kinship and relationships, since those we may be able to recognise archaeologically are not simply those considered important by the ancient Egyptians, but are also a subset of those that have been preserved and recovered archaeologically. However, the existence of several terms in the Egyptian language that refer to larger kin groups implies that kinship was of fundamental importance in social organisation (Campagno 2009).

3.7.1 Age and Social Grouping

Usually when considering social organisation in Egyptian contexts, one discusses the formation of the state and the development of social hierarchy. The potential for their scientific recognition will be considered later, but first it is important to consider the life course and social life. What are the different stages of the human life course and how can these be recognised archaeologically? How was adulthood socially sanctioned and socially recognised?

Indeed, the Greek legend of the Sphinx is very well-known. According to this legend, a question was asked of travellers on the road to Thebes by the Sphinx, with the question being "What is the creature that walks on four legs in the morning, two legs and noon, and three in the evening?" The answer, given by the legend's hero Oedipus, was "man", and apparently led to the death of the Sphinx. This recognition demonstrates the social importance of different parts of the life course at the time. Furthermore, given that the placenta was a requirement for embodiment and for individuals to attain personhood (Meskell 1994), and that skeletal or biological maturity offers no direct indication of social maturity, how can the child/adult threshold be identified in Egypt? Clearly issues arise in ascertaining the status of children when, by implication, the mortuary evidence is primarily of *dead* children and of adults who survived and lived *through* their childhood period.

Circumcision, known to ancient Egyptians as *sebi*, and occurring around the age of nine, may have acted as a puberty ritual defining the transition to adulthood (Janssen & Janssen 1990). Following the circumcision ceremony, depictions of individuals lack the characteristic side-lock of hair used to denote childhood (Strouhal 1992). But, as noted earlier in the section discussing bodily modification, archaeological recognition of circumcision requires the preservation of sufficient mummified tissue.

With the development of more precise and accurate methods for assigning skeletal age to juvenile remains, through the use of dental development, dental eruption, long bone length and epiphyseal fusion, it is possible to assign juvenile skeletal and mummified remains to relatively tight age bands. Combining this with analyses of artefact inventories allows social age groups to be recognised, in which age is defined in terms of categories of people associated with contextually specific objects (Sofaer 2005). The same approach may be used to recognise age categories within adults and the elderly, such as marital status or other categorical groupings.

3.7.2 *Social Status and Ranking*

As noted earlier, during the Predynastic, kinship probably constituted the main axis of social organisation in village communities, with the clustering of burials within cemeteries at sites such as Badari, Naqada and Hierakonpolis being considered to reflect contemporaneous descent groups (Campagno 2009). In contrast, during the Dynastic period, the state formed a new mode of social organisation; however it is likely that kinship was still important in certain situations, such as the organisation of agricultural practices within family units or the management of irrigation (Campagno 2009). Scientific methods of ascertaining kinship and biological relatedness have been evaluated earlier; here, it is the establishment of any potential link between genetic descent and ascription of social ranking that is important.

From the funerary record, social status is primarily ascertained from examination of the broad burial context, including grave goods, tomb structure and decoration or association with other burials and interments. Most of these aspects have been considered elsewhere in this volume, but it is important to link these together to form a cohesive whole. In scientific terms, this involves linking analyses of the burial objects and mortuary art with analyses of the spatial organisation of the cemetery and of the individual body itself. Primarily therefore, scientific analyses of social status comprise identifications of composition in association with assessments of provenance, of the skill and costs involved in production, and in the distribution networks required to obtain or deliver the items.

3.7.3   *Population and Ethnicity*

A further aspect of social organisation is the variation in population composition. Issues relating to migration and mobility were considered earlier (Section 3.6), and so here we discuss the integration of these aspects with race and ethnicity. Modern groups commonly distinguish people on the basis of outward physical characteristics, such as skin or hair colour, but these traits do not necessarily match with ethnic or other group affiliations. Furthermore, ethnic groups have frequently been conflated with races.

Race has been scientifically defined as distinct polygenetically derived species, as discrete mutually exclusive types, as geographically isolated subspecies, or as crosscutting gradients or clines of populations (Harrison 1998). Most human genetic diversity exists as differences between individuals within specific 'populations', whereas only 15.6 per cent can be used to differentiate genetically between the major human 'races' (Templeton 1998: 663). This level of diversity is well below the usual threshold used to identify and distinguish subspecies or races within nonhuman mammals (Templeton 1998). As a result, there is no biological validity for the race construct. Within the folk concept of race, the traits usually used to distinguish races depend upon external and observable features such as skin colour (Shanklin 1998). By contrast, the features used in folk classification of 'racial' characteristics are absent from most literature deriving from the classical period. This implies that the structuration or segregation of a population was undertaken on the basis of other characteristics, such as occupation, kinship, or language, rather than external physical characteristics. For example, it has been noted by Smedley (1998) that Herodotus states that "the Colchians are of Egyptian origin... because they have black skins and wooly hair [but that this is unimportant as this] 'amounts to but little, since several other nations are so too'" (Smedley 1998: 693).

Despite the past use of race and skin colouration within Egyptian archaeology and Egyptology, this volume is not the venue for a prolonged discussion of Afrocentrism and associated issues. By contrast, the concepts of ethnicity and population affinity are more useful for scientific analyses. Ethnicity is a personal construct, and hence is fluid and malleable in nature (Beck 1995; Lucy 2005). Population affinity is simply an evaluation of biological relatedness. It is clear that the ancient Egyptians constructed ethnic groupings as people from different geographic areas who were depicted and illustrated in different forms. For example, by the time of the New Kingdom, Puntite and Egyptian males were usually depicted in similarly reddish skins, whereas Nubians typically had darker skins and Libyans generally had light-coloured or yellowish skin (O'Connor & Reid 2003). It is unclear whether these matched biological reality, or were simply part of the Egyptian artistic canon. Viewing ethnicity as a category of social identity that may be either assigned or self-imposed permits the use of integrative contextual analysis, in the same manner as the social age groups noted above.

In contrast, people may be divided into genetic groups on the basis of their mtDNA, Y or other haplotype. For example, modern Nilotes, such as Sudanese, are usually members of a haplogroup within Y-chromosome clade A (Karafet et al. 2008). These allele groups rarely match with features used in folk race classification systems, such as skin colour. Identification of genetic groupings has been discussed in Section 3.6: Migration and Mobility. Both phenotypic and genotypic similarities between groups may be biologically identified, with historical linkage implied on the basis of these biological population affinities. However, in the same way, lack of biological similarity may indicate biological disjunction or isolation by distance (Zakrzewski 2007).

Box 3.11 The Potential of Genetic Kinship Studies in Ancient Egyptian Human Remains

Abigail Bouwman

Genes are inherited from our parents but DNA is more than just genes. A vast proportion of DNA is not translated into proteins and so changes (mutations) in these sections of DNA are more likely to be passed on to descendants than changes in coding sections of DNA. In addition, mutations in non-coding sections of DNA accumulate at a faster rate than in DNA that is translated into proteins. Kinship studies primarily study these kinds of genetic differences. The two main differences looked at are Single Nucleotide Polymorphisms (SNPs) and Short Tandem Repeats (STRs).

Single Nucleotide Polymorphisms (SNPs) are where an individual base (A, T, G or C) varies. These differences are identified by reading the DNA sequence from an individual and comparing it to either a standard (such as the revised Cambridge Reference Sequence, rCRS, for human mitochondrial DNA [mtDNA]) or to other individuals. These polymorphisms accumulate relatively quickly on the non-coding D-loop region of the mitochondrial genome, which is only inherited from the maternal line. Therefore, SNPs of mtDNA are used to determine maternal kinship.

Short Tandem Repeats (STRs) are areas of the genome which have a short base sequence that is repeated a varying number of times (e.g. ATTG•ATTG•ATTG, ATTG•ATTG). These STRs occur mostly in non-coding areas of nuclear chromosomes. Chromosomes are either autosomal (chromosomes 1–22) or allosomal (chromosomes X and Y). Allosomal STRs on the Y-chromosome are most commonly used to determine paternal kinship. Y-chromosomal SNPs can also be used in the same way as mtDNA SNPs, but the advantage of STRs is that there are multiple alleles (different number of repeats) whereas SNPs tend to only have two possible alleles (e.g. C to T). This means that fewer targets are needed to identify kinship.

Autosomal STRs are often used in forensics (e.g. to compare a suspect with a DNA sample from a crime scene) and in paternity cases (e.g. to determine which man is the father of a girl). Because 50 per cent of an individual's autosomal DNA is inherited from each parent, where there is only one generation between individuals this technique works quite well. In some cases, cousins or grandparents have been compared using autosomal STRs, but the certainty of relatedness is reduced the further apart, in kinship terms or generations, the two individuals are. For this reason, autosomal STRs are rarely used in ancient DNA investigations.

Mitochondrial DNA is most commonly used in ancient DNA because there are many mitochondria in most cells of the body (apart from red blood cells which do not contain mitochondria), and therefore many copies of the mtDNA genome. In contrast, most cells contain only two copies of each chromosome (exceptions include sperm and egg cells). In addition, the non-coding regions of the mtDNA genome are highly variable, and very well studied in modern population genetics, including aspects of varying frequencies of different allele types.

Knowing how common an allele is in the population is very important in determining kinship. For example, assuming that two samples have the same allele, if in the population that allele is only present 1 per cent of the time, this means that there is a higher chance of the two samples being related than if the frequency of that allele in the population is 99 per cent. Multiple variable targets are needed to see if a) the two samples match and b) they are unlikely to match by chance. Due to the difficulties in aDNA analysis of archaeological remains, is it very unlikely that enough individuals will be able to be sequenced or typed to obtain the allele frequency of any population.

A genetic study of the individuals interred in Grave Circles A and B from Mycenae in Greece (contemporary to the Egyptian 2nd Intermediate Period) investigated mtDNA SNPs and Y-chromosomal STRs and SNPs. Sadly, DNA preservation from these individuals was very poor; no DNA could be recovered from Grave Circle A, and only four of the 22 individuals tested from Grave Circle B had amplifiable mtDNA. Although chromosomal DNA was recovered from one individual (Y-chromosomal DNA from individual labelled Gamma 55) this could not be repeated, and so an analysis of this sequence is not scientifically valid. Without the DNA of any other individual with which to compare, kinship relationships could not be postulated. Of the four individuals that yielded mtDNA, two had identical mitochondrial haplotypes, belonging to haplogroup K. Gamma 55 and Gamma 58 are thought to have been buried within a short time of each other within the same tomb (Gamma 58 appears to have been moved to the side of the tomb whilst still articulated in order for Gamma 55 to be interred) and have been identified as being a male and female of similar age. The inference that Gamma 55 and Gamma 58 are maternally related is based on the fact that today, haplogroup K is only present in 5.6 per cent of the European population, and the assumption is that the frequency was similar in the past (Bouwman et al. 2008, & unpublished data).

Bones and teeth appear to be better reservoirs of DNA than soft tissue, and aspects of the mummification process of ancient Egyptian nobility may, in itself, be damaging to DNA (for example, bitumen is known to bind with DNA and thus cause the DNA to become damaged; Schoket et al. 1988). In the above study on Mycenae there was no evidence of mummification in Grave Circle B (1650–1550 BCE), although Schliemann did identify one individual from Grave Circle A (1600–1500 BCE) as being a 'mummy' (Hood 2012). The soft tissue did not survive long after excavation, however, indicating that, if any artificial mummification of these elite individuals had taken place, it was not similar to the processes used in Egypt at that time.

In terms of ancient Egyptian studies, the main limitations are the availability of samples to study and the ability to isolate enough undamaged genetic material from different individuals with which to compare. However, ancient DNA has been recovered from numerous individuals in Egypt (e.g. pathogen DNA from skeletal remains; Zink et al. 2003), including some mummified remains (e.g. Ramesses III; Hawass et al. 2012). With the continuing optimisation of DNA extraction and analysis techniques, the prospects for kinship analysis are constantly improving.

References

Bouwman AS., Brown KA., Prag AJNW. & Brown TA. (2008) Kinship between burials from Grave Circle B at Mycenae revealed by ancient DNA typing. *Journal of Archaeological Science* 35: 2580–2584.

Hawass Z., Ismail S., Selim A., Saleem SN., Fathalla D., Wasef S., Gad AZ., Saad R., Fares S., Amer H., Gostner P., Gad YZ., Pusch CM. & Zink AR. (2012) Revisiting the harem conspiracy and death of Ramesses III: anthropological, forensic, radiological, and genetic study. *British Medical Journal* 345: e8268.

Hood S. (2012) Schliemann's Mycenae albums. In: *Proceedings of the Archaeology and Heinrich Schliemann Conference held in 1990*. Aegeus: Society for Aegean Prehistory. http://www.aegeussociety.org/images/uploads/publications/schliemann/Schliemann_2012_70-78_Hood.pdf

Schoket B., Hewer A., Grover PL. & Philips DH. (1988) Covalent binding of components of coal-tar, creosote and bitumen to the DNA of the skin and lungs of mice following topical application. *Carcinogenesis* 9: 1253–1258.

Zink AR., Grabner W., Reischl U., Wolf H. & Nerlich AG. (2003) Molecular Study on Human Tuberculosis in Three Geographically Distinct and Time Delineated Populations from Ancient Egypt. *Epidemiology and Infection* 130: 239–249.

4 The Biography and Analysis of Objects

The history of an object can be considered as a long chain of events which affect, to a greater or lesser extent, the material component of that object. This chain of events has more recently been described as a biography (Gosden & Marshall 1999). These biographies impact on, and are influenced by, human biographies. They can be relatively simple and short, where an object is simply made, used once and either expended in that use or discarded after use. Modern non-recyclable packaging in western cultures falls into this group. It has long been realised, however, that many objects have very much longer biographies, containing multiple phases of use and re-use. It could be argued that this is especially the case for early cultures, where raw materials were perhaps rather more valuable, objects themselves rarer, and the contrast between very wealthy elites and the poorest in society very obvious. Thus high status ceramics, for example, might be re-used by families of lesser status, broken and then their use changed completely into toys, writing surfaces or building material. Metal tools might be re-sharpened and re-used, or re-melted to become something else altogether, and stone re-used again and again for increasingly smaller objects. In addition, the object's 'use' goes beyond the purely functional. Every object has important symbolic, ritual and other meanings to the culture in which it is created and/or used. These will affect all the stages of the biography in just the same way that the functional uses have an effect, from choice of raw materials, processing, re-use, discard and so on.

Determining the full biography of an archaeological object is a very difficult process and involves a whole range of different disciplines. Analysis plays an important part in this process, but has in the past concentrated on some stages more than others. This is, in part, due to the fact that some aspects of an object's history impact on its material component more than others, for it is the material aspect of the object that is of immediate interest to analysis. However, analysis has also been influenced by the interests and worldview of those carrying out the analysis. Those who, in the past, spent their lives in pursuit of the scientific analysis of objects, not surprisingly tended to place functional uses above others and to view the objects through material/functional lenses. However, increasingly scientists are working alongside archaeologists and anthropologists, and thus a fuller biography of the object is becoming openly questioned by analysis. Those looking for a fuller discussion of theoretical approaches to objects should see Gosden & Marshall (1999) and for Egypt in particular the excellent work by Meskell (2004).

One of the first questions asked of archaeological scientists when discussing an object is 'What is it made of?' This is a fundamental question, and answering it forms the basis of all further questions asked. Being able to accurately describe an object is, of course, fundamental to archaeology and object descriptions have always relied heavily on the experience of objects and their contexts. Recently however, science has played an increasingly important part in the identification of the material component of an object. The second question asked of science is frequently 'Where does the object come from?' This is a far more complex question than the first and even interpreting the question can be difficult. This question can actually be broken down into three more specific questions: 'Where did the raw materials come from?', 'Where was the object made?' and sometimes 'Where was it found?' In many cases, particularly with low value objects, these three places will be proximal to each other; however, often the rarer and more interesting objects tell a more complex story. Tracing an object to where its raw materials might have come from and/or where it might have been made is fundamental in establishing trade and exchange patterns, and forms the basis of perhaps one of the most interesting and widely applicable studies involving science within Egyptian archaeology. The third question to be asked is 'How was it made?' and this brings in the interpretation of both finished objects and rare workshop debris to determine production technologies and facilities.

4.1 Identifying the Material Component: What Is It Made From?

Objects recovered from Egyptian archaeological sites fall into many, many material categories (Lucas & Harris 1962; Nicholson & Shaw 2000). However, these are usually divided by modern analysts into two fundamental groups: organics and inorganics (Brothwell & Pollard 2001). This is a modern division, probably more or less meaningless to the Egyptian producers themselves, but valid to archaeologists if used carefully. Organic materials are all those that stem directly from living processes and therefore contain carbon as the most abundant element in their structure, and include bone, leather, meat, foodstuffs, resins, wood, seeds, reeds, textiles and so forth. They tend to be very rare on most archaeological sites around the world since they are very vulnerable to decay over time. However, the hot and relatively dry climate of Egypt means that greater amounts of these materials than usual are recovered from Egyptian archaeological sites and they form an important part of the record. Even given the relatively high proportion of organic materials that survive on Egyptian sites recorded in published contexts or curated in museums, they are still rare compared with the inorganics, partly due to the fact that they were were sometimes often not the focus of early excavations and overlooked or ignored. Notable early exceptions include the textile analysis by Thomas Midgley in Egypt, including the Badarian region (Brunton 1927: 70; 1937: 92–93, 112, 138, 145). There was particular interest in textiles in Bolton, Lancashire, UK given that it was an important location for textile manufacture through the 19th century AD. Inorganics include all those materials that have been utilised ultimately stemming from a geological source. In the old children's game, where organics are 'animal and vegetable', inorganics are firmly 'mineral'. This group includes rocks and stone directly and all those materials into which either nature or human agency changes those rocks. The Nile and the Egyptian climate weathers the rocks into mud and clays from which ceramics and bricks are created, and into sand from whence faience and glass. Metals come directly from the ore-bearing rocks. Thus, it is the inorganic materials that are most commonly encountered on archaeological sites and found especially in museum collections. The analysis of organic and inorganic materials is carried out in fundamentally different ways. This is primarily because the two material types contain very different elements. As mentioned above, organics contain carbon as the most abundant element, with oxygen, nitrogen and hydrogen alongside other much rarer but important elements. Since organics all contain the same 'light elements', analysis of organics tends to focus on how they are arranged relative to each other, i.e. the structure of the molecules, and the isotopic make-up of individual elements (Brothwell & Pollard 2001). In inorganic materials, while the most abundant element (except in metals) is almost always oxygen, all the other elements are highly variable across the whole periodic table (Brothwell & Pollard 2001). Analysis of inorganic materials therefore tends to focus less directly on structure and more on composition. The two different material groups are discussed here in different sections because the questions asked and techniques used to answer them are so different. What follows here is a description of the analysis of inorganic materials – stone, ceramics, glass and metals.

4.1.1   *Inorganic Objects – HH-XRF*

Identifying the material component of an object relies first on macroscopic observation and a good deal of experience. Often this is enough to judge whether an object is ceramic or glass, stone or metal. However, with corroded and unusual objects there are analytical techniques that can help. One of the most important of these is Hand Held X-Ray Fluorescence (HH-XRF). This is a relatively new development of a rather old technology, XRF. HH-XRF was developed for the routine and rapid screening of scrap metal (see websites for the manufacturers of the equipment such as Bruker, Niton and Oxford Instruments). However, it is increasingly being applied in archaeology to provide analyses of objects. It has the huge advantage of being completely portable, fitting into a small suitcase and weighing only a few kilograms. A related instrument known as a portable XRF is also available. This is slightly larger and more expensive, but can be regarded as essentially the same as HH-XRF. Both are entirely non-destructive, not requiring a sample to be taken. It is therefore possible to take these instruments onto an archaeological site or into a museum and do analyses of objects in situ. It is especially useful for metals, where it can relatively easily identify copper from bronze or other alloys. It is also useful in helping to identify fine grained stones and to distinguish badly weathered faience and glass from stone or ceramic. However, there are certain drawbacks to the technique. Firstly, it analyses only the surface layers of the object, so is particularly sensitive to dirty, weathered or uneven surfaces. Secondly, it is a radiation device that generates quite strong X-rays, and so needs to be treated with care. Finally, there is the question of interpreting the data derived. If HH-XRF is simply used to identify which elements are in the object, then this is relatively simple to interpret. The trouble comes when there is a need to go beyond the simple presence/absence and attempt to quantify the elemental composition. This is a difficult task to accomplish and requires much experience and some good, relevant standards. It is very easy to produce a lot of numbers from HH-XRF that really bear very little relation to reality. Great caution must therefore be exercised with the technique (see Stugar & Mass 2012). However, as a 'triage' to identify groups of ceramics or interesting metals it is both fast and easy to use.

Clays, glass and faience all have silicon as the most abundant material. However, they tend to have different proportions of other elements. Clays are usually characterised by higher levels of aluminium and iron, whereas glass and a faience glaze will have sodium and colouring elements such as copper or manganese. Even when badly weathered, it is usually fairly easy to tell these materials apart by HH-XRF. Stone has a range of compositions, but HH-XRF is particularly good at distinguishing fine grained stones in the field, which can be difficult to examine as a hand-held specimen, especially for those not trained as a geologist. For example, fine grained mid-grey stones could be siltstone or limestone, or even possibly basalt. Each of these is distinguishable by a short non-destructive HH-XRF analysis: limestone for its high calcium content with very little silicon, siltstone for the opposite and basalt for relatively high magnesium (difficult to see by HH-XRF), aluminium, iron and calcium.

Thus, while inorganic objects can be grouped into material type largely on the basis of the way they look, this can be assisted by using portable and HH-XRF both to further subdivide ceramics and metals and to identify interesting and unusual objects. Inevitably, there will be a desire to subdivide these categories even further, confirm the names of the groups and then to say something about where each of these groups might have been made.

4.2 Raw Materials

Where an object might have been made essentially resolves into two related questions, the source of the raw materials and the place where these raw materials might have been put together to create the finished object. Production technologies, which address the question of how an object was made, are discussed below.

Ever since tools were first made, there has been a necessity to locate and exploit raw material sources to create them. In most areas of the world, sources local to human habitation will provide the majority of raw materials for sustaining that society. This is, after all, one of the principal reasons for the sites of towns and cities. For early Egypt, important resources include organic materials, such as straw, papyrus and some wood, and inorganics, such as mud, some stone and clay. Such materials were both widely available and frequently exploited. However, it is where raw materials are rarer that the distance between their source and place of use can get larger.

It is a general rule that the more 'valuable' the raw material, the greater the time, effort and potential risk that will be put into exploiting it. The difficulty with this rather obvious statement is the definition of value. An object or material might be valued for many reasons: physical, emotional and/or spiritual. Usually, the rarer the material, the more valuable it is. But the rarity has to have an obviously visual or otherwise physical component, i.e. it usually cannot just be rare, it must be recognised by other people as being rare. In Egypt, this display property of a rare material is often associated with colour, and so brightly coloured rarities seem to have been particularly prized (Wilkinson 1994: 104–125).

4.2.1 The Provenance Hypothesis

In studies of the sourcing of raw materials, the 'provenance hypothesis' is often discussed (Wilson & Pollard 2005). This lays out the rules by which it is possible to say that the raw material for an object did, or did not, come from a particular source. It is a general theory, and is applicable to all raw material types and most types of analysis. Provenancing aims to identify a particular 'fingerprint' within an object that reflects the source of the raw material used. This fingerprint might be simply the appearance of the material, but more often it is a particular pattern of elements or isotopes within the material. In order to be able to identify the source of the raw material of an object, the first necessity is that the fingerprint characterising the object passes unchanged, or changed in a very well understood way, from the raw material to the object. If the 'fingerprint' changes randomly when an object is made, perhaps in sorting or firing, then it is impossible to use it to trace raw material sources. The best example of a material that passes absolutely unchanged from raw material to finished object is stone, as this is simply shaped. The second requirement for provenancing is that it is necessary to know all the possible places from which the raw material might have been taken and to have analysed samples from all these places. The third condition is related to this: there must be sufficient variation between the sources to allow them to be distinguished from each other, and sufficiently low variation within a given source to ensure that it does not overlap with any other. This can be abbreviated simply as 'inter-source heterogeneity, but intra-source homogeneity'. The final necessity for provenancing is that a sufficiently sensitive analytical technique is available both to measure all the fingerprints of the different sources and objects, and to ascribe reliably the correct fingerprint to each.

Finding a system or material type that fulfils all these criteria is actually very rare. While it is relatively common to be able to measure compositional fingerprints in objects in a reliable way and to be relatively sure that they have passed unchanged from raw material to finished objects, the real problems come in determining all the possible raw material sources. Even if a fingerprint has identified one potential raw material source for an object and distinguished it from other sources which have different fingerprints, it is often impossible to know that there is not another source with exactly the same fingerprint which has not yet been identified. This has very important implications for provenance studies. It means that when, as is often the case, it is impossible to be sure that all the possible raw material sources for an object have been analysed, it is not possible to positively ascribe a source for an object. All it is possible to do is say that a source is consistent with the material that an object is made from, and more importantly other sources are not consistent. This means that provenancing turns into a negative procedure – it is much easier to identify a source that was definitely not used in its manufacture rather than a source that definitely was.

4.3 Distribution and Consumption

The plotting of patterns of trade and exchange is, along with the description of technologies, one of the most important aspects of the analysis of objects. As discussed above, the question 'Where does this come from?' is one of the most commonly asked of the archaeological scientist specialising in objects. In an ancient society, the vast majority of objects found will be locally produced, and a rare import has and gives value and prestige. It is a fact that external trade routes are often controlled almost exclusively by elites. This means that these foreign objects tend to be concentrated in temples, palaces, high status houses and elite graves. This gives them an interest in their own right, but there are further reasons why these objects are important: they give evidence of foreign contact and they provide important 'pins' for cross dating between different states.

From the Neolithic onwards, there is much evidence for contact between Egypt and its southern and northern neighbours. These contacts are ongoing through the Old and Middle Kingdoms (Tallet & Marouard 2012: 40–43), with a particular upsurge in trade within the eastern Mediterranean in the New Kingdom. Reliefs of the New Kingdom show expeditions abroad, such as the one to Punt depicted in the Temple at Hatshepsut at Deir el Bahari (Kitchen 1971). There are many examples of foreigners bringing gifts to Egyptian rulers as tribute. Some of the most famous of these include the tribute given to Tuthmosis III at the Temple of Karnak following his military victories in Syria, depicted in the Hall of the Annals at Karnak. Here there is a scene showing foreigners coming to Tuthmosis III's court and presenting gifts in the Tomb of Rekhmire at Thebes (TT100) (Davies 1935; 1943). Some of the earliest and most frequently repeated depictions of foreigners are the large numbers of images of captured soldiers, bound and prostrate, which are shown associated with many objects belonging to the King. This occurs most notably in the goods from the Tomb of Tutankhamen (KV62), where it is one of the more common images. However, this last point shows a problem with the interpretation of these images and scenes. It is certain that the Egyptian kings and queens, some more than others, used conventional scenes of captured and cowed foreigners to bolster their image at home. How much actual contact there was, friendly or otherwise, is therefore very difficult to tell. One of the key ways that this problem may be resolved is to attempt to identify the foreign objects themselves in Egypt and, alongside stylistic assessment, science has a big part to play in this process.

The identification of foreign objects has a further important role as they can provide dating evidence. How the Egyptian chronology was constructed and the importance of king lists, imported goods and scientific dating techniques is discussed in detail in Section 2.1: Time. Suffice it to say here, the identification of foreign objects in closely dated Egyptian contexts has been vital in the construction of foreign chronologies. This has been particularly the case in Mesopotamia, but also in the Mediterranean and Greece, and the reverse is also true.

Imported objects fall into several categories. The simplest is where a raw material is transported into Egypt to be made into a finished object in Egyptian workshops. These tend to be valuable raw materials like semi-precious stones and are discussed below. The second category is where a raw material is worked into a finished object abroad and then imported into Egypt as such. This might be because the finished object itself was of substantial value, or it might be a container for other valuable commodities.

Ceramics fall into this category, as they have value both intrinsically and for the goods they hold. The Canaanite Amphorae project (Box 4.1) is a good example of the sort of research undertaken on ceramics of this type. Finally there is a more difficult, although perhaps the most interesting, category. This is where a raw material is converted into a 'medium of exchange', typically an ingot, which is then transported. The ingot is then worked into a finished object in Egypt. There are two major material types that follow this method and/or medium of exchange: metal and glass.

4.3.1 Chaîne Opératoire?

The assessment and delineation of production technologies has, over the years, been one of the most important and widely studied areas of archaeological science. This has involved attempting to draw up a *chaîne opératoire* of the different stages involved in the process that converts a raw material into a finished usable object. In order to do this, careful examination of many finished objects is carried out, looking for evidence that might give information about the production technique. In addition, production areas are sought and studied, and production debris, in the form of tools, remnants and waste products, are examined at length. This is often put together to build up a scientifically based *chaîne opératoire* for a particular object or technology. The validity of this approach has more recently been challenged on several grounds. Using models drawn from anthropological theory and ethnographic studies, it has been pointed out that many, many stages in the production process can be missed in a simple physical, scientific assessment of an object (see Gosselain & Livingstone Smith 1995; Livingstone Smith 2000; Sillar & Tite 2000, for more details and discussion). This is because the scientific analysis of the object can only detect stages that result in a measurable physical change. Certain acts, such as saying an incantation over an object, might be (in the mind of the producer) a vital stage in its production, but, since there is no scientifically detectable physical change in the object, there is no way of detecting and reconstructing it later.

The second weakness that has been pointed out in the *chaîne opératoire* model is its attempt to try to establish why a certain methodological decision has been made. Until relatively recently, most of the work on the analysis of archaeological objects was carried out by people trained mostly or entirely in the physical sciences and they have, whether consciously or subconsciously, reconstructed ancient technologies following their own world view. For example, in the analysis of a ceramic vessel, if large quantities of shell are found incorporated into the clay, then it might be interpreted that this has been deliberately added for some reason. So far, there would probably be little debate. However, it is in the why this shell might have been added that problems occur. In the the past, it has been common to reconstruct production decisions of this type as

Box 4.1   The Canaanite Amphora Project: Study by Thin Section Petrography

L.M.V. Smith

Introduction

The case study given here forms part of the Canaanite Amphorae Project, which studies the contexts, shapes, fabrics and contents of transport amphorae used to import mainly organic products into New Kingdom Egypt (Bourriau et al. 2001; Smith et al. 2004). It was recognised that these transport amphorae, found throughout the Aegean, eastern Mediterranean, Egypt and Nubia, could offer important insights into trade networks and ancient economies. The project is based on material from the Egypt Exploration Society's excavations at Memphis (Kom Rabi'a) and Amarna, together with comparative material from sites in Israel, Syria and Cyprus.[1]

The main aims of the project are three-fold: 1) to establish a visual classification system for the fabrics used for the amphorae, thus enabling their identification to be secured wherever they occur; 2) to determine their regions of manufacture; 3) to identify the commodities they carried, whether in primary or secondary use.

Extraction and identification of organic residues from the sherds by gas chromatography/mass spectrometry showed that the main commodities carried were different types of resins and oils (Serpico et al. 2003; Serpico & White 2000a, 2000b). The provenance study of the vessels required a methodology that could allow the identification of source areas for the pottery despite the absence of identified production/kiln sites for the amphorae at the time the project was initiated.

Methodology

The first step was to substantiate and, where necessary, refine the existing classification, based on descriptions of sherd sections at 25× magnification using a binocular microscope. These descriptions included the type, relative abundance and size-range of main inclusions, the degree of fabric porosity, hardness, and colour according to the Munsell Colour Chart (Munsell Colour Company 1975). The procedure adopted for the petrological study entailed making thin sections from chips of the sherds using standard techniques (Gribble & Hall 1992; Kerr 1977). The thin sections were then examined using a polarising microscope at between 50×

and 400×. Presence/absence of constituents was recorded, as well as the sorting, shape and sphericity of the major types, which were classified on scales derived from Pettijohn et al. (1973: 585–6) using a set of standard comparator charts. Also noted were the degree of *sorting* (the extent to which inclusions are of uniform size or otherwise) and the degrees of *sphericity* and *roundness* (the extent to which grains are spherical or elongated in shape and their smoothness or roughness). Quantitative data were obtained from a point counter attached to a stepping stage. A total of 200 points was counted for each section using the 'multiple intercept' method (Middleton et al. 1985), as previously described (Smith et al. 2000). The petrographic data provided a basic grouping of fabrics. It was considered more rigorous, however, to use more than one method of analysis, so the petrographic groupings were tested, where possible, by chemical analysis using Inductively-coupled Plasma Emission Spectroscopy (ICP-AES) and Neutron Activation Analysis (NAA).

The provenance study followed. The sections were first compared with the collection held at Tel Aviv University, courtesy of Professor Yuval Goren. Sections representative of local geology and comparable with this series were selected and recorded using the methods applied to the Egyptian examples. Two workshop sites considerably later than the Bronze Age were included because of evidence for continuity of resource exploitation (Y. Goren, personal communication 1998). For comparison, examples of local pottery from Ras Shamra (Schaeffer excavations) were obtained from the Louvre. The petrographic groups were then related to the geological information on the locations where particular combinations of mineral and rock types could be found, to give the areas from which the amphorae originated (Figure 4.1).

Group 1: characterised by a high proportion of quartz sand with sand-sized limestone grains, including calcitic microfossils, and basalt fragments. The most probable area for the provenance of this group is the seaward portion of the Jezreel Valley, to the north-east and south-east of the Carmel Ridge, northern Israel.

Group 2: characterised by a moderate proportion of quartz, limestone and coarse to very coarse chalk inclusions. Potential areas for provenance can be limited to two possibilities at present: the zone where the coastal plain meets the hinterland between northern Israel and southern Lebanon, or the coast to the south of Haifa, the latter being more likely since the proportion of quartz is similar to deposits on the Carmel coast.

Group 3: includes andesine basalts, limestone, chalk and chert, with some microfossils (foraminifera). Comparison with thin

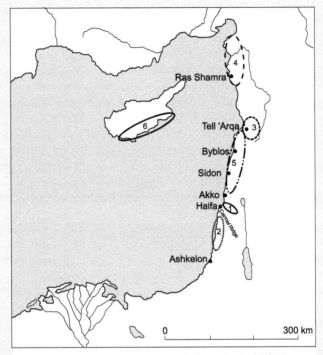

Redrawn by the author from original by M. Ownby using 'Professional Draw' by Gold Disk Inc.

Figure 4.1 Map of the Eastern Mediterranean showing the most probable source areas for the fabric groups on the basis of geology

Key: 1 = Group 1; 2 = Group 2; 3 = Group 3; 4 = Group 4; 5 = Group 5; 6 = Group 6.

sections from sites in the inland region of the Jezreel Valley with a basaltic fabric indicated that the latter differed in the type of basalt present. Other probable areas of origin where this combination of constituents is present are northern Israel, in the Galilee region, or southern Syria/northern Lebanon, in the region of the the 'Akkar Plain. The types of foraminifera are more consistent with the Lebanon coast, so the latter, around the site of Tell 'Arqa, is the more likely.

Group 4: characterised by igneous rock fragments, some showing alteration to serpentinite, together with quartzite, schist and replacement and radiolarian chert. These constituents indicate a source where components of an ophiolite complex exist, the most probable being the Baër-Bassit in northwest Syria. Examples of fabrics considered local to the site of Ras Shamra within this locality appeared very similar to Group 4 in thin section.

Group 5: basically contains microfossils of coastal types, sometimes coastal limestone (beachrock), chalk and chert, usually with a low or very low quartz content, and sometimes with rare serpentinised inclusions. The widest range for the origin of the group as a whole is along the coastal zone between about Akko, Israel, and the southern 'Akkar Plain area, Lebanon.

Group 6: having a similar set of inclusions to Group 4, but exhibiting a higher proportion of metamorphic rock fragments and quartzite, and a virtual absence of radiolarian chert. It is considered that this group originated in Cyprus because of the relatively high proportion of weathered metamorphic fragments, and the fact that the radiolarian chert has been removed by erosion from the ophiolites present on Cyprus.

Conclusions

Initially, a fabric classification has been established in which petrographic analysis is integrated into a visual classification, so that all sherds, whether thin-sectioned or not, can be classified and the material for statistical analysis in further research can be extended. Using the thin section petrography combined with comparison with other pottery representative of the geology of local areas and examination of geological maps allows regions of origin for all six fabric groups to be suggested. It should be noted that the success of the method depends substantially on the degree of variation in geology within the overall region of interest (effectively the eastern Mediterranean coast and immediate hinterland). The combinations of mineral and rock inclusions in some groups (such as Group 1) could only be found in a limited area so that a closely-circumscribed source region could be proposed. In contrast, the combination of inclusions in Group 5 can be found in numerous locations along the eastern Mediterranean coast, so that only a much broader source area could be proposed. Despite the variation in the degree to which the source areas can be specified, the study permits patterns of trade to be seen. All the fabric groups from New Kingdom Memphis and Amarna, except Group 2, originate in the northern Levant and Cyprus.

Note

1 The Egyptian material was exported with the permission of the Supreme Council for Antiquities, Egypt. Comparative samples were collected in Israel by permission of the relevant excavators and the Israel Antiquities Authority; those from Ras Shamra by permission of the Department

of Near Eastern Antiquities, the Louvre, Paris; those from Cyprus, by permission of the Director of Antiquities and Ian Todd and Alison South, the excavators of Kalavassos. Funding was provided by the Egypt Exploration Society, the McDonald Institute for Archaeological Research, Cambridge, the Wainwright Fund at the University of Oxford, the British Academy and the Society of Antiquaries of London. Thanks are due to Prof. Y. Goren, Tel Aviv University, for the identification of the source areas of the Fabric Groups, to Dr J. Bunbury, Department of Earth Sciences University of Cambridge, for advice on the mineralogy, to Dr J.N. Walsh, Royal Holloway (University of London), for the ICP-AES analysis and to Dr M. Hughes for collaboration on the statistical analysis.

References

Bourriau JD., Smith LMV. & Serpico M. (2001) The provenance of Canaanite Amphorae found at Memphis and Amarna in the New Kingdom. In: Shortland A. (ed.) *The Social Context of Technological Change: Egypt and the Near East 1650–1550 BC*. Oxford: Oxbow Books. pp. 113–146.

Gribble CD. & Hall AJ (1992) *Optical Mineralogy: Principles and Practice*. London: UCL Press.

Kerr PF. (1977) *Optical Mineralogy*, 4th edn. London: McGraw-Hill.

Middleton AP., Freestone IC. & Leese MN. (1985) Textural analysis of ceramic thin sections: evaluation of grain sampling procedures. *Archaeometry* 27: 64–74.

Munsell Colour Company (1975) *Munsell Soil Colour Charts*. Baltimore: Munsell Colour Co.

Pettijohn FJ., Potter PE. & Siever R. (1973) *Sand and Sandstone*. New York: Springer-Verlag.

Serpico M. & White R. (2000a) The botanical identity and transport of incense during the Egyptian New Kingdom. *Antiquity* 74: 884–897.

Serpico M. & White R. (2000b) Oil, fat and wax. In: Nicholson P. & Shaw I. (eds) *Ancient Egyptian Materials and Technology*. Cambridge: Cambridge University Press. pp. 390–429.

Smith LMV., Bourriau JD. & Serpico M. (2000) The provenance of Late Bronze Age transport amphorae found in Egypt. *Internet Archaeology* 9. http://intarch.ac.uk/journal/issue9/bourriau_toc.html

Serpico M., Bourriau J., Smith L., Goren Y., Stern B. & Heron C. (2003) Commodities and containers: a project to study Canaanite Amphorae imported into Egypt during the New Kingdom. In: Bietak M. (ed.) *The Synchronisation of Civilisations in the Eastern Mediterranean in the Second Millennium B.C. II*. Vienna: Verlag der Österreichischen Akademie der Wissenschaften. pp. 365–375.

Smith LMV., Bourriau JD., Goren Y., Hughes MJ. & Serpico M. (2004) The provenance of Canaanite Amphorae found at Memphis and Amarna in the New Kingdom: results 2000–2002. In: Bourriau J. & Phillips J. (eds) *Invention and Innovation: The Social Context of Technological Change II, Egypt, the Aegean and the Near East, 1650–1150 B.C.* Oxford: Oxbow Books. pp. 57–77.

almost entirely the result of a technological decision aimed at improving the physical functionality of the vessel – making it survive firing, making it stronger, more versatile, more heat resistant, etc. In other words, the ancient potter is working with a rational scientific mind-set in the modern sense to improve his or her product. However, ethnographic studies have shown that this is very rarely the reason why additives of this type are used (Smith 2000 and others). Other reasons overwhelming dominate; these are largely symbolic and religious. This has led some scholars to largely dismiss the study of ancient technology as a useful pursuit. More recently, the focus has shifted once again to a less bipolar view and something of a compromise has been reached (see Jones 2004). Those studying ancient technologies have started to realise the limitations of their techniques, and, on the other side, there has been a realisation that the material component of an object is also an important aspect of the object itself.

The production process can be defined as the method by which raw materials are transformed into a usable finished object. This will involve several stages that vary considerably depending on the type of material used and the object that is to be produced. The first stage comprises the transport of the raw materials to the site of production. In this case, the phrase 'raw materials' should be taken in its broadest sense, in that it will involve not only the material components of the object, but all the other components that are needed in that process. These might include fuel, tools, material to make specific special structures such as kilns, and so on. The production site might be very close to the source of the raw material, but sometimes it will be located somewhat distant because of these various factors.

The second stage is the production itself. In Egypt many different technologies were used, from ceramics, glass and metal to resins, food and detergents. One of the problems in discussing Egyptian technologies is that it is uncertain how they were organised into groups, i.e. whether certain technologies co-existed, perhaps with shared resources and/or workers. In modern texts it is common to talk about 'pyrotechnologies', that is to say those technologies that utilise high temperature processes, for example ceramics, glaze and metals. These tend to be discussed separately from, for example, 'artists' technologies', which involve the use of pigments, sculpture and inlay. It seems, however, that the Egyptians did not see it that way, since at Amarna there is apparent evidence of glaze workshops for the production of faience being situated within artists' compounds, most notably, the House of the Sculptor Tuthmose (Arnold 1997; Shortland, 2000a). Equally, colorants used in glaze and glass are also present in artists' paints, so some crossover must be present here. In grouping technologies following a modern pattern, information on the ancient activities is being lost and modern 'prejudices' introduced.

This tends towards thoughts of *etic* and *emic* categorisation (Arnold 1971) since, as archaeological scientists, we need to place the etic categories and decisions made that are recognised archaeologically into the emic categories that are theorised for the ancient Egyptians. Past manufacturers of objects, such as ceramic vessels, clearly had sophisticated understandings of the raw material from which they made their objects. This has several potentially major implications for Egyptology: the cultural patterns manifested in the artefacts may be cognitive; it may be possible to recreate and reconstruct these emic categories from the physical (and etic) characteristics; itnis likely that the local community of producers used these discrete ethnosystems and hence this may enable archaeological understanding of aspects of the local cognitive structuring system (Arnold 1971).

While the range of Egyptian technologies is very large and covers all aspects of Egyptian life and society, scientific analysis of production technology has been used mostly for ceramics, glaze, glass and metal, and to a lesser degree on the production of organic compounds (e.g. resins, perfumes, etc.) and pigments. This may well change as analytical tools continue to improve, but it also reflects the relative amounts of evidence available. Ceramic vessel fragments are by far the most common remains on almost all Egyptian sites, probably followed by glazed objects or faience. In terms of the remains of production sites, ceramic kilns are again the most common, alongside bread and cooking ovens, which have been somewhat less studied. The following will consider the role of science in the study of these main technologies, with notes on others.

4.3.2 Transport and Supply

Analysis has been very successful at identifying the rare material sources for a wide range of ceramics, stone and metals. As has been discussed, some of these sources are very close to their point of use, whereas others are very distant. For the distant sources, the next question to be determined is how the materials were transported to where they were needed. This includes the form in which they were transported, by whom, the route taken and the method of transport itself. As discussed above, one of the limitations of the scientific analysis of objects is that it can only detect stages in a process that result in a measurable physical change. One of the problems when trying to apply scientific techniques to determine how a raw material has been transported is that the transport process rarely results in any measurable change in the material. Whether the material was transported by water or land, or by which route, leaves very little 'signal' for the analyst to detect and interpret. There are much better ways of working on this problem, be they iconographic, textual or ethnographic, but science generally has little input. Transport and trade in finished objects, as opposed to raw materials, is discussed at length later.

4.3.3 Production Evidence from Finished Objects

The great majority of information on the production technologies is drawn from the examination of finished objects. This is partly because production sites, which would in theory give the best evidence, are often unidentified, poorly preserved, indifferently excavated and/or the studies unpublished. This is further discussed below.

4.4 Egyptian Materials

This section aims to give some background on the materials used in ancient Egypt, and how they were manipulated and preserved. However, it is particularly concerned with how science has been applied to each of the materials to address the questions discussed above. The discussion presented here begins with stone, the least physically modified of all the material types, and then goes on through clays to the highly modified and rather complex systems of metals, glass and glaze.

4.4.1 Stone

Stone was extremely important in Egypt for the building of prestige buildings, especially temples. It also featured in high status and high visibility areas of palaces and important houses, such as door frames and lintels (Lucas & Harris 1962: 50–63). Away from building work, stone vessels are relatively common, especially in the early periods, and the use of stone in jewellery and ritual contexts continues throughout Egyptian history (Lucas & Harris 1962: 406–428). The ability to manipulate stone also played an important part in the excavation and construction of tombs. Egypt is particularly rich in the quantity and quality of the stone available to its population.

The population was largely restricted to the Nile Valley floodplain, with the surrounding desert rising up on both sides, and other fertile areas such as the desert Oases and the Fayum. This desert provides ample stone suitable for construction (Aston et al. 2000: 8–15), and stone is both a heavy and fragile material to have to move around in any quantity. So the fact that a lot of the stone was available locally to the population of the Nile Valley meant it was relatively easy to use in large quantities in temples, tombs and pyramids. The Delta is quite different, with some stone types being available in the vicinity of modern Cairo (e.g. limestone and quartzite) and the Fayum (basalt), but with other hard stone types only coming from the Upper Egyptian Nile Valley (e.g. granite), although a combination of local and non-local stones were used together in mortuary structures in the Memphite Region from the Old Kingdom onwards (Klemm & Klemm 1993; 2008). This contrasts strongly with the very wide fertile plains of the Tigris and Euphrates occupied by the contemporary Mesopotamian states, where stone was only rarely locally available (Moorey 1994: 21). This led to a much wider use of mud and glazed ceramic in prestige building projects in Mesopotamia in situations where stone would have been the material of choice in Egypt (Paynter & Tite 2001).

Stone can be divided into two loose groups, 'soft' and 'hard'. 'Soft stones' are usually sedimentary stones such as limestone, sandstone and siltstone. They vary from bright white limestones such as the Tura limestone (Lucas & Harris 1962: 53–54), used in many areas of tombs, temples and pyramids, especially in the Old Kingdom (Aston et al. 2000: 40–42), through fine grey siltstones and shales to red, brown and buff sandstones (Aston et al. 2000: 54–56). However, it is immediately obvious to the archaeologist walking around any of the temples in the Nile Valley that, while these soft stones make up the majority of the stone in almost all cases, other stones were frequently used, particularly in areas of prestige or high visibility. These are mostly 'hard stones' and include banded metamorphic rocks and igneous rocks such as basalt, dolerite and granite. In most areas, igneous rocks are almost always more rarely exposed on the Earth's surface than sedimentary rocks (Press & Siever 1997: 60–72). In Egypt, hard rocks are found mostly in the south of the country (Said 1990). The nearest hard rocks to the temples of the Nile Valley, certainly in terms of journey time and effort, are perhaps the granites at Aswan. Since, as the term 'hard stone' suggests, they are more durable than soft stones, they tend not to naturally weather as fast, meaning that they form hills and mountain ranges. The first cataract at Aswan is formed by the Nile flowing over the relatively weather-resistant granites of the area. Apart from their rarity, there appear to be two main reasons why hard rocks such as granites were chosen. Firstly, since they are more durable and more resistant to wear, they can be carved more sharply. Secondly, they tend to be more strongly coloured and often have multiple colours, particularly the coarse granites such as the Aswan granite (Kozloff & Bryan 1992: no. 62, plate 36) and the banded diorites (see Box 4.2) such as those found in the Western Desert (Hayes 1953: 70, 197; MMA 1999: nos. 61 & 99; Shaw & Bloxam 1999). It should be noted, however, that many of the constructions made of hard stone appear to have been painted, thus covering up the (to the modern eye) beautiful coloured stone beneath. This makes the simple argument that hard stones were used for their colour more difficult to make and suggests that there were other more complex ritual reasons for the use of these stones in these areas (see Aufrere 2001 and many of the papers in Davies 2001).

Box 4.2   Egyptian Quarry Landscapes
Patrick Degryse & Tom Heldal

Ancient quarry landscapes, though often neglected, are important parts of our cultural heritage, providing insight into aspects of daily life throughout human history. Apart from showing layer upon layer of quarrying a stone resource, quarries are also testimony to secondary resources and a chain of production, and often show remains of complicated logistics, and bear evidence of social infrastructure. In particular, ancient Egyptian quarries contain multiple layers of information, from the Palaeolithic well into the Roman and later periods, displaying different uses and contexts of the same resource through millennia. As such, quarries can even be indicators of events or changes in society.

Quarrying techniques and stone working have long fascinated researchers, particularly in connection with the great monuments of antiquity (Heldal 2009). In Egyptian archaeology, this includes the works of Petrie (1883) on the pyramid sites, Clarke & Engelbach's (1930) interpretation of the unfinished obelisk in the quarry in Aswan, the discovery and description of Chephren's Quarry (Engelbach 1933; 1938), the survey of the quarries in the Faiyum depression (Caton-Thompson & Gardner 1934) and the quarrying of the Aswan granite (Röder 1965). Later, the formation of ASMOSIA (Association for the Study of Marble and Other Stones In Antiquity) has made significant contributions to the understanding of ancient quarrying and in developing methods for linking quarries to their places of use (Herrmann et al. 2002; Herz & Waelkens 1988; Lazzarini 2002; Maniatis et al. 1995; Schvoerer 1999; Waelkens et al. 1992). The excavations of two Roman quarry sites in the Eastern Desert of Egypt run by the University of Southampton (Maxfield & Peacock 2001; Peacock & Maxfield 1997) demonstrated definitively that quarry sites can contain a rich archaeological record. Similarly, depth of archaeological record was noted during several seasons of surveys and excavation at Chephren's Quarry in southern Egypt (Bloxam 2000; 2003; 2005; Harrell & Brown 1994; Shaw & Bloxam 1999; Shaw & Heldal 2003). Surveys of several quarries in the Faiyum area (Bloxam & Storemyr 2002; Harrell & Bown 1995; Heldal et al. 2009) continued the work initiated by Caton-Thompson & Gardner (1934). Survey work carried out by Harrell et al. (1996), Harrell (n.d.; 2002), Harrell & Storemyr (2009) and Klemm & Klemm (1993; 2008) of quarries in Egypt have added significant knowledge to understanding stone procurement in ancient Egypt, and clarified and removed numerous misinterpretations regarding stone sources and their use.

Furthermore, the latter research work and active cooperation from Harrell made it possible for the Egyptian heritage authorities to make a comprehensive national inventory of ancient quarries, through the QuarryScapes project (Heldal 2009).

In spite of the positive contributions of the last decades, research into quarries or production sites often still remains compartmentalized into studies of either single periods (i.e. Neolithic, classical antiquity, Pharaonic Egypt, medieval), commodity (i.e. grinding stones) and/or geographical region (i.e. Egypt). As ancient quarries rapidly disappear through modern development or through the lack of registration of such sites as cultural heritage (Storemyr 2009; Storemyr et al. 2007), the need to raise awareness of these important places was tackled by the QuarryScapes consortium (http://www.quarryscapes.no) through designing tools for a better characterization and valorization of quarry sites as cultural heritage sites (QuarryScapes 2009). By using a characterization method that is standardized enough to allow comparison between different places and different periods and yet open enough to take account of the individual characteristics, the evaluation of quarry landscapes would be made easier and more targeted. The method of describing and characterizing quarry features by grouping them into four main elements (the resource, the production, the logistics and the social infrastructure) provides a method of connecting physical remains in quarries by micro-level analysis to processes, technology and organization. The term "quarry complex" was introduced as a necessary tool of interpretation, in between a quarry and a quarry landscape. The characterization of the resources and the material remains together with interpretation of production techniques and logistics leads to the identification of connected activities and finally to a re-interpretation of the quarry landscape. This, in turn, then impacts on how the significance of such landscapes from a purely elite stone procurement is evaluated, and leads towards viewing the entirety as a composite system of quarrying through time. When deconstructing the quarry landscape into its complexes, it is also easier to point out particularly important places for future conservation that are good informational projections of the different activities (Heldal 2009).

References

Bloxam E. (2000) Transportation of Quarried Hard Stone from Lower Nubia to Giza during the Egyptian Old Kingdom. In: McDonald A. & Riggs C. (eds) *Current Research in Egyptology 2000*. BAR International Series 909. Oxford: Archaeopress. pp. 19–27.

Bloxam E. (2003) *The Organisation, Transportation and Logistics of Hard Stone Quarrying in the Egyptian Old Kingdom: A Comparative Study.* Unpublished PhD dissertation. Institute of Archaeology. London: University College London.

Bloxam E. (2005) The Organisation and Mobilisation of Old Kingdom Quarry Labour Forces at Chephren's Quarry (Gebel el-Asr), Lower Nubia. In: Cooke A. & Simpson F. (eds) *Current Research in Egyptology II.* BAR International Series 1380. Oxford: Archaeopress. pp. 11–19.

Bloxam E. & Storemyr P. (2002) Old Kingdom Basalt Quarrying Activities at Widan el-Faras, Northern Faiyum Desert. *Journal of Egyptian Archaeology* 88: 23–36.

Caton-Thompson G. & Gardner EW. (1934) *The Desert Fayum.* Vols I & II London: Royal Anthropological Institute.

Clarke S. & Engelbach R. (1930) *Ancient Egyptian masonry. The building craft.* Oxford: Oxford University Press.

Engelbach R. (1933) The Quarries of the Western Nubian Desert. A Preliminary Report. *Annales du Service des Antiquitiés de l'Egypt* 33: 65–80.

Engelbach R. (1938) The Quarries of the Western Nubian Desert and the Ancient Road to Tushka. *Annales du Service des Antiquitiés de l'Egypt* 38: 369–390.

Harrell JA. (n.d.) *Research on the Archaeological Geology of Ancient Egypt.* http://www.eeescience.utoledo.edu/Faculty/Harrell/Egypt/AGRG_Home.html

Harrell JA. (2002) Pharaonic Stone Quarries in the Egyptian Deserts. In: Friedman R. (ed.) *Egypt and Nubia: Gifts of the Desert.* London: British Museum Press. pp. 232–243.

Harrell JA. & Bown T. (1995) An Old Kingdom Basalt Quarry at Widan el-Faras and the Quarry Road to Lake Moeris in the Faiyum. *Journal of the American Research Center in Egypt* 32: 71–91.

Harrell JA. & Brown VM. (1994) Chephren's Quarry in the Nubian Desert of Egypt. *Nubica* 3/1: 43–57.

Harrell JA. & Storemyr P. (2009) Ancient Egyptian quarries—an illustrated overview. In: Abu-Jaber N., Bloxam EG., Degryse P. & Heldal T. (eds) *QuarryScapes: Conservation of ancient stone quarry landscapes in the Eastern Mediterranean.* NGU Special Publication 12. Trondheim: Geological Survey of Norway. pp. 125–154.

Harrell JA., Brown VM. & Masoud MS. (1996) Survey of ancient Egyptian quarries. *Egyptian Geological Survey and Mining Authority Paper* 72.

Heldal T. (2009) Constructing a quarry landscape from empirical data. General perspectives and a case study at the Aswan West Bank, Egypt. In: Abu-Jaber N., Bloxam EG., Degryse P. & Heldal T. (eds) *QuarryScapes: Conservation of ancient stone quarry landscapes in the Eastern Mediterranean.* NGU Special Publication 12. Trondheim: Geological Survey of Norway. pp. 125–154.

Heldal T., Bloxam E., Degryse P., Storemyr P. & Kelany A. (2009) Gypsum quarries in the northern Faiyum quarry landscape, Egypt: a geoarchaeological case study. In: Abu-Jaber N., Bloxam EG., Degryse P. & Heldal T. (eds) *QuarryScapes: Conservation of ancient stone quarry landscapes in the Eastern Mediterranean*. NGU Special Publication 12. Trondheim: Geological Survey of Norway. pp. 51–66.

Herrmann J., Herz N. & Newman R. (2002) (eds) *ASMOSIA 5, Interdisciplinary Studies on Ancient Stone – Proceedings of the Fifth International Conference of the Association for the Study of Marble and Other Stones in Antiquity*. London: Archetype Publications.

Herz N. & Waelkens M. (1988) (eds) *Classical Marble: Geochemistry, Technology, Trade*. NATO ASI Series E, Applied Sciences, Vol. 153. Dordrecht: Kluwer Academic.

Klemm R. & Klemm D. (1993) *Steine und Steinbrüche im Alten Ägypten*. Berlin: Springer-Verlag.

Klemm R. & Klemm D. (2008) *Stones and Quarries in ancient Egypt*. London: British Museum Press.

Lazzarini L. (2002) (ed.) *Interdisciplinary Studies on Ancient Stone – ASMOSIA VI, Proceedings of the Sixth International Conference of the Association for the Study of Marble and Other Stones in Antiquity*. Padova: Bottega d'Erasmo.

Maniatis Y., Herz N. & Basiakos Y. (1995) (eds) *The Study of Marble and Other Stones Used in Antiquity*. Archetype Publications: London.

Maxfield V. & Peacock D. (2001) *The Roman Imperial Quarries Survey and Excavation at Mons Porphyrites 1994–1998. Vol. I: Topography and Quarries*. London: Egypt Exploration Society.

Peacock D. & Maxfield V. (1997) *Mons Claudianus Survey and Excavation 1987–1993. Vol. I. Topography and Quarries*. Fouilles de l'IFAO 37. Cairo: Institut Français d'Archéologie Orientale.

Petrie WMF. (1883) *The Pyramids and Temples of Gizeh*. London: Field & Tuer.

QuarryScapes (2009) *QuarryScapes: Conservation of ancient stone quarry landscapes in the Eastern Mediterranean*. Abu-Jaber N., Bloxam EG., Degryse P. & Heldal T. (eds) NGU Special Publication 12. Trondheim: Geological Survey of Norway.

Röder J. (1965) Zur Steinbruchgeschichte des Rosengranits von Assuan. *Archäologischer Anzeiger* 3: 467–552.

Schvoerer M. (1999) (ed.) *Archéomatériaux – Marbres et Autres Roches*. Bordeaux: Presses Universitaires de Bordeaux.

Shaw I. & Bloxam E. (1999) Survey and Excavation at the Ancient Pharaonic Gneiss Quarrying Site of Gebel el-Asr, Lower Nubia. *Sudan and Nubia* 3: 13–20.

Shaw IM. & Heldal T. (2003) Rescue work in the Khafra Quarries at Gebel el-Asr. *Egyptian Archaeology* 23: 14–16.

Storemyr P. (2009) Whatever else happened to the ancient Egyptian quarries? An essay on their destiny in modern times. In: Abu-Jaber N., Bloxam EG., Degryse P. & Heldal T. (eds) *QuarryScapes: Conservation of ancient stone quarry landscapes in the Eastern Mediterranean*. NGU

Special Publication 12. Trondheim: Geological Survey of Norway. pp. 105–124.

Storemyr P., Bloxam E. & Heldal T. (2007) *Risk Assessment and Monitoring of Ancient Egyptian Quarry Landscapes.* QuarryScapes report. Trondheim: Geological Survey of Norway www.quarryscapes.no

Waelkens M., Herz N. & Moens L. (1992) (eds) *Ancient Stones: Quarrying, Trade and Provenance – Interdisciplinary Studies on Stones and Stone Technology in Europe and Near East from the Prehistoric to the Early Christian Period.* Acta Archaeologica Lovaniensia 4. Leuven University Press: Leuven.

4.4.1.1 Analysing Stone

What stone is it? Where was it mined? These are the two questions regarding stone objects most commonly asked of the archaeological scientist. Stone objects are essentially made of virtually unaltered (other than shaped) rock, so the internal structure and characteristic minerals of stone would be recognisable to any geologist used to dealing with the identification of rock outcrops. The first approach that a scientist would use with a stone object would therefore be almost exactly that of the geologist in the field, examining the surface of the object by eye and with a hand lens. The word 'almost' is used advisedly, since a geologist in the field would want to examine a fresh, unaltered surface of the rock, so would therefore take a hammer and knock a piece off. This is very rarely allowed with archaeological objects, despite the fact that it is often the broken edge of such an object that gives the most detail to the trained eye, rather than the polished surface. This is because different minerals break in characteristic ways, and this can only be seen in the broken edge, not on the polished surface. Geologists would tend to use other destructive tests too, such as scratching to test for hardness and dilute hydrochloric acid to detect rocks high in calcite. Once again, these tests are rarely permitted for archaeological materials. Despite this, examination by the trained eye and hand lens will be enough to tell apart most rock types and show from what the object is made.

Ascertaining where the stone came from can be more complex. It is certainly true that some rock types are limited to one source and so can easily be identified and provenanced. However, many areas produce rocks that look very similar to the eye and lens. This is especially true of medium and fine grained rocks, whether they are soft (limestone, siltstone) or hard (basalts and rhyolites). There are usually multiple possible sources for these rocks and, to tell them apart, it is necessary to look at them in more detail. Once again the techniques of the geologist are used – thin section petrography. 'Petrography' is the description and classification of rocks and can be applied directly to stone objects (Nockolds et al. 1978). It is the primary analytical technique of the archaeological scientist who works with both stone and ceramics (discussed below). To do this, a sample of the stone must be taken. This can be problematic, since the sample must be representative of the stone as a whole. This means that the sample usually needs to be at least approximately the size of a finger nail and often larger. This may be possible with large broken objects, such as fragments of colossal statues, but is obviously impossible with smaller, complete objects, such as stone vessels. The sample has a flat surface cut on it and this is mounted on a slide and polished down until it is only 0.03 mm thick. By this point, almost all of the minerals in the sample have become translucent (Ehlers 1987). This is known as a thin section. The thin section can then be examined under a polarising microscope, with special filters allowing the various minerals to be differentiated (MacKenzie & Guilford 1980). Petrography therefore allows the minerals and other features of the stone, perhaps fossils, to be determined. This can differentiate not only different rock types, but potentially different sources of the same rock type. These might be distinguished through the presence of either low abundance minerals ('accessory minerals') that vary from location to location within the rock type and are characteristic of one particular source, and/or fossils which give a geological age to a limestone, thereby distinguishing different outcrops of varying ages (Adams et al. 1984; MacKenzie et al. 1982).

Thin section petrography is thus a very important and useful tool for the provenancing of stone, but its major disadvantage is that it is destructive. Obtaining samples large enough to use this technique from objects, especially those in museum collections, is becoming increasingly difficult. Another issue arises with very fine grained or glassy rocks as either these have minerals too fine to see under the microscope or there are no minerals present. In this case, chemical techniques, such as inductively coupled plasma atomic emission spectroscopy (ICP-AES) or inductively coupled plasma mass spectrometry (ICP-MS), can be used. These distinguish the different rocks not by their mineralogy, as thin section petrography does, but by their bulk chemical composition. These techniques are discussed more fully under ceramics, since they are much more frequently employed for ceramic objects than rocks, but it is useful to look at a couple of examples here. Again, the technique is limited because it requires a sample, and a reasonably large one that is representative of the rock as a whole, rather than just a microsample. One of the most important rock types which is routinely analysed by chemical means is obsidian.

Obsidian is naturally occurring volcanic glass that forms in lava flows associated with certain volcanoes, both modern and ancient, and is an important, if rare, material that was transported over very long distances (Dixon et al. 1968). Although a stone, since it is glassy with few obvious structures or minerals, a lot of obsidian looks very similar, even under the petrological microscope or scanning electron microscope (SEM). It can be compositionally distinguished and much research has been undertaken into the compositions of different possible volcanic sources of obsidian (Brown & Harrell 1995; Lucas 1942; 1947), most of which occur in the Aegean and further east into Anatolia. Indeed some of the earliest provenancing work performed in the 1960s was carried out on obsidian from the Eastern Mediterranean (Dixon et al. 1968; Renfrew et al. 1965; 1966). However, relatively few Egyptian obsidian objects have been analysed. Again, this is mostly due to the techniques' requirements for samples to be taken, thereby rendering them problematic for rare objects.

The final provenancing technique for stone tends to be used when all the others discussed above have failed. Some rock types look very similar not only under the hand lens and in thin section, but are also chemically very similar. The best example of this is marble, which was used extensively throughout the Mediterranean in the Ptolemaic and Roman periods both as a building material and for statues and other objects (Herz 1992). Marble was available from a wide range of quarries throughout the Eastern Mediterranean, with especial foci on the islands. However the various quarry sources usually cannot be uniquely distinguished by any of the techniques discussed above. A further technique has therefore been attempted – isotopic analysis. Marbles are essentially calcium carbonate, and the isotopes ratios of carbon ($\delta^{13}C$) and oxygen ($\delta^{18}O$) in the carbonate of the marble in the different quarries have been determined (Brown and Harrell 1995). This showed that there was some variation in the two isotope ratios, so some quarries could be distinguished. But there was also a huge overlap between the quarries. At worst, some $\delta^{13}C$ and $\delta^{18}O$ values are shared by at least eight different quarries from widely different areas of the Eastern Mediterranean, and that were in use at very different times (Herz 1992). Isotopic analysis of marble, therefore, has to be supplemented by the (not terribly useful) petrographic information that is available and any historical sources that might help identify and distinguish the most likely quarries. Isotopic work is rarely attempted on other stone types in Egypt because of the good results that can be obtained from other less destructive and less expensive techniques.

4.4.1.2 Working Stone

Huge quarries where stone was extracted can still be seen in many areas of Egypt. Soft stones were relatively easy to work, with large pieces being broken off by driving wooden wedges into artificial or natural cracks and joints. These blocks could be roughly shaped by chipping, using hard stone or metal axes, or cut with copper and bronze saws, albeit with some difficulty (Aston et al. 2000; Stocks 2003: 25–99). Fine details were carved using a range of copper and stone tools, and finished to produce a beautiful polish. Huge hard stone quarries can still be seen where

Box 4.3   Provenancing Stone and the Red and Black Granites in the Egyptian Antiquity Museum of Turin

Andrew Shortland

Stone statues are some of the most striking and most widely collected objects from Egyptian antiquity. Large numbers were transferred to western museums during the nineteenth century and form the backbone of major museum collections in London, Berlin, New York and so on. These statues are frequently of hard stone. Attempts to trace the quarry from which the stone was extracted have been going on almost as long as the statues have been collected. Amongst the many papers on this subject, the work of James A. Harrell at the University of Toledo, Ohio, stands out. He has published an extremely wide range of articles and books on the subject of quarries and stone in Egypt. As in Box 4.2, here we highlight a website constructed by him: http://www.eeescience.utoledo.edu/Faculty/Harrell/Egypt/AGRG_Home.html.

Entitled *Research on the Archaeological Geology of Ancient Egypt*, the website represents an exceptional resource for the student and specialist interested in the use of stone in Egypt. The website includes a useful introduction to the rocks and minerals used in Egypt, the use of stone in Ottoman mosques and a discussion of an early papyrus dealing with quarries. It is, however, the survey of Egyptian quarries that presents the most useful resource. Here Professor Harrell has presented maps that show all the most important Egyptian quarries. In addition to this, he has divided them into sources of hard stones, soft stones and gemstones. For each quarry, the stone is identified and images are presented representing the texture of rocks found. This therefore acts as a textbook for the visual identification of Egyptian rock types and is a valuable resource for the student and professional.

Inspired by Professor Harrell's work, Serra et al. (2010) undertook an excellent example of the classic approach to the provenancing of hard stone. This research project examined seven hard stone objects in the Turin Museum and attempted to answer the question that is usually asked – where was the stone mined? The objects comprised seven "masterpieces of New Kingdom ... art" (Serra et al. 2010: 964), including the lid of the sarcophagus of Nefertari, a statue of Hathor and three of Sekhmet, a ram-headed sphinx and a statue of Ramesses II. Sampling was allowed and, as always, they were "trying to find a balance between the need to represent the compositional variability of plutonic [therefore coarse grained] rocks in Egyptian sculpture and the obvious limitations of conservation" (Serra et al. 2010: 969). Samples were removed from cracks or uncarved areas. A reference set of material was gathered from fieldwork in eight quarries in the Aswan region of Egypt, *red granites* from the region of the Unfinished Obelisk, and *black granites* from near Gebel Ibrahim Pasha. The two sets of samples, museum and quarry, were compared by subjecting them to petrographic examination using a polarising light microscope in conjunction with SEM with EDS analysis. The polarising microscope enabled the minerals of the samples to be identified. The red granite quarry samples contained a main mineralogical assemblage of alkali–felspar, quartz, plagioclase, biotite and amphibole, with accessory apatite, ilmenite and others. This was very similar to the sarcophagus lid, both macroscopically and microscopically, but different from the sphinx, which, while made from red granite, lacked amphibole and had other inconsistencies. The black granite quarry samples were technically identified as granodiorite and tonalite, with the same major mineral constituents as the red granites, but in different proportions and with contrasting textures. Two of the Sekhmet statues were very similar to the granodiorite, but one, whilst still being granodiorite, had higher biotite and other differences. The statues of Hathor and Ramesses II were tonalites and broadly similar to the samples from the Aswan quarries. The SEM-EDS was used to analyse the composition of individual mineral grains in the samples, specifically the amphibole, biotite and plagioclase. This work supported the microscopic conclusions that the while the sarcophagus was consistent with the Aswan quarries, the sphinx statue was not. The paper speculates that the sphinx statue is more similar to High Dam granites further south. The SEM-EDS of the other statues again support the optical work, although one of the Sekhmet statues is slightly different from the other two, suggesting it may have been made at a different time, but from the same rock source.

This study shows the process of stone provenancing and the interpretation of the rock and mineral textures and their compositions. It is an excellent example of taking a few important archaeological objects and determining their provenance with useful historical implications.

References

Serra M., Borghi A., D'Amicone E., Foiroa L., Mashaly O., Vigna L. & Vaggelli G. (2010) Black and red granites in the Egyptian Antiquity Museum of Turin: a minero-petrographic and provenance study. *Archaeometry* 52: 962–986.

granite and other stones were extracted. Particularly interesting are the granite and porphyry quarries at Aswan and in the Eastern Desert (Aston et al. 2000: 35–37). The Aswan granite is a course-grained rock with prominent lath-shaped grains ('phenocrysts') of pink potassium feldspar. It is probably the most widely used hard stone in Dynastic Egypt, and the quarries at Aswan provided stone for such projects as sarcophagi, the lining and facing stones for pyramids (especially the King's chamber and Grand Gallery in Khufu's pyramid and the casing of Menkaure's pyramid at Giza), colossal statues such as that of Ramesses II in the Ramesseum, and obelisks such as those at Karnak (Aston et al. 2000: 35–37). An unfinished obelisk, broken during its excavation, still lies in one of the quarries (Arnold 1991; Aston et al. 2000; Engelback 1922). Much later Roman quarries still survive at Mons Claudianus and Mons Porphyrites in the Eastern Desert, where stone was extracted for columns, flooring, vessels and other uses in Roman temples and houses (see Peacock and Maxfield 1997; 2007 and other volumes).

Whilst soft stone could, in some cases, be cut with metal tools, hard stones present a very different problem. They are so hard that the copper-based metal tools of the period would not even scratch them. To work such stones, a combination of different techniques needed to be used. To extract large pieces from outcrops of rock, artificial grooves were carved into the rock using stone tools made of granite or even harder stone (see Arnold (1991) for an excellent account). Wedges of wood were driven into the grooves, just as with the soft stones, and these were soaked in water (Aston et al. 2000 and references). Wood swells, and the forces produced are so great that they can crack granite, which, while being very hard, is also very brittle. This technique, however, is only able to remove blocks up to a certain size, and only if outcrops of the right shape exist. For larger pieces, such as obelisks, this technique was not possible. The only mechanism available to the ancient Egyptians was to extend the technique used in making the initial grooves for wedging. Spherical dolerite pounders were used, some of which can still be found around the unfinished obelisk at Aswan. Dolerite is finer grained than granite and tends to be harder, although not by a great deal. The balls were hammered and ground to create deep trenches around the obelisk. The bottom side of the unfinished obelisk still exhibits the characteristic curved shapes left by the balls. Even when the object was roughed out, shaping, polishing and finishing of granite requires considerable effort. Once again, stone grinders were probably used.This process could well have been assisted by using abrasives made of quartz sand and for softer stones at least, copper tools (Stocks 1993). It is perhaps fortunate that quartz, the most common mineral on the Earth's surface and the most abundant of all Egyptian sands, is also one of the hardest, being rated at 7 on the Mohs scale. This is harder than almost all other stones and minerals in Egypt. The process, however, would still have been hugely laborious and required great skill.

4.4.1.3 Minerals and Semi-precious Stones

So far we have discussed those stones that are used in large quantities in buildings and statues, but which are less commonly used in smaller objects such as amulets and jewellery. However, there are other stones too rare and perhaps too valuable or otherwise unsuitable for use in these larger projects. This is a rather disparate group that includes semi-precious stones, rarer rock types and minerals (Lucas & Harris 1962: 386–405). They are grouped together here because they are broadly treated the same way by scientific analysis, but their uses to the ancient Egyptians were hugely varied: as jewellery, pigments, medicine, inlays, and other ritual objects amongst others.

'Semi-precious stones' is a rather ill-defined modern term which is widely used to describe the sort of minerals and stones found commonly in ancient Egyptian jewellery. Precious stones are better defined. This term relates to the very highest value gemstones, specifically diamond, ruby, emerald and sapphire. Some definitions would include other gemstones as precious – topaz and tourmaline, for example. None of these precious stones is known to have been used in Dynastic Egypt, although there are emerald mines in the Eastern Desert that were exploited in the Roman and Byzantine periods (Shaw et al. 1999). Semi-precious stones fall into an ill-defined group of gemstones which, in modern times, are not as valuable as precious stones. Perhaps the most important and most used in Dynastic Egypt would include: lapis lazuli, malachite, quartz (amethyst, rock crystal and other types), chert and flint, felspar, garnet and a number of others (Aston et al. 2000). Some are locally available in Egypt; others were transported very long distances to arrive in Egyptian contexts. Related to semi-precious stones are a whole series of minerals and rocks that are sometimes used in jewellery, in pigments and in metallurgical or other 'industrial' processes. These include ore minerals such as galena and malachite, calcite (as alabaster and other forms), and carvable stones such as steatite etc.

Some semi-precious stones are the very rarest of materials and have very high values. This tends to mean that they are incorporated into the most significant, often royal, objects. It is worth considering an example – lapis lazuli, perhaps the most interesting, as well as one of the most important to Egypt. Lapis lazuli is a rock containing significant quantities of the deep-blue mineral lazurite, often with secondary pyrite (Aston et al. 2000: 39–40). This creates a blue rock speckled with 'gold'. It was particularly prized by the Egyptians who were greatly influenced by symbolic associations with the colour blue (Wilkinson 1994: 107). As such, it is often found in royal regalia and in other high profile roles. Its position on offertory lists shows that it was the most valuable of all the stones and second only to precious metals (Sherratt & Sherratt 1991). Lapis lazuli is very rare, formed only under very special circumstances. The only known cluster of sources that was available to the cultures of Egypt and Mesopotamia is located in the Badakhshan Province of Afghanistan (Lucas & Harris 1962: 398–400; Wyart et al. 1981). It lies high in the Western Hindu Kush and is only accessible for a few brief months in the summer. This source was exploited from the Predynastic Period onwards and all lapis lazuli found in Egypt was mined there. This makes lapis one of the easiest materials to provenance. Specimens can be relatively accurately identified by hand, since it is a bright blue and the only material with which it could probably be confused is cobalt blue glass. The latter was probably fabricated to resemble lapis as closely as possible. However, specimens, once identified as lapis, must have travelled thousands of miles to the Nile from Afghanistan. As such, it represents one of the earliest examples identified of long-distance transport of materials and one of the longest examples of 'down-the-line' exchange in the ancient world. Both these aspects emphasise the value placed on this stone by the Egyptians.

Questions asked of the scientist about semi-precious stones are the same as those for building stone: what is it and where did it come from? Semi-precious stones present special challenges for the scientist, largely stemming from their value. The techniques described above for building stone required a sample to be taken. This is virtually never possible for semi-precious stones – they are too small, too rare and (even if sampling was allowed) often too hard to sample easily. Therefore, non-destructive techniques almost always have to be employed with semi-precious stones. The risk of damage to the stone, or perhaps to its mount, also usually rules out any technique that might involve putting the object in a vacuum chamber (such as SEM). This means that most analyses have to be carried out in air or helium atmospheres.

Two approaches are possible, and these can be used in combination. The first is to attempt to determine the chemical composition of the stone. The first technique for this is XRF, with the advantage being that it is completely non-destructive and, for almost all stones, will not cause damage. The disadvantage is that airpath XRF will not detect (or detect only poorly) light elements. Unfortunately these elements, for example sodium, magnesium, aluminium and silicon, make up the majority of most semi-precious stones. To partially combat this, analysis is sometimes carried out in a helium atmosphere, which, being less dense, allows through more of the low-energy X-rays from the lighter elements. Even this technique, however, will still struggle with some elements. This means that a second approach, looking at the structure of the gem rather than its chemical composition, is often preferred. The first of these structural methods is Fourier-Transform Infra-Red spectroscopy (FT-IR). This compares the absorption of different wavelengths of light by the stone to a series of reference spectra, with a match confidently identifying the stone type. The second is X-ray diffraction (XRD), where X-rays are diffracted by the crystal planes in the gem in characteristic ways depending on the spacing and order of the planes. Once again, the resulting spectra are compared with reference libraries. Modern XRD machines can carry this out entirely non-destructively, whereas older ones tend to require a sample. Using a combination of the techniques described above, especially if it is possible to combine structural analysis with chemical analysis, it is usually possible to identify the stone type.

Going beyond simple identification of type to determine where the stone might have been mined adds a huge extra level of difficulty. On the very rare occasions that destructive sampling might be allowed, this is sometimes possible. One of the few examples where sampling was permitted was, once again, for lapis lazuli. Analysis has been undertaken of lapis lazuli objects where it was suspected that the object might be a late copy of an original. This is because, in later periods, the Afghan mine was no longer regularly producing significant quantities of lapis (Lucas & Harris 1962: 398–400). Instead, large deposits had been discovered in the Chilean Andes with smaller deposits found elsewhere in the world. These deposits, however, have significantly different chemical signatures, thereby allowing them to be distinguished from the Afghan source. Late copies made of Andean or other lapis can hence be distinguished from original Dynastic Egyptian lapis objects (Schmidt et al. 2009).

The success of provenancing other stone types very much depends on the material concerned. Very simple and very pure stones, such as quartz (e.g. amethyst, milky quartz or rock crystal), chert (e.g. jasper) or chalcedony (e.g. onyx, citrine), are extremely hard to provenance. Throughout the world, these stones have, to all intents and purposes, identical structures and compositions. There is simply not enough variation in the stone itself to allow different sources to be characterised. However, other stones have more promise. Some stones exhibit what is known as solid solution series (varying levels of a range of elements substituting for each other in the same site in the crystal structure). These substitutions affect both the structure and chemical composition of the stone and can make them quite characteristic of a source. A good example of this is garnet, which has a complex series of substitutions possible in its structure and so has been successfully provenanced non-destructively to various areas and periods (Bimson et al. 1982).

4.4.2 Clay and Ceramics

Most raw materials used in Egypt originated within a few kilometres of the place where they were used. This is the case for most ceramics, building materials, and general household materials (mats, beds, fuel, etc.). The Nile silt and associated marls and other clays were easily available and widely exploited. They are also remarkably uniform for the length of the Nile, making it rather difficult to state with certainty where, within the Valley, a particular ceramic vessel might originate. In general, however, clays can be very variable and the nature of the clay fundamentally affects its working and firing properties (Henderson 2000: 112–115; Whitbread 2005). By far the greatest focus over the last hundred years in the examination of objects by scientific methods has been dedicated to the study of ceramics. This is unsurprising as ceramics form the bulk of the archaeological material recovered from most sites. This is particularly so in Egypt, where ceramics are so abundant that, on many sites, it is impossible not to walk on fragments of broken vessels that lie scattered everywhere. This frequency, along with the use of ceramic sequences in the dating of sites and levels, has made them very important for archaeologists.

Ceramic analysis uses techniques largely derived from geological sciences to trace the clay of a ceramic back to its source. The most basic and rapid use of this technique is the initial ceramic sorting that goes on at every archaeological site. As mentioned above, for most Egyptian periods, ceramics are extremely abundant. This creates a problem that is quite unusual for most archaeological sites throughout the world – too many finds. The ceramic finds need to be processed, such that information can be extracted from them without having to draw, photograph and describe what can be literally tens of thousands of sherds. Thus ceramics are sorted firstly by whether they are 'diagnostic' or not (whether they have sufficient preservation, usually a rim or base, to show the size and shape of the vessel) and secondly by fabric. The fabric of the ceramic is the appearance of the cross-section of the sherd, most clearly seen on fresh breaks and is caused by a combination of the raw materials used and the production technologies employed (Bourriau et al. 2000: 129). It includes the number, size and composition of any particles in the clay, the colour and characteristics of the two surfaces and the presence or absence of porosity, organic material, mineral grains and so on. In the field, ceramics can be rapidly sorted into fabric groups that look similar, and then samples can be taken of each of the groups, which can then be further analysed to determine where and how they might have been made. Each of the groups of non-diagnostic fragments can then (at least) be weighed in order to give some idea of the relative incidence of the different fabrics on the site.

After representative samples of the different fabric groups have been taken, ceramic petrography proper starts. 'Ceramic petrography' is simply the description and classification of ceramics, essentially man-made rocks, using exactly the same tools that have been described above for the analysis of stone (Whitbread 2005). It is the primary analytical technique of the archaeological scientist who works with ceramics (Bourriau et al. 2000). The technique relies on taking a sample of a sherd of the particular fabric type, mounting it on a slide and polishing it down to create a thin section (see Section 4.4.1.1 *Analysing Stone)*. The thin section is examined under a polarising microscope, and, through their distinctive characteristics, the different minerals, clasts, voids and other phases of the ceramic are identified. The characteristics can then be compared with those in possible clay sources, and a conclusion drawn as to which is most similar and therefore which is most likely to be the source. Petrography therefore allows the mineralogy and structure of the ceramic to be determined. However, it can also give important and interesting information on the technology used in the ceramic. This can be of several different types and is discussed under production in the next section.

Thin section petrography is often supplemented by techniques designed to determine the chemical composition of the ceramic in terms of elemental abundance. Various techniques have been used to do this, including wet chemical analysis, neutron activation analysis (NAA), inductively coupled plasma atomic emission spectroscopy (ICP-AES) and inductively coupled plasma mass spectrometry (ICP-MS) (Hatcher et al. 1995; Velde & Druc 1999: 259–284; Whitbread 2005). ICP-AES and ICP-MS are now the standard techniques for this research as they give the best range of elements and precision for the lowest cost. They involve taking a sample of the ceramic and dissolving it in acids before measuring the elemental abundance of the sample. They are 'bulk' techniques; they give an average analysis of all the components in the ceramic. These can be particularly valuable when the clay is fine and does not contain large mineral grains. In these circumstances, optical techniques, such as petrography, are less applicable because the grains are too small to be easily identified. However, where large mineral grains or temper/grog have been added (see next section), then the analysis will reflect the addition of extra material in addition to the clay, and so will not be the same as the original clay analysis. A compromise solution that offers both the possibility of obtaining mineralogical information alongside composition is to mount fragments of the ceramics in polished blocks and to examine them using SEM (Tite et al. 1982a; 1982b; Tite & Maniatis 1975). While having many advantages, the drawback of this is expense, as much machine time is needed to go through the number of ceramics that would be required in any sort of significant study. This technique is therefore often used to give additional information on questionable samples or to highlight non-local clays within a mass of local material.

There are many studies of Egyptian ceramics in the literature, with a good summary of techniques and findings in Arnold et al. (1993), but often it is the attempt to spot rare imports and determine their provenance that is the driving force behind them. Particularly interesting in this regard are the Canaanite amphorae project (see Box 4.1) and the work carried out on the Amarna letters (Goren et al. 2004). More standard applications of ceramic analysis can be seen in many excavation reports from sites throughout Egypt, especially the larger ones such as Amarna and other major sites (Arnold 1988; Hope 1979; 1980; 1981; Nicholson & Rose 1985).

4.4.2.1 Identifying Production Technology

Analysis of Egyptian ceramic objects, to interpret both their provenance and their technologies, has been going on since the beginning of the twentieth century. The techniques used in the determination of the clay source are discussed above. This identification and grouping of clays in Egypt has been codified into what has become known as the Vienna system (see Box 4.4). This is a set series of descriptions used by ceramic specialists. Many of these same techniques are used for examining production technology, but two dominate (thin section petrography and SEM). This is because most production technologies do not result in chemical changes within the clay, but instead show microstructural changes. In order to detect them, therefore, it is necessary to be able to image the ceramic. Usually the imaging is of a cross-section. Several stages of ceramic production can be recognised (Bourriau et al. 2000; Henderson 2000: 115–130). Firstly, the raw clay can often be tempered. This means that another material is introduced into it. Various tempers can be recognised, including sand, shell, straw and chaff, either through their direct identification (in the case of minerals) or by identifying the shape of the void they have left (in the case of burnt-out organic components). The other possible addition is grog. This is the deliberate addition of small fragments of another clay or fired ceramic to the clay before shaping.

There could be several reasons for the addition of tempers and grog. In modern ceramics, they are added to change the thermal properties of the vessel, so as to make them more resistant to thermal shock (rapid changes in temperature). This could be both in the firing process and in subsequent use, for example, if the vessel is used in cooking (Tite & Maniatis 1975; Tite et al. 1982b). However, this may not have been in the mind of the ancient potter, who, viewing his world in a different way from the way we do now, may have had other ideas. Certainly, the ritual addition of materials to ceramics is extremely important ethnographically and it is difficult to believe that it would not have been practised, at least to a limited extent, in Egypt (Livingstone Smith 2000; Sillar 2000; Sillar & Tite 2000). Caution, therefore, has to be taken in interpretation.

The second stage of the ceramic production process is shaping. This can be done by hand modelling, but frequently involves a potter's wheel, allowing the process of making a vessel to be easier and quicker. The use of a wheel can leave uniform and characteristic striations on the surfaces of the vessel, which are indicative of its forming process. The faster the wheel spins, the easier it is to create a pot, and so fast wheels were a definite technological development. In fast-thrown pots elongate particles or minerals in their fabric may take on a preferred orientation which can be seen in thin section or SEM, and is indicative of this type of manufacture. The first wheel-thrown pottery came into common use around 2250 BC, during the Middle Bronze Age in Mesopotamia (Henderson 2000: 119).

Box 4.4   The Vienna System
Janine Bourriau

The 'Vienna System' was devised as a visual classification for Egyptian pottery fabrics used from the end of the Old Kingdom to the end of the New Kingdom. The aim was to provide a terminology and a methodology for the recording of pottery fabrics in the field. The Vienna System was intended to be a framework for further research and not to become a fixed system, but, to avoid confusion, it is necessary to identify fabrics or groups added later. The authors were a group of archaeologists working in Egypt in the 1980s: Dorothea Arnold, Manfred Bietak, Janine Bourriau, and Helen Jacquet-Gordon working with Hans-Åke Nordström whose microscope descriptions were the starting point for discussion. The full description was published in 1993 (Nordström & Bourriau 1993) and there have been several later studies using petrological and chemical analyses to test the groupings. These are listed at the end of this box.

Pottery is a man-made artefact so can lead directly to an understanding of the processes by which it was made. The characterisation of the raw material, the prepared clay, may be the first step in identifying the source, date, and function of a vessel. The recording is done by examining a fresh section cut through the vessel wall and orientated so that inclusions line up parallel to the rim. Using a binocular microscope at 20´, a basic description of the clay matrix, inclusions (size and quantity), sorting, porosity, hardness, wall-thickness, firing and colour is made. Quartz and felspar (‹sand› using the terminology of archaeologists), plant remains, limestone, mica, shell and grog (pottery dust) can be identified with confidence (Bourriau 2010: 17–21). At this level, additional mineral inclusions can only be described by colour, quantity, size and shape. Nevertheless, with the help of a digital colour image for reference and a set of fabric samples kept on site for consultation, more consistent descriptions of pottery fabrics can be made.

The fabrics are sorted into Nile and Marl groups (Bourriau & Nicholson 1992), as follows:

- **Nile fabrics**: These alluvial clays are divided into five groups, Nile A–E according to the relative quantity and size of their most common inclusions, sand and plant-remains, which relate in turn to properties of porosity etc., see above. Nile D and E are further distinguished by conspicuous, abundant inclusions of limestone and rounded sand grains (Bourriau 2007: 142–143), respectively.

Each letter represents a group of fabrics, not a single fabric, and the demarcation between them is sometimes difficult, such as between Nile B2 and Nile C.

- **Marl fabrics**: These fabrics derive from calcareous and calceo-ferruginous clays containing varying amounts of sand but few plant remains (except for Marl E). Fabrics show a wide colour range, sometimes on the same vessel, ranging from pink/yellow through grey to olive green. Marl fabrics are usually harder and denser than Nile alluvial fabrics and show a greater variety in quantity and size of mineral inclusions. They have been divided into Marl A–E.

In what ways does this classification tool aid the archaeologist to understand methods and networks of pottery production? The classification makes it easier to compare pottery from different sites and so generate hypotheses for mechanisms of production and distribution. For example, it can be shown that Marl C vessels found from the Old Kingdom to the beginning of the 18th Dynasty have a special technology of manufacture. Vessels, usually for storage, were thick-walled and hand-made by coiling and turning. The combination of technology, fabric and function means that sherds are easy to identify. They have been found on routes to the Oases, to Sinai, to the Red Sea and to Nubia – that is on expeditions organised by officials from the Middle Kingdom capital, at Lisht, south of Memphis. This suggests that at least some workshops were under the control of royal officials. Similarly, the decline of Marl C production at the beginning of the New Kingdom in the Memphite area is noted, and the introduction at the same time of vessels in fabrics with previous histories in Upper Egypt, such as Marl B, Marl A2 and Marl E, suggests that a re-alignment of pottery production took place alongside political and cultural change (Bourriau 2010: 65, 81–87).

The study of New Kingdom transport amphorae has been deepened by analysis of the fabrics of which they were made (Aston 2004; Hope 1989). Firstly, although amphorae are made in Nile clays, Marl clays and Mixed clays, it can be suggested that some, at least, were made in the same workshops because the same material, Marl D, was used as a covering slip and for the handles for some amphorae of all three clay groups. Secondly, Marl D amphorae themselves often carry royal stamps on the handles and dockets in hieratic, giving date by regnal year, contents, quantity, place of origin of contents and donor. This indicates that, like the Marl C production of the Middle Kingdom, some amphorae production was under royal control. Using this information, including shape and context, the archaeologist has a powerful tool for an integrated study of pottery and commodity distribution in the Nile Valley.

There is more work to be done to establish the production area of fabrics. A large neutron activation analysis (NAA) study of samples classified by the Vienna System (Bourriau et al. 2004; 2006) has shown that Marls A, B, C and D each had a different composition and that samples of Marl A3, B, C and D, regardless of provenance, generally grouped together. This suggests, firstly, that most Marl fabric groups have a distinct region or regions of origin, and secondly, that the Vienna System may provide a link to allow the results of analysis of a sample to be related to all the Marl pottery recorded in the same way. Since only a few hundred sherds, at most, can be analysed at one time, this is extremely useful. For Nile silt fabrics, the situation is different. At the level of the Vienna System, it is difficult to distinguish samples from different sites. This requires the more powerful compositional analysis which chemical or petrological examination provides (Bourriau 1998; Bourriau et al. 2000: 5–12). The Vienna System, however, does allow the experienced ceramicist to distinguish classes of Nile silt fabrics, Nile A, B1–2, C, D and E from the same site. This is because visual criteria of porosity, sorting, quantity of inclusions and colour are used in addition to composition. Analytical work, based upon the Vienna System, has led to a general hypothesis that Nile silt vessels usually have a local source whereas Marl clay vessels are more likely to be distributed beyond their production centres. This needs further testing, but from the ceramicist's point of view, it gives an added significance to the identification of Marl clay fabrics in any assemblage.

The identification of fabrics made of a mixture of Nile and Marl clays was not possible when the Vienna System was developed although it was recognised that they must have existed. A start has been made on this (Bourriau et al. 2000: 19–28) but further work will need to be done as the periods after the New Kingdom are studied more closely.

References

Aston D. (2004) Amphorae in New Kingdom Egypt. *Ägypten und Levante* 14: 175–214.

Bourriau J. (1998) The Role of Chemical Analysis in the Study of Egyptian Pottery. In: Eyre C. (ed.) *Proceedings of the Seventh International Congress of Egyptologists*. Peeters: Leuven. pp. 189–199.

Bourriau J. (2007) The Vienna System in retrospect: How useful is it? In: Hawass Z. & Richards J. (eds) *The Archaeology and Art of Ancient Egypt: Essays in Honor of David O'Connor*, Vol.1 *Annales du Service des Antiquités de l'Egypte* Cahier 36. Cairo: Supreme Council of Antiquities. pp. 137–144.

Bourriau J. (2010) *Kom Rabia. The New Kingdom Pottery. Survey of Memphis IV*. London: Egypt Exploration Society.

Bourriau JD. & Nicholson PT. (1992) Marl Clay Pottery Fabrics of the New Kingdom from Memphis, Saqqara and Amarna. *Journal of Egyptian Archaeology* 78: 29–91.

Bourriau JD., Smith LMV. & Nicholson PT. (2000) *New Kingdom Pottery Fabrics: Nile Clay and Mixed Nile/Marl Clay Fabrics from Memphis and Amarna*. London: Egypt Exploration Society.

Bourriau J., Bellido A., Bryan N. & Robinson V. (2004) Neutron activation analysis of Predynastic Pottery from Minshat Abu Omar, Hemamieh and Armant. In: Hendrickx S., Friedman R., Ciałowicz KM. & Chłodnicki M. (eds) *Egypt at its Origins: Studies in memory of Barbara Adams*. Orientalia Lovaniensia Analecta 138. Peeters: Leuven. pp. 637–663.

Bourriau J., Bellido A., Bryan N. & Robinson V. (2006) Egyptian Pottery Fabrics. A comparison between NAA groupings and the "Vienna System". In: Czerny E., Hein I., Hunger H., Melman D. & Schwab A. (eds) *Timelines. Studies in Honour of Manfred Bietak. Volume III*. Orientalia Lovaniensia Analecta 149. Peeters: Leuven. pp. 261–292.

Hope CA. (1989) Amphorae of the New Kingdom. In: Hope CA. (ed.) *Pottery of the Egyptian New Kingdom: Three Studies*. Victoria College Archaeology Research Unit Occasional Paper 2. Burwood: Victoria College Press. pp. 87–126.

Nordström H-Å. & Bourriau JD. (1993) Ceramic Technology: Clays and Fabrics. Fascicle 2. In: Arnold D. & Bourriau JD. (eds) *An Introduction to Ancient Egyptian Pottery*. Mainz: Von Zabern. pp. 144–190.

Ceramics, especially low value utilitarian wares, can be left undecorated, but many, if not most, ceramics in Egypt had some form of decoration applied after the vessel had been made, but not yet fired (Arnold et al. 1993; Bourriau et al. 2000; Henderson 2000: 122–127). Perhaps the simplest way this can be done is by burnishing, which is where a smooth object, such as perhaps a pebble, is rubbed over the external surface of the vessel to create a smooth finish. This can be identified macroscopically from the surface appearance of the vessel. An alternative (or an addition) to this is where patterns are incised into the surface of the vessel with a tool, which may be either single pointed (such as a wooden needle or twig) or multiple pointed (a comb-like tool). Close examination of the outer surface of these vessels through an optical microscope can reveal the type of tool used. It is also possible to attempt to determine whether multiple vessels were made with one tool, and therefore made in one workshop or even by one potter.

The next possibility is to add material to the surface to create a decoration. There are three ways this might be done: a slip, a paint or a glaze. A slip is a fine-grained clay applied to the outside of a vessel to change its appearance. This might be the same clay as that from which the ceramic is made, or another. Thin section or SEM analysis can tell these apart. Paints were often applied to Egyptian ceramics, particularly red, black and brown. Small samples taken from the paints and analysed by SEM-EDS (Scanning Electron Microscopy – Energy Dispersive Spectroscopy) can identify the pigment used. Red and brown pigments tend to be iron-rich compounds such as ochres. Black colours can also be iron-rich, but an alternative is the use of manganese compounds. Occasionally other colours appear in the palette of Egyptian ceramics. Perhaps the most striking is the use of cobalt blue pigments in the 18th Dynasty, in what is known as palace ware. This cobalt blue pigment is the same as that used in glassmaking and is discussed below. (For more detail on surface treatments on Egyptian ceramics, see Arnold et al. 1993: 85–102.) A glaze is a glassy layer that is fired onto a ceramic (Henderson 2000: 123–127). Glazes were not applied to Egyptian clay ceramics until the late first millennium BC, in stark contrast to Mesopotamia where they were used from around 1500 BC (Paynter & Tite 2001). Glazes were used in Egypt, but only in faience (discussed in the next section).

Although sometimes paints are applied after the ceramic has been fired, usually firing is the last stage. This converts the clay from a more-or-less water soluble, semi-rigid state to a permanently stable, usable ceramic. There are several possible firing methodologies with varying degrees of complexity, from bonfire firing through simple to more complex multi-chambered kilns (Bourriau et al. 2000; Henderson 2000: 135–42; Tite et al. 1982b; Velde & Druc 1999: 169–176). Different types of kilns can produce different types of firings. There are three variables that are key to the understanding of firing processes: temperature, time and the amount of oxygen in the kiln.

Close examination of the microstructure of a ceramic can give evidence as to the temperature to which it has been fired and potentially for how long it has been fired. As firing temperatures get higher and higher, ceramics react more and more during firing. This involves the breakdown of less stable minerals within the ceramic and the conversion of the clay minerals into glass (Henderson 2000: 132–135). If the temperature is high enough for long enough, the entire ceramic, except for a few very stable minerals, can turn into molten glass. This usually results in the collapse of the vessel. This process can be seen microscopically in the SEM or optical microscope, where the increasing areas of glass can be picked out. It can also be seen macroscopically in that the ceramic pores (natural generally elongated spaces in the clay) become more pronounced and more rounded. They eventually join together to form 'bloating pores' which are large bubbles on the surface and in the interior of the ceramic (Shortland 2000b: 31–42; Tite et al. 1982a; 1982b). The starting material in each case was part of the same lump of Nile silt. Shortland (2000b, Figure 4.12) clearly shows the breakdown of the various minerals as the temperature gets higher and the development of more and more glass. Also clear is the development of more and more pores, coalescing to create larger and rounder bloating pores. Using a series of images such as this for Nile silt, it is possible to compare fired ceramics to the image sequence, and thus obtain an estimate of the firing temperature that might have been used for that particular vessel. It also helps to fix a maximum temperature that could have been used as, up at around 1200°C and higher, the vessel becomes unstable and will collapse and flow under its own weight.

Controlling the amount of oxygen in a kiln, or parts of the kiln, can have a dramatic effect on the colour of the vessel (Velde & Druc 1999: 122–126). The most common clay in Egypt, Nile silt, naturally contains relatively high levels of iron (typically 11–12 per cent FeO). In normal conditions, where there is plentiful oxygen in the kiln, this results in a reddish or brownish pot. However, if the oxygen in the kiln is removed, perhaps by sealing the kiln and through the burning of charcoal inside it to convert the remaining oxygen to carbon dioxide, then the ferrous red $Fe3+$ iron cannot obtain enough oxygen to form, and instead the ferric black $Fe2+$ form is created. This means that the surface of the vessels (including at times the interior) turns black. Sometimes part of the vessel was buried in ash so that only part of the vessel turned black. The best examples of this are the famous Predynastic 'black-topped redwares', which were created by just this process. Experimental replication of these vessels, backed up by SEM analysis, has demonstrated that this is the case (Lucas & Harris 1962: 379–381).

4.4.3 Metals

In Egypt, the metals of importance can be divided into 'base metals' and the 'noble metals'. The base metals are more abundant and formed the main constituents of metal tools and implements. Noble metals were almost entirely restricted to jewellery, decoration and ritual activities (Lucas & Harris 1962: 195–269). The most important base metal is copper as it was widely used in both its base form as copper and as the main constituent of copper alloys (Craddock 1985). Copper was used throughout Egyptian Dynastic history, and formed the main component of all metal tools until the first millennium BC. Three elements were commonly alloyed with copper: arsenic, tin and lead (Ogden 2000).

Arsenic minerals are common accessories to copper ores, so 'natural' or accidental traces of arsenic are found to be present in many Egyptian copper objects. However, when arsenic is found to be present within the alloy at the level of a few percent (and this is found commonly in a certain range of objects), this appears to have been deliberate addition (Ogden 2000: 152–153). Arsenic has the dual properties of reducing the viscosity of the copper melt, so that it flows more easily into a mould, and of greatly increasing the hardness of the finished object. The latter is obviously an important characteristic for tools and weapons. However, it has the very great disadvantage of being extremely toxic. The addition of tin to copper, to create tin-bronze, has many of the same effects as arsenic (lower viscosity, greater hardness), but without the problem of slowly poisoning the metalworkers. Levels of tin in excess of 1 per cent are usually taken as evidence of deliberate addition of this metal (Ogden 2000: 153). Levels of tin in bronze objects are usually around 5–10 per cent, as this level creates the optimum working properties without actually using up large amounts of tin, which was presumably rare and expensive (Ogden 2000: 153–155). There are, however, occasional bronze objects which have up to 20 per cent tin. These seem to have been created for their colour, which turns more silvery and begins to look very like electrum (see below). Apart from some very rare finds of tin beads and rings, tin is not found on Egyptian sites as a metal on its own. There also are rare finds of the tin ore cassiterite, but the use of tin seems to have been almost entirely confined to alloying with copper (Lucas & Harris 1962: 253–257). Arsenic compounds were used as the pigment orpiment (Lucas & Harris 1962: 349–350). Arsenical coppers tended to be used in early Egyptian history, with tin-bronzes more common later. During the New Kingdom, both were used contemporaneously to create objects.

The final metal to be routinely alloyed with copper is lead. Once again, this can be a common trace element in copper ores, but in substantial quantities it can affect the working properties of the melt (Cowell 1987). Its primary effect is the lowering of melting and working temperatures of the metals, and a large decrease in its viscosity, making it much easier for the melt to flow into a mould. Copper alloys with 1–2 per cent lead were present in the New Kingdom but were rare during the Middle Kingdom, however high lead copper alloys (around 5 per cent) started to be used in the Late New Kingdom (Ogden 2000: 154–155). Very high lead alloys (20–30 per cent) were common in the Late Period and later, and were frequently used in the casting of small statues. Other elements are occasionally found, apparently deliberately added to copper alloys, including antimony and zinc. The presence of significant quantities of these two elements is very rare in Egypt and their deliberate use in copper alloys is either questionable or possibly occurs only very late. Lead objects themselves are known, especially from the New Kingdom onwards, when lead was sometimes used for fishing net weights (Lucas & Harris 1962: 243–244).

For most of the Dynastic period, copper was the most important metal in terms of both volume and spread of use. Iron, whilst the most important metal very late in the Dynastic period and into the Roman period, was virtually unheard of earlier (Lucas & Harris 1962: 235–243; Ogden 2000: 166–168). There are a few, rare exceptions, especially iron objects from incontrovertible contexts. There are indications in the Amarna letters that iron weapons were given as gifts between kings, and an iron dagger was found by Carter in the tomb of Tutankhamen (Lucas & Harris 1962: 239). It has been proposed that several earlier finds of iron were telluric in origin (from the exploitation of iron-rich meteorites). The presence of high nickel levels in iron is also indicative of telluric iron, but might simply be present in certain iron ores, and hence the use of telluric iron in Egypt is still a subject for debate (Craddock 1995; Craddock & Lang 2003; Lucas & Harris 1962: 237–238).

The noble metals include gold, silver and their alloy electrum (Lucas & Harris 1962: 224–234; Ogden 2000: 161–166). The normal definition of electrum is a gold–silver alloy with less than 75 per cent gold (Gale & Stos-Gale 1981). 70–85 per cent gold was the norm for the gold composition of ancient Egyptian gold objects of the Middle and New Kingdoms. High levels of silver were very common, but the presence of copper, iron, tin and platinum group metals was also common. Indeed, the gold content of gold–silver objects seems to have varied down to only a few percent gold (often termed aurian silver). High purity gold objects (greater than 85 per cent gold) were really only common in the Late Period and later, when the refining of gold to improve its purity had been perfected (Gale & Stos-Gale 1981). The silver for many silver objects is thought to come from the same mines as the gold, as they also contained a few percent gold. This fits into a variable source, stretching from gold with a few percent silver to silver with a few percent gold. However, some silver seems to have had very low percentages of gold, and, since the Egyptians found it impossible to remove the gold from the silver, this must mean it comes from another source (Gale & Stos-Gale 1981). Lead, copper, tin and a range of other trace elements are commonly present in this ancient silver and may hint at its source, but problems of mixing and re-use are always going to make this very difficult to rely on.

The questions asked of scientists concerning metal objects are slightly subtler than those asked when dealing with stone or ceramics. The classification of an object into mostly gold, mostly silver, copper alloy, iron or lead, etc. is fairly simple and can be carried out usually by visual inspection. However, what is obvious from the brief summary discussion above is that the compositions within these broad groupings offer valuable information on the technology and date for an object. Knowing the precise composition therefore becomes very important. If sampling is allowed, it can be done by drilling the object and extracting a small amount of metal powder, which can then be analysed by ICP-AES, ICP-AAS or ICP-MS amongst several other techniques. There has been, however, an increasing move away from destructive techniques towards analysis without taking a sample. For metal objects, this has often been carried out by airpath XRF. In addition to not requiring a sample to be taken, unlike when XRF is used on stone and semi-precious stone, a further advantage is that almost all the relevant metallic elements have relatively high atomic masses, and give off relatively energetic characteristic X-rays. This means that they can be seen and usually measured in an airpath XRF (in a way the light elements that make up the bulk of stone compositions cannot). Thus airpath XRF can indicate whether an object is copper, arsenical copper or tin-bronze and, if properly calibrated, has the potential to give percentage concentrations of all the elements discussed above. In the past, benchtop-mounted XRF were common in large museums and much research on metal objects was undertaken using this technique. More recently, the invention and fairly widespread introduction to museums and the archaeological world of handheld XRF has meant that it is now possible to analyse more objects.

Several words of warning must be said about this type of work. The first is that while it is relatively easy to analyse an object and detect which elements are present within it, it is actually extremely difficult to quantify such analyses. This is because there is a very wide range of factors that all play a part, and there is frequently a lack of good, relevant standards with which to compare and calibrate the XRF. However, even if the calibration and standards were perfect, other issues are even more significant. The most important of these is the state and preservation of the surface. All metals, except gold, are very vulnerable to surface corrosion. XRF is a surface technique. Unlike drilling, it analyses only the first few microns of the surface of an object (Meeks 1986; Mohamed & Darweesh 2012). If the surface is corroded, then XRF will not define the composition of the bulk of the object. If corrosion is minimal and the item is curated in a museum, one immediately suspects that it might have been cleaned early in its museum history. This again can affect the surface composition. Even if corrosion really is minimal, it is known that some processes concentrate certain elements into the surface layers. Tin enrichment of surface layers in bronzes is one of the most important (Meeks 1986). In these situations, once again, analysis of the surface will not give a result that reflects the composition of the object as a whole. Great care has to be taken therefore when using XRF analysis and comparing XRF data with that derived using other techniques.

4.4.3.1 Metal Provenancing

This section will concentrate on the trafficking and exchange of metals, which were frequently traded in the form of an ingot. Two fundamentally different material types, metals and glass, were transported in the form of ingots into and out of Egypt. Tracing the source of ingots, or fragments of them, is one of the most difficult of the import questions for the archaeologist or art historian to answer, since metal or glass from Egypt looks identical to metal or glass deriving from anywhere else. Scientific analysis, therefore, is usually the only hope for determining from where the material might have come.

While some stone and clays are local and others come from more exotic sources, metals are almost always distantly sourced. The reason for this is not always the rarity of the element in nature (aluminium and iron, for example, are the third and fourth most abundant elements in the Earth's crust after silicon and oxygen); it is the rarity of the element both in an exploitable form and in sufficient concentrations. A mineral containing a metallic element that can be exploited is known as an ore, and ores tend to be concentrated in certain types of geological terrain.

Tracing the source of the ore that was used to create base metals is fraught with difficulties for copper and (especially so) for tin. The reason for this is that the chemical process that converts the mineral ore to the metal fundamentally changes almost all the properties of the material. The process of turning an ore into a metal involves a long and complex procedure of crushing, sorting, heating (often several times, and frequently to temperatures over 1000°C) and controlling the amount of oxygen in the furnace (Henderson 2000: 220–231). These processes result in the mineralogy of the original ore being almost completely removed and the chemistry being radically altered in order to concentrate the metallic element. The challenge to the analyst is, therefore, to find some property of the metal that reflects the ore in some way, and thus allows the source to be narrowed down. In an ideal case, the property might remain constant through the entire process, but properties that change in a predictable and repeatable manner are also usable.

Early work in this field concentrated on the use of trace element analysis, using such techniques as wet chemical analysis, NAA and ICP-AES. The theory was that almost all ores have traces of other elements with the main metal desired, such as, for example arsenic, lead, zinc, antimony, or iron in copper. These additional metallic elements (or their combinations) might be unusual enough to distinguish a particular ore. Alternatively, the particular ratios between the additional elements and the main metal might be sufficiently characteristic as to be identifiable. If the rare elements pass into the final metal and the ratios remain similar, this might give a traceable signature to the ore source. This certainly seems to work in some cases. A good example of this is the presence of trace levels of arsenic in copper objects of the early Dynasties, indicating the use of copper from an arsenical copper ore being used prior to the New Kingdom (Cowell 1987; Kaczmarczyk & Hedges 1983: 73). However, there are several problems. Firstly, when copper starts to be deliberately alloyed with arsenic, tin, and later with lead and zinc, it becomes very difficult to say whether the arsenic (for example) is a trace in the ore or a deliberate addition from another ore altogether. Additionally, when alloying starts, the alloy metal(s) introduce their own range of trace elements, and so discovering which trace element is incorporated with which metal becomes much more complicated. Even more problematic is the suspicion that, in many cases, the trace element content of the metal is significantly changed by the production process, so the technique may produce misleading results.

A huge jump forward in the sourcing of metals came when, rather than using the ratios of trace elements associated with the metal (which are known in many cases to change), isotopes were used instead. The most successful work to date has used lead isotope analysis (LIA). Lead is a common, if not particularly abundant, element in copper ores. The theory that LIA relies upon is that, while the absolute amount of lead may vary between the ore body and the metal produced from it, the ratios of the various lead isotopes within that lead do not change (Gale & Stos-Gale 1981; 1993). There is good reason to believe that in most, if not all, cases this is true. LIA therefore provides a way of characterising the ore source of copper, and has been applied to Egyptian copper, silver and lead objects (Brill et al. 1974; Gale & Stos-Gale 1981; Shortland et al. 2000). However, this promising technique has suffered from several significant problems. One of the most serious is the realisation that a number of significant ore sources have very similar lead isotopic characteristics, which make them rather difficult to tell apart (this is the same issue as stable isotope analysis of marble, discussed above). Another problem seems to be that some copper objects apparently do not match any known ore source, suggesting perhaps that the ore was completely exploited in antiquity or that mixing from different sources might have occurred (Gale 2009; Pollard 2009). However, despite these and other methodological difficulties, the technique has shed light on the sources of Egyptian copper (see Box 4.5), showing that much of it came from such distant sources as Cyprus and Lavrion in Greece (Stos 2009).

The initial success of LIA led to hopes that other isotopes, particularly those of copper and tin, might also be used to directly source these metals rather than relying on the passing on of trace amounts of lead from the ore. However, very significant technical and chemical problems have meant that copper and tin isotopes have never been successfully applied on any scale (Gale 1997; Gale et al. 1999; Stos 2009). Both of these problems are now being re-examined and it is possible they may soon produce some exciting new results (Haustein et al. 2010).

Attempting to determine the source of gold and silver throws up another problem, touched on above, that may be also a problem for base metals – the problem of mixing. Gold and silver are rare and valuable; they therefore tend to be reused over and over again rather than thrown away and discarded. This means that if multiple ore bodies are known, there is the potential that the signature of these ores, whether it be trace element or isotopic, will increasingly become mixed in the finished object as they go through multiple cycles of use and re-use. This is a problem that in most cases has never been solved. The provenancing of Egyptian gold by analytical

Box 4.5   Provenancing Oxhide Ingots
Andrew Shortland

Metals and glass seem to have been transported in the form of ingots – usually substantial quantities of the raw material were shaped to provide a robust way of transporting and stacking the material, sized for carrying at an appropriate weight. One of the most iconic ingot types found in the Late Bronze Age is the copper oxhide ingot. As this modern name suggests, the ingot is shaped like the hide of an ox, flat with four lobes, one on each corner of a quadrilateral, representing or at least resembling the legs of the ox. Other names for these ingots include talenta, kissenbarren and double-axe ingots. They typically weigh 20–30 kg. They are restricted to the Late Bronze Age, and found throughout the lands that border the Mediterranean and the islands within it, from Cyprus in the east to Sardinia in the west. The earliest oxhide ingots are found on Crete and date to Late Minoan I (about 1550–1450 BC, the early 18th Dynasty). Most of the finds, however, are somewhat later and are concentrated (although not exclusively) on three islands in the Mediterranean, Cyprus, Sicily and Sardinia, and two famous wrecks, Cape Gelidonya and Ulu Burun (see Box 4.6). Both of the wrecks contained significant quantities of ingots, especially of copper oxhide ingots.

Some debate has occurred as to whether oxhide ingots were true ingots, designed as a method of transporting raw copper for melting down to make tools and other objects, or whether they might have had other uses, as a form of currency or in ritual exchange. Oxhide ingots are relatively frequently shown in Egyptian tomb scenes, such as the Tomb of Rekhmire (TT100), where they are depicted being carried by foreigners from the land of Retjenu, Syria, making offerings to Tuthmosis III. Other scenes show them stacked in temple storage rooms. The provenance of such ingots is therefore of great interest and considerable analysis has been carried out.

Elemental analysis has shown that almost all of the oxhide ingots are pure copper, with less than 1 per cent composition of other trace elements. While these trace elements might be used as an indicator of provenance in particular ideal circumstances, this can often be difficult as the ratio of one to the other has the strong possibility of changing, or "fractionating", during smelting and/or further heating. This has meant that the main work on the copper oxhide ingots has been isotopic, specifically lead isotopes, where fractionation is usually not thought to be a significant issue. Samples have been taken from many ingots found both on the shipwrecks and throughout the

Mediterranean (especially Cyprus and Sardinia). These were then compared with the existing databases for lead isotopic characteristics of copper ores from throughout the Mediterranean and beyond. This comparison is done quite carefully, as for certain lead isotope ratios there are substantial overlaps between different ore bodies. This means that, for some lead isotope values, there are several ore source possibilities. In addition, another potentially significant problem is possible mixing. Copper alloys were a valuable commodity in the ancient world and, as such, would not be thrown away when they ceased to be useful, but instead would be remelted. The remelting of metals from different ore bodies results in the mixing of their lead isotope ratios, thereby making tracing their ore sources very difficult indeed. If, however, the oxhide ingots were made from raw copper straight from the smelting at the mine, and either not intended for functional "use" or represent an early stage in use, then such mixing may not be quite so much of a problem as might be supposed.

Numerous lead isotope analyses on oxhide ingots have been undertaken, with, for example, over 300 from the various types of ingots on the shipwreck of Ulu Burun alone. The Ulu Burun ingots all appear to be isotopically uniform and the different shapes of the ingots are not reflected in different isotopic signatures (Stos 2009). Instead, all ingots seem to derive from one geologically uniform deposit, probably on the island of Cyprus. Research on oxhide ingots from sites other than the Ulu Burun wreck, such as those found in Cyprus, Greece and the islands of the Western Mediterranean, has also included lead isotopic analysis (Pollard 2009; Stos 2009; Stos-Gale et al. 1997). No matter from where the oxhide ingots are found archaeologically, the great majority of oxhide ingots, if not all, have lead isotopic signatures matching those found in copper mines on the island of Cyprus. Indeed, it has been suggested that many of them might come from one particular mine, Apliki, from the Solea region close to the northwest coast of Cyprus. It is clear that without analysis, and especially the widespread use of lead isotopic analyses, this would never have been discovered and the importance of this area to broader regional trade might not have been fully recognised.

References

Pollard AM. (2009) What a long strange trip it's has been: lead isotopes and archaeology. In: Shortland AJ., Freestone I. & Rehren T. (eds) *From mine to microscope: advances in the study of ancient technology*. Oxford: Oxbow Books. pp. 181–189.

Stos ZA. (2009). Across wine dark seas... sailor tinkers and royal cargoes in the Late Bronze Age Eastern Mediterranean. In: Shortland AJ., Freestone I. & Rehren T. (eds) *From mine to microscope: advances in the study of ancient technology*. Oxford: Oxbow Books. pp. 163–180.

Stos-Gale ZA., Maliotis G., Gale NH. & Annetts N. (1997) Lead isotope characteristics of the Cyprus copper ore deposits applied to provenance studies of copper oxide ingots. *Archaeometry* 39: 83–123.

means has therefore been fraught with difficulties. Egypt, however, has large gold reserves in the Eastern Desert which were exploited throughout antiquity (Said 1990). It is highly likely that these reserves provided the vast majority of the gold used. Throughout the Dynastic period, the relative purity of Egyptian gold objects improved, suggesting that the gold was being refined before being used in objects (Ogden 2000: 161–162). Gold can be found in a range of geological environments, of which two of the most important are vein gold and placer gold. Vein gold is the original geological deposition of gold in quartzites and other rock types. When this weathers, the gold accumulates as nuggets in stream beads, known as placer gold. The nuggets also include elements from the weathering of other minerals, especially platinum group elements (PGEs). The finding of PGEs in a gold usually means that it must derive from a placer deposit. It is usually thought, however, that the signatures are not specific enough to locate the exact deposit (Meeks & Tite 1980). Silver, which is rarer than gold in Egyptian objects of most periods, is usually found associated with lead sources, such as the galenas of the Eastern Desert (Gale & Stos-Gale 1981; Lucas & Harris 1962: 245–253). Aurian silver (silver with a significant gold content), which makes up a significant proportion of Egyptian silver objects, seems to be derived from gold mines. However, LIA of purer silver objects suggests that the silver sources lay outside the borders of Egypt (perhaps in Greece; Gale & Stos-Gale 1981).

4.4.3.2 *How Was It Made? Workshops and Workshop Debris*

The process by which an ore is converted into metal has been briefly mentioned above. The investigation of the processes involved in fabrication is complex and technical. Three major lines of evidence are used: analysis of workshop and mine debris, analysis of the finished object and analysis of experimentally reproduced materials. This applies to all pyrotechnologies. It is especially important for metals and glass because they retain relatively little evidence in their structure of the process by which they were made. This is particularly true when comparing them with ceramics. It is therefore worth taking a step back and considering workshops in general, and how they can be identified and distinguished. This is often not such an easy task in the field.

Examination of production sites and their debris can give valuable insights not only into production processes, but also into the conditions the workers operated under and their general day-to-day life. One problem has been the lack of workshop evidence caused by a number of factors. Firstly, early excavators generally were not very interested in the industrial areas of the cities they were excavating, preferring instead the palatial and temple contexts. Perhaps the first person to take notice of proper workshops as interesting and valuable site types for study was W.M.F. Petrie, who recorded workshop areas in a number of the sites that he excavated (Petrie 1894; 1909a; 1909b; 1911). Even into the 1950s and later, this was not common practice. Workshop debris tends to be scattered, equivocal, not aesthetically appealing and considered the realm of the specialist. As a result, it was not accorded the priority it perhaps deserved. Over the last couple of decades this focus has changed, and there are now excavations that specifically target workshops and industrial areas.

'What might have been produced here?' This is usually one of the first questions asked by archaeologists when a workshop is uncovered. In a well-preserved site, there may be kilns or furnaces, tools (broken crucibles, moulds, stands or miscellaneous copper objects) and debris (ash, broken or fused clay, wasters, etc.). The presence of high-temperature kilns or furnaces can usually be determined fairly easily. Work on examining kiln walls by SEM and thin section has shown that when the kiln is fired, the heat penetrates into the wall causing microstructural changes resulting in vitrification of the kiln itself (Maniatis & Tite 1979; Tite & Maniatis 1975). The higher the temperature and the longer the heating, the more the changes are evident and the further into the wall those changes penetrate. This type of study can distinguish pottery kilns operating at below 1000°C from glass, glaze and metal kilns operating at this temperature and well above, and shows that a site might have been involved in one of the pyrotechnologies. By far the most common of these is pottery production. This can go on at a domestic scale, within a house or garden, or at semi-industrial scales in temples and palaces. Identifying any industrial debris tends to be a matter for the specialist, although pottery kilns can usually be identified by their relatively low firing temperatures. The low firing temperature means that the wall of the kiln is not as highly vitrified as it tends to be after operating at higher temperatures. Furnaces used in metal production can also be determined fairly easily. Metal production tends to be a rather messy process, and small quantities of the metals often become assimilated into the debris created, which might include crucibles, furnace walls or the high-temperature fused clay that results from the firings of the furnace. These small spherical metal inclusions are called 'prills' and their presence in number is a clear indication of metalworking. They can sometimes be seen in the field with the naked eye, but more often are detected in subsequent analysis of the debris by SEM or thin section. Where copper working is evident (by far the most common type of metalworking in Egypt), these prills are often green in colour due to weathering of the copper metal.

Fundamental to the study of the technological processes in workshops is the analysis of their debris. Although it had a slow and sporadic start at the end of the nineteenth and beginning of the twentieth century, there is now a thriving field of analysts studying the debris of pyrotechnologies. Workshop debris may be classed into two types. The first is the evidential remains of successful operation of the furnaces and facilities, including the furnaces themselves, tools for both operating the furnaces and manipulating what is being fired, ash from fire boxes and slag. The latter is the molten remains of the rocky part of the ore that has neither volatilised nor ended up in the metal itself (McDonnell 2005). The second is those remains of unsuccessful attempts that are preserved. These would include collapsed furnaces, batches of slightly the wrong composition or temperature which have failed, and broken and discarded remains called 'wasters'. Very little of this material is of huge aesthetic value, so, even if it does end up in a museum, it is unusual for it to be displayed. Given this, and its usually frequent abundance on a site and its normally fragmentary form, sampling this material usually presents fewer problems than finished objects. Most commonly, samples are taken for SEM. This is because both the composition of the slag or furnace wall and the phases within it, plus the microstructure and the inter-relationship of the phases within the furnace wall, provide valuable information about the processes used. The samples are mounted in resin blocks and polished to give a flat surface and are then inserted into the SEM. A range of different structures can often be seen, indicating from which pyrotechnological process the slag derives. Different phases in the slag will form and be stable at different temperatures, thereby giving information on the temperature reached in the process, potentially also the time for which it was held at that temperature and indications of the oxidation states of the furnace.

For metallurgical debris, a second technique is often used and can be used complementarily to the SEM. The metallurgical microscope has many similarities to the polarising light microscopes used in ceramic and stone petrography. Again it consists of a precision microscope with a range of magnifications and filters. The metallurgical microscope, however, is used primarily in reflection. It is illuminated from the top and the light reflects off the specimen. This contrasts with light shone through the specimen (transmission), which is the norm for microscopes used for petrography. Because many of the most interesting phases in metals and ores are opaque, light will not pass through them and hence reflectance must be employed. Similar information can be derived as by SEM, although ideally they are used in tandem.

'How is the metal made into that object?' Analysis of finished metal objects can answer questions as to the processes involved, such as metalworking, that lead to the conversion of a metal ingot into the finished object. A wide range of non-destructive techniques can be used, but if sampling is permitted, then the potential information gain can be far greater. How the object was made is very important as varying techniques can be used at different stages in the manufacturing process. Casting was common with many metals, and various techniques can be identified, such as one-part or multipart moulds, lost wax, hollow or solid casts in one piece or multiple pieces that were later joined. Hammering, annealing and cold working were also used to modify the shape or working properties of a metal. All the above processes affect the internal structure of the metal and hence are more-or-less detectable if a sample is taken. Objects in some periods were often made as multiple separate pieces, which then had to be joined together (Ogden 2000: 155–160). Solder was frequently used to do this, and can be analysed by the techniques described above. Different solders (and their compositions) tend to be indicative of different temporal periods. Other joining techniques include rivets, pegs and dowels. Surface finishes may also include burnishing and polishing, the application of gold and/or silver leaf, or decorative inlays.

Much discussion and analysis has been devoted to trying to determine what the original colour of bronze ritual objects might have been. Colour played such an important part in both Egyptian life and ritual that objects might be interpreted completely differently if they were of different colours. Over time, copper and copper alloy objects weather to greens and blacks in certain conditions. It is, however, widely thought that the surface of some copper alloy objects was deliberately altered, during manufacture, to turn it from a metallic copper colour to a black colour (Craddock 1998; Mohamed & Darweesh 2012 and their references). This effect is often seen in objects inlaid with gold and silver, where the black provided a contrast, thereby allowing the inlay to be more clearly seen. This effect seems to have required the deliberate addition of a silver-rich gold to the alloy. In addition, the surface of the resulting cast needed treatment with an acidic solution to create a thin, but resistant, black copper oxide layer on the surface. These objects are sometimes termed 'black bronzes' and are probably linked to the Egyptian phrase *hsmn-km* ('black copper'). This phrase is found in offering lists from the 18th Dynasty onwards (Ogden 2000: 160).

The final source of evidence for process is experimental archaeology linked with analysis. Analysis of an object may reveal a possible way that it might have been made, such as the brief description of the manufacture of a black bronze above. Experimental work tends to fall into a spectrum of different approaches, the two end members of which might be characterised as 'laboratory' and 'field'. The laboratory approach tends to use relatively pure reagents, a modern high temperature, very controllable kiln, and employs multiple runs to change the various factors in the process. The advantage is that it is relatively easy to undertake, relatively inexpensive and can give a good overall idea of the science behind (probably) a small part of the process. An example might be the addition of various amounts and compositions of gold to copper alloy in the making of a black bronze. This tends to be the approach that a scientist wishes to use, as they are generally well trained in limiting variables and tightly controlling experiments in order that concrete conclusions can be drawn about the physics and chemistry operating. The disadvantage is that this approach does not replicate very well the complex, dirty and impure world of an ancient workshop. The opposite extreme, here labelled 'field' experiments, addresses this issue. Following this methodology, the ancient processes are replicated or duplicated as closely as possible. Actual raw materials are used that mimic the ancient. A furnace is built mirroring excavated examples, and fuelled with whatever material or fuel is thought to have been used. Every attempt is made to make the whole operation as authentic as possible. The advantages of this approach are that there is more direct relevance to the ancient technology. Additionally, the whole manufacturing process tends to be replicated (rather than just some part). There are, however, several disadvantages. Firstly, the techniques for the whole operation of an ancient kiln are hardly ever known; many assumptions have to be made in the building and running of the experiment. Perhaps more importantly, the raw materials can almost never be guaranteed to be the same as the ancient ones; they might be very similar, but small variations can be very important in unpredictable ways. It is also much more difficult to get a good idea of the underlying physical processes of the technology as there are so many variables involved. This is important, because it means that it is difficult to extrapolate from the specifics of one experiment to the generality of many experiments and ancient firings. However, the field experiments do make very good teaching tools, and increasingly good television, so there are other positive and beneficial factors affecting their use.

The two extremes, 'laboratory' and 'field' experiments, form end members of the experimental archaeology continuum. Many experimental projects fall somewhere in between, perhaps using modern raw materials in an 'authentic' kiln, for example. This is often because the ancient raw materials are unknown or unavailable, or the original furnaces were in a country (and therefore a climate or locale) that is now difficult to visit or work in. The products of the experimental work, regardless of the experimental conditions under which they were created, are analysed in the same way as the ancient samples. The results are then compared and, if discrepancies in composition or microstructure are found, an attempt may then be made to account for these, such as considering the way the chemical and physical processes might have created the effects on the raw materials used. The initial model for the technique, as set up in the initial experiment, can then be subtly altered, and, ideally, the experiment re-run under these new conditions. This will hopefully create a result closer to the ancient. Various factors may be changed in this way (such as the proportions of raw materials, temperature, oxidation state, heating time, etc.) and their effects observed. This use of multiple runs with small changes is much easier to do in a modern laboratory furnace than using replicas of ancient models where the precise control of temperature and heating time is much more difficult. As a result, the very precise examination of the processes of production almost always is carried out in a laboratory. For further information on the design, running and criticism of experimental techniques, see Ferguson (2010) and Millson (2010).

While this discussion and evaluation of experimental replication falls here in the metals section, the very same techniques have been used to investigate the production processes of glass, glaze and (perhaps to a slightly lesser extent) ceramics.

4.4.4 Glass

Glass (apart from the natural volcanic glass, obsidian, discussed above) is a man-made material, made from the firing of pebbles or sand with a flux, such as ash or mineral soda. As discussed above, glass is an end product that looks identical to the eye even though it may have been produced in a whole range of different places. The material itself has virtually no characteristics visible either by the naked eye or under an optical microscope which might give a hint as to its provenance. In this respect, it is very similar to metal and unlike ceramics.

The questions asked of the scientist about glass are essentially the same as those asked about metals: the origin and the production technology. To understand fully the techniques for provenancing glass, it is very important that the raw materials and the method of fabrication of the glass are understood. As a result, the production of glass is considered here first, before discussing questions of provenance.

The first glass in the archaeological record appeared around the 16th century BC and persisted in Egypt until the end of the New Kingdom (Lilyquist & Brill 1993; Nicholson 1993; Peltenburg 1987). There then appears to have been a hiatus in both glass production and use until the middle of the first millennium BC. The use of glass really expanded around the time of Christ, when the invention of glassblowing meant that the production of vessels became much cheaper and therefore more widespread (Degryse & Shortland 2009; Freestone et al. 2000; Jackson 2005a). There are therefore two temporal phases for glass production: the Late Bronze Age and the Iron Age to Roman period. Similar techniques are used for each, but they tell rather different stories.

Most glass has very little internal structure and, even for those that do, that structure is of little help in determining provenance. The analysis of glass, therefore, relies almost entirely on compositional and, increasingly, on isotopic techniques. All the glasses found in Egypt are soda-lime-silicate glasses (Caley 1962; Turner 1956a; 1956b). The main component is silica (as SiO_2) which forms the network or structure of the glass. This network is modified by the presence of soda and lime. The former lowers the melting and working point of the glass, and the latter (as calcium) adds stability. Analysis of these and other major elements in glass is normally carried out by either SEM with an attached EDS, or by microprobe. This is discussed more in the production section below.

When glass was first regularly produced in the early New Kingdom, it usually appeared in bright opaque colours; blues were almost always dominant, but yellow, white, red, purple, green and black were all present (Lilyquist & Brill 1993). It was made into jewellery, amulets, inlays and glass vessels (Nolte 1968; Shortland 2000b). The scientific study of glassmaking started with the analysis of the glass objects themselves, with the aim of identifying their raw materials and production process. Very precise analyses of the glass, which is usually fairly homogeneous, were required. The dominant technique for this work has been SEM-EDS (Lilyquist & Brill 1993). As mentioned earlier, SEM can give both compositional information and images of structure. This is very useful in glasses as it was quickly realised that whilst translucent glasses were fairly free of structures, the opacity in the yellow and white (and their related green and turquoise) glasses was caused by the presence of deliberately added, or created, particles (Shortland 2002; Weyl 1951).

While SEM-EDS was extensively used in the past – and is still used to great effect when very carefully calibrated – increasingly more precise and accurate values have been desired for glass compositions. To accomplish this, SEM-EDS has been to a certain extent superseded by the microprobe, i.e. SEM with wavelength-dispersive spectroscopy (SEM-WDS), which has lower detection limits and is relatively free of the overlapping peaks problems that can plague EDS (Henderson 1988). Within an ancient glass, typically 20 or more elements are analysed by microprobe, thereby providing a very good idea of the composition (Shortland & Eremin 2006). Interpretation of the results has proved more difficult than might be expected, although certain conclusions are clear. There are two major compositional groups of glass found in Egypt. In the New Kingdom, the glass was a soda-lime-silicate with relatively high levels of magnesium and potassium (Lilyquist & Brill 1993). These two elements are thought to be linked to the sodium in the glass, and represent the fluxing agent used to lower the melting temperature of the glass batch when the glass was made. A lot of research has been undertaken to identify this fluxing agent, and it is generally agreed that it must be the ash of a plant that is itself rich in sodium and these other elements (Barkoudah & Henderson 2006; Tite et al. 2006). This means that it must be a halophytic plant, living either on the edge of the desert or in brackish conditions close to the sea. This is thus known as plant ash glass, made from combining quartzite pebbles with soda-rich ashes. Later in

Egyptian history, primarily from the Late period and through into Roman times, with occasional cases from as early as the Third Intermediate period, analysis shows that the glass being produced had a different composition. It had much lower levels of both magnesium and potassium. These levels are thought to be too low to be plant ash glass, and thus another relatively pure source of sodium was used as a fluxing agent (Turner 1956a; 1956b). This was probably natron, carbonates and other compounds of sodium found in evaporitic lakes such as the Wadi Natrun. There was thus a switch in the composition of glass at the end of the New Kingdom or later to a new form made from sand and natron (Shortland et al. 2005; Turner 1956a).

'How is the glass coloured?' Egyptian glasses are almost always coloured, and the colouring elements and raw materials have also been identified (Weyl 1951). The most common colours (Shortland 2000b) are dark blue (coloured with cobalt) and turquoise blue (with copper from copper metal or bronze scale). Yellow and white were both created from antimony compounds; lead antimonate for yellow and calcium antimonate for white (Shortland 2002). In addition to this analysis of finished objects, analysis has been carried out on the debris from glass and faience workshops, especially at Amarna and Qantir (Nicholson 1995; 2007; Pusch & Rehren 2007; Rehren & Pusch 2005). Mostly using SEM-EDS, research has identified the intermediate stages in the production of glass, as these may show how the raw materials were combined (Rehren & Pusch 1997; 2005; Shortland 2008; Shortland & Tite 2000). Were they subjected to one round of heating or two? Or many? Was colourless glass transported around between workshops and coloured locally, or was the form of exchange glass that had already been coloured? Examination of the microstructure of crucibles and melting vessels, backed up by experimental laboratory replication, has shown that the temperatures required to create the glass are high. These are almost certainly above 1000°C and perhaps as high as 1150°C if a shorter firing time is desired (Shortland 2000b). Field replication of furnaces (Nicholson & Jackson 1998) in Egypt has shown that a relatively simple furnace, fuelled with locally available materials, was capable of reaching such temperatures.

4.4.4.1 Provenancing of Glass

'Where was the glass made? Has it come a long way?' One of the most interesting questions concerns the provenancing of glass. When glass first appears in the archaeological record, it appears to do so almost simultaneously in Egypt, the Near East and perhaps Greece (Peltenburg 1987). The obvious question to ask is whether all this glass represented the production of one workshop, several workshops in one state, or multiple workshops throughout the East Mediterranean and the Levant. Stylistic approaches to examining the objects were inconclusive (Nolte 1968; Oppenheim et al. 1970). The spread of glass over such a wide area certainly appears as if it should be the result of multiple production sites. However, analysis of the glass by SEM-EDS and microprobe has indicated that it was compositionally rather homogeneous (Brill 1999; Henderson 1998; Lilyquist & Brill 1993). There were some possible variations, but they were by no means certain, and difficult to attribute to one particular area. An exception was the glass coloured with cobalt. This has proved to have one particular compositional signature: when cobalt was present, the glass always had lower levels of potassium and greater quantities of aluminium, manganese, nickel and zinc relative to other glasses (Kaczmarczyk & Hedges 1983). This gave a signature for the cobalt ore, which has been traced to alums in the Western Oases of Egypt (Kaczmarczyk 1986). The fact that cobalt blue glass was made from a cobalt colorant apparently from Egypt, while it is also one of the most common colours found in Egypt (but very rare in the Near East), suggests that this colour glass was the product of Egyptian workshops. Work on lead antimonate yellow glasses in Egypt, using lead isotope analysis, has suggested that glasses of this colour from the Amarna period were made using a source of lead that probably originated from the galena mines of the Eastern Desert of Egypt. They also were probably locally made, although the very first yellow glasses in Egypt may have come from the Near East (Lilyquist & Brill 1993; Shortland et al. 2000). All other colours, however, were much more difficult to provenance, leading to the conventional belief that in the Late Bronze Age "glass from Egypt and Mesopotamia cannot be unequivocally distinguished on the basis of their chemistry" (Jackson 2005b: 1750–1752).

It took the use of a relatively new technique in the analysis of Egyptian glass to clarify the situation. Just as the microprobe has lower detection limits than the SEM-EDS, ICP-MS offered even lower detection limits (Gratuze 1999). This means that, instead of analysing and quantifying some 20 or so elements, as is carried out by microprobe, ICP-MS can do the same for two or three times that number, with detection limits many orders of magnitude better. If the ICP-MS has a laser ablation system attached (LA-ICP-MS), then the very same resin-mounted samples that were previously analysed by SEM and microprobe can also be analysed by LA-ICP-MS. The result is the compositional equivalent of increasing the magnification on an optical microscope – suddenly much more detail can be seen. Most significantly, when a whole range of elements are plotted, but perhaps especially titanium, chromium, lanthanum and zirconium (only one of which could be routinely quantified in these glasses by microprobe), there is a clear distinction between glass found in the Near East and that found in Egypt in the Late Bronze Age. This shows that there were at least two production sites, one in each region (Shortland et al. 2007). When contemporary Greek glass was analysed, the plant ash glass found was compositionally very similar to either the Near Eastern or the Egyptian glass (Walton et al. 2009), suggesting that significant quantities of Egyptian glass were exported into the Aegean world (see Box 4.6).

Box 4.6 The Ulu Burun Wreck – Exchange in Action
Andrew Shortland

At the extreme south western tip of Turkey is a sheer promontory known as Ulu Burun. In 1982, Turkish sponge divers working in the sea off this point found a mass of ancient ingots. Lying on a slope dipping down over 50 metres below the surface were the remains of a Late Bronze Age shipwreck. This wreck was excavated systematically between 1985 and 1994 by a specialist team from the Institute of Nautical Archaeology at Texas A&M University, led by George Bass and Cemal Pulak. The finds were conserved and displayed in the specially created Bodrum Museum of Underwater Archaeology.

The excavations showed that the ship was a 15-metre-long vessel made of cedar wood which could have been propelled by both sails and oars. A full-size experimental replica has been built and is now displayed in the Bodrum Museum. Radiocarbon dates on short-lived material, such as olive stones and small branches carried for firewood, have given a date of 1342–1314 BC at 95 per cent probability, dating the wreck to the last years of the 18th Dynasty, coincident with the reigns of Akhenaten, and his immediate successors Smenkare, Tutankhamen, Aye and Horemheb (Pulak 1995; 2005: 90).

The vessel could have carried up to around 20 tons of cargo, and much of this cargo was recovered from the seabed. So undisturbed was it that it was still possible to tell how the cargo was stacked and whereabouts within the ship each load was carried. The whole wreck was initially announced in some early general publications (Bass 1987) but details have now been superbly published in a volume that accompanied a travelling exhibition of some of the most important finds that took place at the Deutsches Bergbau-Museum in Bochum (Yalcin et al. 2005). The ship contained a wealth of objects from around the Mediterranean, with over 18,000 individual objects catalogued. These included about 10 tons of copper oxhide ingots, more than 40 tin ingots, 175 glass ingots, beads of faience, agate, stone, ostrich shell and rock crystal, gold, silver and shell jewelry and rings, other objects of silver and gold, stone anchors for use and exchange, precious woods, ivory of both elephant and hippopotamus, amber, weights in metal and stone, a ton of resin for perfumes, and foodstuffs for trade (acorns, almonds, figs, olives, pomegranates). Many of the latter were stored in Canaanite jars. There were also vessels of faience, ivory and ceramic including Cypriot ring base bowls, white slip bowls, wide-mouthed jugs, lamps and pithoi, plus tools and weapons.

The finds from the wreck have been subjected to perhaps the most intensive and wide-ranging study of any site of its size from the ancient world. Firstly, the techniques employed have included absolute and relative dating to date the ship and its contents, especially dendrochronology and radiocarbon dating. The results of these have been compared with the relative dates developed based upon analysis of the shape and decoration, especially of the ceramics and rare items (such as a gold scarab inscribed for Nefertiti). A whole range of compositional analysis has been carried out on metals, ceramics, glass and glaze. These range from simple identifications of the metals of each of the ingots, through major and trace-element analysis of glass and ceramic to lead isotopic provenancing of lead and copper objects, and of the ingots in particular. Some of the first experimental work to explore the possibilities of tin isotopic provenancing was carried out on the tin ingots. The organic residues from the inside of the many ceramic vessels have been examined to reveal their contents. The rock types of the stone objects have been identified and their sources sought.

Much of this analysis has been aimed at answering some very fundamental questions that directly affect the interpretation of the wreck. While it is, of course, clear where the wreck was found, where the ship was coming from and to where it was going is far less certain, but both are extremely important. Looking inside the structure of the ship, anything that may have belonged to a member of the crew might suggest the nationality of the ship. The source of the cargo might give a hint as to where the ship's last voyage originated. In the simplest option, the cargo might all come from one place where it was loaded for transportation elsewhere. The interpretation, however, has proved much more complex. The ship carried objects characteristic of the Greek world, Cyprus, the Levant, the Near East and Egypt. Lead isotope analysis suggests the copper ingots seem to have originated in Cyprus (see Box 4.5 on oxhide ingot provenancing), but other cargo came from a whole range of different places. One interpretation is that the ship was sailing anti-clockwise around the Eastern Mediterranean (the winds dictate the direction) stopping off at different ports and "exchanging" or "trading" objects. The last port perhaps was in Cyprus, where the tons of copper were loaded. However, another interpretation is also possible. Looking at the contemporary value of the cargo, it seems to be far above what one might expect for a simple trading vessel. This presents the possibility that it is not a trading vessel at all. It could be argued that the only people who would have access to such wealth would be the very top of the administrative hierarchies, specifically the kings. If this is the case, it makes it more likely that the cargo is a royal gift, offered by one king to another. This exciting interpretation could mean that the Ulu Burun wreck is physical evidence of the intensive gift-giving activity discussed in several ancient texts, including parts of the Amarna letters (Moran 1992).

References

Bass GF. (1987) Oldest known shipwreck reveals splendours of the Bronze Age. *National Geographic* 172: 692–733.

Moran WL. (1992) *The Amarna Letters.* Baltimore: Johns Hopkins Press.

Pulak C. (1995) Das Schiffswrack von Uluburun. *Antike Welt* 23: 43–58.

Pulak C. (2005) Das Schiffswrack von Uluburun. In: Yalcin U., Pulak C. & Slotta R. (eds) *Das Schiff von Uluburun: Welthandel vor 3000 Jahre.* Bochum: Deutsches Bergbau Museum.

Yalcin U., Pulak C. & Slotta R. (2005) (eds) *Das Schiff von Uluburun: Welthandel vor 3000 Jahre.* Bochum: Deutsches Bergbau Museum.

Egyptian glass production also seems to have been important in later periods. Analysis of glass from the first millennium BC shows that it was made with a mineral soda flux, natron, rather than a plant ash flux (Caley 1962; Turner 1956b). In this respect, it is similar to contemporaneous glass found across the Greek and Roman worlds. The source of this natron is almost certainly one or more evaporitic lakes. Egypt has some rather large examples in the Wadi Natrun, located west of Cairo on the edge of the Western Desert (Shortland 2004; Shortland et al. 2005; Tahar 1999). Furthermore, significant glass factories have been found near the lakes, suggesting that they were exploited and that production was in the form of large tank furnaces. These were often used in the Roman period and later (Freestone & Gorin-Rosen 1999; Nenna 2000). Analysis has been carried out on the deposits of the lakes. These analyses demonstrate that their evaporitic products would have had to have been gathered at one time of the year and that they probably would have required some processing and refinement to make them useful in glass (Shortland 2004; Shortland et al. 2010; Tahar 1999). Research has also demonstrated the relative importance of the Egyptian sources of natron for the huge amount of glass production occurring during the Roman period (Turner 1956a; 1956b).

Analysis of rare Egyptian glass of the first millennium BC has followed the same series of techniques as for Late Bronze Age glass, with the addition of some new analytical techniques. Isotopic techniques have been described under the metals sections (where lead isotope analyses have been used extensively). This technique has been employed to look at the source of lead in lead antimonate yellow glass (Lilyquist & Brill 1993; Shortland et al. 2000), but other isotopes also have been applied to glass. Oxygen isotopes were tried early (Brill 1970), but with only limited success. After the development of techniques for strontium and neodymium isotopes, significant results were produced. Both these isotopes provide an indicator of the location from which the sand component of the glass might derive. Since the sand is not thought to have been transported far, these isotopes provide some idea of the provenance. They have shown that the simple theory for Roman glasses, that all the Roman glass was produced in the Levant (using both fluxes from the Wadi Natrun and perhaps other natron), is probably too simplistic (Degryse & Schneider 2008). This work has recently been extended to Late Bronze Age glass, but there are perhaps still too few analyses to show clearly the patterning for this period (Degryse et al. 2010; Henderson et al. 2010).

4.4.5   Glaze

'How are objects glazed? What really *is* the 'glaze'?' Glazed objects are linked to glass as a glaze is essentially a glass that has somehow been applied to the surface of another material. Some of the first glazed materials in Egypt are glazed stones, such as steatite and quartz (Tite 1987). The stone was shaped into a small object, and then covered with a liquid glaze slurry and fired. The objects tended to be small (often zoomorphic) amulets and scarabs. The glaze was usually green or turquoise and copper coloured (Tite 1987). Glazed stone objects tend to date from the early periods as the technology of glazing stone was, to a certain extent, superseded by faience (see below). An exception to this, however, is glazed steatite, and especially steatite scarabs, which are found throughout Egyptian Dynastic history (Lucas & Harris 1962: 155–156). Steatite is an unusual stone, and is sometimes known as soapstone or talc-schist. It contains a significant quantity of the mineral talc, which is one of the softest known minerals (hardness 1 on the Mohs scale). Thus steatite can be easily carved. Upon heating, most probably during the glazing process, the mineralogy of the steatite changes. It becomes much harder, changing into predominantly enstatite, developing a hardness of around 6 on the Mohs scale. This property seems to result in glazed steatite persisting in Egypt when other glazed stones do not.

As early as the beginning of the Dynastic period, a new technology was used in the creation of glazed objects. Instead of a large crystal of quartz being glazed, crushed quartz was used. Given an appropriate medium to make it less thixotropic (it flows slowly, but breaks if forced to flow faster – like toffee), it could be formed into a 'paste' which could then be relatively easily moulded into amulets, figurines and vessels and glazed.

This material, known as faience, went on to become one of the most common finds on Egyptian sites after ceramics (Friedman 1998). It was used in a whole range of objects from tiny beads only 2 mm across to major architectural or ritual pieces over a metre in length.

Analysis by SEM-EDS has revealed that the raw materials of the glaze were similar to that of glass as plant ash and a colorant was used (Tite & Bimson 1986; Vandiver 1982; 1983). The most common colorant by far in faience was copper blue. This created a range of colours from pale blue-greens through to deep blues, sometimes with additional manganese black decoration (Kaczmarczyk & Hedges 1983; Shortland 2000a). In the New Kingdom, at the same time as the creation of the first glass, new colorants were used on faience for the first time to create yellow, white and red. How this faience was made has been the subject of some research using the imaging capabilities of SEM on polished sections through objects (Tite et al. 1983; Vandiver 1982; 1998). The profiles through the glaze and into the body of different types of faience vary: some have thick glazes, others are thin, some have glass in the core of the objects others not and so on. Experimental replication of these patterns described by SEM has allowed three distinct methods for the production of faience to be hypothesised: efflorescence, cementation, and application (Nicholson 1998; Nicholson & Peltenburg 2000; Tite et al. 1983).

Efflorescence glazes are created using a natural process whereby the water of a drying clay or sand carries some dissolved salts to the surface, from where it then evaporates. This can often be seen on relatively modern bricks, where a white powder appears on their surface. In efflorescence glazing, this physical process is deliberately employed. The glaze powder is added to the crushed quartz and/or sand core which is wetted and moulded as it dries. Some of this glazing powder is dissolved in the water and comes to the surface of the object as the core dries, precipitating on the surface as a powder. On firing, this forms a glaze. Experimental replication of this glazing technique has shown that significant amounts of the glazing powder remain in the middle of the object, the core, where it forms a glass on firing. This means that objects glazed by efflorescence tend to have a lot of glass in their cores compared with faience objects made by other techniques. The glaze also tends to be thickest where the rate of efflorescence was quickest, which is where the drying was quickest. Drying is quickest in high surface to volume areas, such as the rims of vessels, so relatively thick glazes in these areas are often taken as indicative of efflorescence glazing (Tite et al. 1983; Vandiver 1983).

Cementation glazes utilise another process. The object is placed in a box of glaze powder. On heating to just the correct temperature, the glaze powder begins to fuse. Importantly, the first point of fusing is on the boundary between the quartz-rich core and the powder. This means that a glaze forms on the surface of the core while it is still within the glazing powder. Excess powder can then be knocked or brushed off when cool.

This is an ideal way of making large numbers of very small objects, such as tiny beads. It results in a very even glaze thickness over the objects, and so this even nature is taken to be indicative of this type of glazing (Tite et al. 1983; Vandiver 1983).

The third type of glazing is application. This is perhaps the most obvious of all the techniques. It involves the painting of a slurry of glaze powder, in some form of suspension, onto the surface of the object. This is left to dry and then fired. The key recognisable features that appear to be indicative of application glazing are drips and runs of the glaze. These are caused both during the painting process and during firing if the slurry has been thickly applied or fired at slightly too high a temperature. Application is one of two ways of creating a polychrome object. If the decoration is polychrome and looks like it has been painted on, then it is most probably an application glaze. The second way of creating a polychrome glaze is to draw in miniature on a technique that is used widely in furniture making and architectural elements – inlaying. When using an inlay method, tiny grooves and depressions are cut into the surface of the faience core while it is damp. Into these cavities, small amounts of faience paste in contrasting colours can carefully be inserted.

4.4.5.1 Production Evidence from Faience and Glass Workshop Debris

The most difficult types of furnace to identify are those associated with glazing and glassmaking. These often present as high-temperature furnaces without any obvious markers within the kiln itself. In these cases, identification rests on the finds associated with the kilns. Almost all high-temperature kilns are found with large quantities of black vitreous 'slag'. This is the result of subjecting Nile silt to temperatures well in excess of 1000°C. Broken kiln furniture is also common, such as crucibles and other vessels, including moulds and stands. Copper tools themselves, as would have been used to lift hot material in and out of the fire, are very rare. This is presumably because, even when broken, they could be melted down and re-used. Perhaps some of the most interesting and useful finds for interpreting the processes operating in a workshop are wasters. These are items that have failed in the production process for some reason. This may be because the temperature of the kiln was too hot, or a shelf inside collapsed, or the clay or glaze mixture was incorrect. They are frequently found in ancient kilns, particularly in the production of pottery and glaze. The Memphis glaze workshops (see Box 4.7), first studied by Petrie, are a good example, with abundant wasters and other working debris. SEM analyses of the wasters can indicate the temperatures to which the kilns were fired, and may also explain the reasons behind some of the failures within the kiln.

While a lot of material (in terms of wasters, moulds and kiln furniture) can often be recovered from glaze workshops, the identification of an

Box 4.7   The Technology of Faience and Glass
Andrew Shortland

Objects of faience, a glazed ceramic with a crushed quartz core, are some of the most common finds on Egyptian sites, particularly in New Kingdom and later periods. Faience is commonly blue, but often features black decoration. It is a vitreous material, which means in composition it is closely related to glass, and it is certainly possible that it was produced by the same workshops and/or furnaces as glass. The first glass appears at the beginning of the New Kingdom and the development of new colours in faience occurs just as these same colours are being used in the new glass technology for the first time. A great deal of analytical time has been spent investigating the technology of faience and glass (see Shortland 2012; Tite & Shortland 2008; Tite et al. 1983; Vandiver 1983). This work has included the identification of the raw materials of the technology, the processes involved and the final shaping and making of finished products. Analysis of finished pieces of faience and glass by SEM and microprobe has given a good idea of the composition of these pieces, more than 99 per cent of which are usually made up of the same 10–12 elements. From these analyses and some experimental refirings, it has been determined that the raw material for both these technologies was either quartz or quartz sand, combined with a plant ash based alkali source, a "flux", which lowers the melting temperature of the components. To these are added the colouring elements – copper and cobalt for blue, manganese for purple, antimony with or without lead for white and yellow and so on. It is clear that Egyptian manufacturers of glass and faience used very similar raw materials and required similar facilities, which leads to the conclusion that these production systems were closely linked. The technologies needed a supply of these raw materials. Some of the raw materials probably derived from fairly close to the factory sites as they are relatively common in Egypt, such as, for example, fairly pure quartz sands and desert plants for alkali-rich ashes. Others, particularly the colouring elements, must have come from further afield. These sources have been tracked using trace element and/or isotopic provenancing to either Egypt (cobalt-bearing alums from the Western Oases, lead from the Red Sea coast) or outside (copper, perhaps from the Mediterranean, antimony potentially from the Caucasus).

It is, however, the study of debris from glass and faience workshops that reveals most information about the processes of glass and faience making. Two sites predominate here, Amarna and Qantir, and it is from these two sites that most of our information about Late Bronze

Age vitreous materials derives. Both have been excavated well over a series of campaigns and both have had scientific input into both the excavation strategy and the interpretation of results (Nicholson 2007; Pusch & Rehren 2007). Careful excavation of suspected vitreous material workshop sites has revealed a wealth of debris associated with production. This includes crucibles for melting and potentially moulding glass, hundreds of moulds for small faience objects, glass rods and spills from glassworking, primary uncoloured glass (potentially an intermediate stage in the glass production process), as well as finished and semi-finished objects. Furnaces have also been uncovered that are probably connected to the production of such vitreous materials. Meticulous excavation in association with analysis has shown how all these different items connect together and were used in the ancient workshops. Having firmly established production on these sites, work has continued in trying to track the finished products through the Mediterranean and Near East, and to find comparative workshops in other countries. Once again, analysis has played a big part in this and the careful interpretation of results has pointed to production site(s) in the Near East, at least one in the north and probably one in the south of Mesopotamia. The work on the elemental analysis of glass has been supplemented by more and more isotopic analysis, particularly of strontium and neodymium isotopes, which impart the geological "fingerprint" of the area from where the raw materials were obtained to manufacture the raw glass and hence to the finished product. Similar work is now being carried out with faience and it is likely that this will also reveal interesting patterns of production and exchange.

References

Nicholson PT. (2007) *Brilliant things for Akhenaten*. London: Egypt Exploration Society.

Pusch E. & Rehren T. (2007) *Hochtemperatur-Technologie in der Ramses Stadt*. Hildesheim: Gerstenburg.

Shortland AJ. (2012) *Lapis Lazuli from the kiln: glass and glassmaking in the Late Bronze Age*. Leuven: University of Leuven Press.

Tite MS. & Shortland AJ. (2008) *Production technology of faience and related early vitreous materials*. Oxford: Oxford University School of Archaeology.

Tite MS., Freestone IC. & Bimson M. (1983) Egyptian faience: an investigation of the methods of production. *Archaeometry* 25: 17–27.

Vandiver P. (1983) Egyptian faience technology, Appendix A. In: Kaczmarczyk A. & Hedges REM. (eds) *Ancient Egyptian Faience*. Warminster: Aris & Phillips. pp. A1–A137.

equivalent for glass has been problematic. Glass production was a two-stage process: glassmaking where the raw materials discussed earlier were transformed into glass, and glassworking where the raw glass was made into finished objects. These two types of production seem to have occurred at separate sites in Egypt during the Dynastic Period, and indeed in the rest of the Near East. Glassworking sites tend to have small finds of glass used in the making of vessels and jewellery, such as glass rods, glass drips and runs, and even the occasional glass waster. They may also have vessels with glass teared down them. These were used for holding molten glass while it was heated. Perhaps the best studied glassworking sites have been found at Amarna, initially studied by Petrie, with extensive later research by Nicholson (summarised in Nicholson 2007). The difficulty is identifying glassmaking workshops. The raw materials of glassmaking, crushed quartzite and plant ash, are not easily identifiable as such during excavation. Furthermore, 'wasters', in the form of failed batches of glass, seem to be very rare. The only really good example so far published is from Qantir (see Box 4.7), and is a mass of semi-fused crushed quartz still in its melting crucible. Study of all of the glassmaking and glassworking debris has been principally carried out by SEM-EDS and SEM-WDS. The combined opportunity to obtain images of the microstructures alongside the ability to get very good compositional analyses has been the focus of much recent glass research.

4.4.6 Pigments

A final group of materials must be considered when discussing the analysis of Egyptian inorganic materials – pigments used in painting. This group includes minerals, rocks and man-made compounds. As a result, pigments fit into both the stone and vitreous materials categories. The techniques used in their identification are very similar to those discussed above, but analysis of pigments is so widely conducted that it is worth a separate mention here. Colours were applied to soft and hard stones, plaster, wood, textile and papyrus (Lee & Quirke 2000). In addition, a wide variety of binders, grounds and other preparations were used. These can be investigated using the same techniques as the pigments themselves.

The earliest pigments used include charcoal black and ochre reds. Their use stretches back into the Palaeolithic. By the Predynastic period, the colours used seem to have included black, red, yellow, white and perhaps blue and/or green; however, due to their rarity and the value of the objects, no modern analyses of any of these colours are available (Lee & Quirke 2000). From the 4th Dynasty onwards, a full range of colours was used, a simplified summary of which is in Table 4.1 (adapted from Lee & Quirke 2000; Lucas & Harris 1962).

Several techniques are used in the analysis of pigments. As usual, wherever possible, non-destructive techniques are used. These can include

Table 4.1 Colours used in pigments

Colour	Pigment
Black	carbon from soot or charcoal
Brown	red ochre iron oxides with *Black*
Red	iron oxides ochres realgar vermillion red lead
Orange	*Red* with *Yellow*
Yellow	yellow ochre orpiment
White	calcium carbonate, in the form of chalk or whiting gypsum huntite
Blue	cuprorivaite in man-made 'Egyptian blue' azurite (?)
Green	wollastonite and copper glass, man-made 'Egyptian green' copper ores copper chlorides (?)
Pink	*Red* with *White*

XRF and XRD. XRF, either hand-held or bench-top, can give qualitative compositional analyses for elements heavier than potassium. If helium is introduced between the object and the X-ray detector, this reduces the density of the atmosphere and means low-energy X-rays generated in the object can be detected, so that sodium and other heavier elements may be identified. It is, therefore, a very valuable, quick and easy technique for surveying large numbers of objects and colours to see if any appear to be compositionally unusual. XRD can be carried out non-destructively on small and/or flat objects. This gives structural information, and hence can identify the lattice parameters of the mineral phase(s) of the pigment, as these are unique to one particular mineral. It is more powerful in some ways than XRF as it can identify a particular phase. For example, XRF of a red pigment might reveal that it contains iron. From this, one may deduce that it was an iron oxide (as opposed to a carbonate, for example), but one could not tell whether it was an anhydrous iron oxide (such as haematite) or a hydrous one (i.e. ochre). In the right conditions, however, the XRD would not only be able to do this, but also demonstrate which oxide or ochre is present, or a mixture, and even potentially give some idea of the proportions of each.

While non-destructive analysis is a good first step, unlike many other materials, pigments are often sampled for analysis. There are several reasons for this. Firstly, pigments tend to occur in large areas of the objects, and often have areas of damage that allow small pieces to be taken with little effect on the object as a whole. Secondly, with large painted objects, such as walls, coffins and statues, there are often small pieces that have already become detached from the object. These, as long as they can be positively identified as belonging to the object, can be used for analysis. Thirdly, several of the techniques used only become possible if samples are taken. Destructive techniques used include the standard SEM-EDS (discussed many times previously in this section). In addition, however, two techniques can be used that, in general, tend not to be used for ceramics, metals and glass: FT-IR (Fourier-Transform Infra-Red) and Raman spectroscopy. FT-IR spectroscopy works by passing an infra-red beam through a tiny sample of the pigment. The molecules in the pigment either absorb or transmit the different wavelengths of the infra-red. These results are then presented as a spectrum. This spectrum can be compared with a large database of reference spectra, and thereby an identification for the compound is obtained. The big advantage of the technique is that it provides a compound identification, in a similar way to XRD, rather than merely elemental composition. Raman spectroscopy is in some ways similar. Using this method, a laser is used to illuminate the sample, and photons from the laser light are absorbed and re-emitted. The re-emitted light is shifted up or down from the original laser frequency. This is a phenomenon known as the Raman effect. The shift provides information regarding the molecules present. Once again, identification can be achieved by comparison with a database of reference samples. Both techniques require only the smallest of samples, and can provide very valuable detail about pigments, binders and grounds.

Box 4.8   Red-shroud Mummies: Links and Sources
Marc Walton & Andrew Shortland

There is a group of mummies dating to the early Roman period known as "red-shroud mummies". Only small numbers of these mummies are known, spread around museums in the USA and Europe. Many of these have been subjected to a series of studies designed to answer questions about the traditions surrounding Romano-Egyptian mummification, the preparation and use of Roman pigments and about their long-term preservation. One of the most interesting aspects of these projects was the investigation of the red pigments that they all share. This was led by the Getty Conservation Institute Museum Research Laboratory and the Getty Museum Antiquities Conservation Department (Svoboda & Walton 2008; Walton & Trentelman 2009).

The J. Paul Getty Museum has in its collection one of these mummies (91.AP.6). The mummy has the name "Herakleides" written in a painted inscription on the feet of the shroud, hence the mummy is usually known by this name. It has a highly accomplished mummy portrait of a man using encaustic techniques on a wooden panel. The shroud of the mummy is covered with red paint decorated with protective and other magical deities and symbols. Herakleides was subjected to a number of investigations including computer-aided tomography (CT) scanning, carbon-14 dating, and multispectral imaging. Details of some of this work are available online from the Getty (http://www.getty.edu/museum/programs/mummy.html). However, one of the major thrusts of the analysis was to develop an understanding of the red surface of the mummy. Red-painted mummies are relatively rare, although the colour is thought to have been considered magically protective by Romano-Egyptians (Corcoran 1995). Most red painting in Egypt was carried out using ochres, red hydrated iron oxides. However, Herakleides' shroud was covered in another red pigment – red lead, Pb_3O_4. The project was greatly enhanced by being able to draw comparative samples from another red-shroud mummy in the J. Paul Getty Museum, Isidora (81.AP.42) and from mummies in the Brooklyn Museum of Art, the Fitzwilliam Museum, the Ny Carlsberg Glyptotek, the Antikenmuseum Basel und Sammlung Ludwig and the Spurlock Museum. Seven samples of red lead could therefore be compared.

The techniques used were a combination of inductively coupled plasma time-of-flight mass spectrometry (ICP-TOF-MS) and Raman microspectroscopy. Time-of-flight detection for mass spectrometry is a variation on the normal ICP-MS, allowing the complete mass range of major and trace elements in a sample (from lithium to uranium) to be determined. This minimises the signal noise, thereby giving better precision to the determination of isotope ratios for some elements. Raman microspectroscopy was used to characterise single pigment grains and so assess minor mineralogical phases that might otherwise be below the detection limits of conventional mineral identification techniques (such as powder X-ray diffraction).

ICP-TOF-MS showed that the samples of red pigment were composed primarily of lead (83–92 per cent by weight), with tin (Sn) present at concentrations ranging from approximately 1900 ppm to 19,900 ppm in the solid material ($\sim$ 0.2–2.0 per cent by weight). Each was associated with a range of trace elements, which the low detection limits of the ICP-TOF-MS could accurately characterise. The study normalised these values to the average upper continental crust composition. This is a standard technique which allows those elements naturally or artificially enriched or depleted to be easily identified. All the samples showed a strong compositional similarity and show elevated levels of many trace elements, especially Ag, Au, Cd, In, Sn, Sb, and Bi. This is consistent with the red lead being made from a litharge (PbO) and the elevated silver and gold content suggests that the parent lead ore was rich in these elements – probably rich enough for the gold and silver to be exploitable. The red lead therefore seems to stem from a metallurgical process. This interpretation is supported by the presence of tin, which is not found in lead ores, and so must have been incorporated into the pigment artificially, almost certainly in a metallurgical context. The trace elements therefore suggest that the lead in red samples came from metallurgical processing of a silver or gold bearing ore body. In order to determine which ore body, the lead isotope ratios of the lead were determined with the ICP–TOF-MS. Lead isotope ratios vary according to the age and geological terrain of the ore. Comparing the results from all seven samples with ore bodies from Egypt, the Mediterranean and the Near East, the isotope composition of them all most closely matched lead-rich slags, debris and silver from mining and processing at the early-Roman site of Rio Tinto in Spain.

Raman microspectroscopy of individual pigment particles in the samples confirmed the major phase to be red lead (Pb_3O_4). However, spectra of the yellow area of the pigment indicated the presence of lead stannate (Pb_2SnO_4) with the red lead. Lead stannate is an entirely artificial product produced by heating mixtures of lead and tin oxides to temperatures between 650° C and 800° C. Lead stannate was not known to have been produced for use as a pigment until the fourteenth century when it was termed *lead-tin yellow*. It was, however, detected, together with red lead, in all of the red pigment samples examined in this study.

This study therefore managed to show that the seven mummies examined were not only similar in their date and macroscopic appearance, but were painted with pigments that probably came from the same ore source and used the same production technology. Both trace element and lead isotope data point to this source being the silver production sites at Rio Tinto, Spain. This suggests that the pigment found on Romano-Egyptian mummies is evidence of trade in either pigments or raw litharge between Spain and Egypt during this period.

References

Corcoran LH. *(1995) Portrait mummies from Roman Egypt (I–IV centuries A.D.) with a catalogue of portrait mummies in Egyptian Museums.* Chicago: The Oriental Institute of the University of Chicago.

Svoboda M. & Walton M. (2008) Materials investigations of the J. Paul Getty Museum's red shroud mummy. In: Dawson J. & Pancalcdo S. (eds) *Decorated surfaces on ancient Egyptian objects: technology, deterioration and conservation.* London: ICON. pp. 148–155.

Walton MS. & Trentelman K. (2009) Romano-Egyptian red lead pigment: a subsidiary commodity of Spanish silver mining and refinement. *Archaeometry* 51: 845–860.

4.5　Use and Re-use of Objects: The Lifecourses of the Objects

What was the object actually used for? How was it used? Some items, such as stone or metal tools, can be analysed through use-wear methods to see what they were employed to process. In addition, surfaces can be tested to assess what materials they have been in contact with. Examples include lipid analysis of the internal surfaces of ceramic sherds or immunological analysis of the surface of stone or metal tools. Although not yet studied fully in Egypt, due to the prohibition on the export of samples at the time of writing, ELISA has been used to identify haemoglobin and myoglobin on the surface of stone tools in many archaeological contexts (Lombard 2014; Stodder et al. 2010). These studies can identify exactly which animals were processed with specific tools. Similarly the analysis of vessel residues can show what was stored or cooked in them (see Section 3.4: Diet and Subsistence). While many, if not most, objects found in the archaeological record have been manufactured for a single use and then discarded, it is worth bearing in mind that at least some objects may have had several uses through their history. It could be argued that this might be especially prevalent in ancient societies where raw materials may have been especially valuable. The simplest example is where objects have been broken, damaged or blunted. Here re-use consists of repair or resharpening, both of which can often be identified through careful analysis of wear marks.

There are some very clear examples of objects being reused in a completely different way after their first mode of use is over. One of the best examples of this from Egypt is ostraca. Ostraca are broken fragments of ceramic vessels or stone, presumably the waste of vessel breakage of production, which have been written or, more rarely, drawn on. They seem to have been mostly used for texts that had a short use time, for example brief messages, notes and student exercises. They are a well-known and very interesting source for those who study Egyptian texts. However, students of Egyptian material culture should constantly be aware that objects might be used and reused for other reasons than those initially intended. One major agent for this, understudied and often ignored as they are, would be children. Children have the imagination to press the most unlikely objects into service as toys, and it seems inconceivable that they would not have done so with discarded material in Egypt. One example of this can perhaps be seen at Amarna, where small faience objects were made within baked clay moulds. These moulds were found clustered around kiln sites, suggesting that these kiln sites were used in the manufacture of faience. However, the moulds are also scattered widely across the Amarna landscape, many houses containing one or two. This might have been caused by normal taphonomic processes, but it might also result from children having reuses these cheap and useful moulds as toys in their games. More work might be devoted to looking for such secondary uses.

4.6 Conservation and Display

Whilst considered here in this section about objects, conservation and display issues also apply to human and animal remains. This is especially so, of course, for Egyptian archaeology, where funerary archaeology, including mummies, is such an important and popular part of museum displays. However, in terms of sheer numbers of items, inorganic objects are by far the most common museum items in galleries and storerooms with Egyptian material. The care of the object, from the time it is excavated to its eventual storage, either in a magazine in Egypt or elsewhere, or in a museum display, is a complex and lengthy process. More and more, science is playing a significant role in this process, whether it be in direct intervention for an object, or in the monitoring of conditions and treatments that an object might have undergone. An excellent account of the problems faced has been published by the Museum of Fine Arts, Boston (Gänsicke et al. 2003a; 2003b). It is from these two papers that this section is largely drawn.

Conservation science is the name usually given to the interdisciplinary field using scientific methodology and analysis for the preservation of historical objects. The term has slightly different meanings in the USA and the UK. In the UK, the term is relatively restricted and applies really only to work involving stabilising and conserving objects and the research connected with that. In the USA however, the phrase is more akin to archaeological science in general, and expands beyond 'conservation' *sensu stricto* to include the more theoretical considerations of provenance, dating, materials analysis and technological inferences. Since these more general fields have been covered at length above, here the application of science in conservation itself is briefly covered.

4.6.1 Field Conservation

Conservation of an object should start even before it is removed from the ground. Indeed the process of conservation now starts before an object is even found. Good, well-organised excavations have dedicated conservators on the staff at all times and are prepared with the facilities and treatments that might be needed for the object types they are likely to recover. Conservators can give advice on the exposing and lifting of an object that can make the subsequent work in the conservation laboratory easier and vastly reduce the risk of damage to the object. Science plays a relatively minor role here, and it is still unusual for scientific analysis to be involved at very early stages. Once the object has been removed from the ground and is available for work and treatment in the conservation areas of the excavation, or slightly later in the storerooms, it is the first concern of the conservator to address the difficulties caused by the removal of the object from the soil or sand matrix in which it has rested. The object might be damp, or perhaps more usually in Egypt, very dry. It is now exposed to different and differing humidities and temperatures for perhaps the first time in its recent history. For some material types, such as hard stones, this is unlikely to cause too much of a problem. Others, for example, glass and ceramics, can deteriorate, but the circumstances would probably have to be extreme or the object particularly vulnerable. However, there are other materials, most especially organic materials such as human and animal tissue, textiles and wood, but also metals, which are especially vulnerable to early degeneration and disintegration if not treated with care and stored in the correct environments. The role of analytical science in this work is pivotal, but currently is unfortunately not usually employed in the field. The role here is the development of conservation techniques in the laboratory which can then be used in field conservation. This work is largely experimental, the approach employed is to test a series of techniques on the problem that needs to be solved, with each iteration of the testing learning from the results of the previous experiment so that the results continue to improve.

In addition to the conservation of portable objects, one type of conservation that has to be done in the field is the conservation of buildings. In the Egyptian context, this includes dwellings, but conservation problems on a far greater scale are generated by temples and tombs. The huge increase in visitor numbers in the late 20th century through to the 21st has led to new problems in both of these areas. Key amongst them, particularly in the tombs, is the problem of humidity. Even in Egypt, deep tombs would naturally usually be cold and dry. However, large visitor numbers, accompanied by problems with ventilation for very long tombs, means the temperature and especially the humidity increases dramatically. This can have a serious effect on paint and plaster layers of the tomb, leading to mould and salt crystallisation, both of which can result in the loss of surface layers. Once again, science plays a leading role in understanding the processes and advising on the best ways of combatting the problem (see both Smeaton & Burns (1988) and Wüst & Schlüchter (2000) and their references for examples). Similar effects can be seen on soft stones (see Bradley & Hanna 1986; Shoeib et al. 1990).

Initial conservation in the field is often intended to be the temporary consolidation of an object to prevent further deterioration until proper work can be carried out at a later date. This later work includes reassembling objects and giving them treatments to counteract further deterioration. In the past, this work has included the restoration and/or the return of an object to something like its state when it was in use, or at least to what is interpreted as that state. This might have included the removal of patina or damaged surface layers and the addition of fills suitably disguised so as to look like the original object. Whether such work should be done is a matter of taste and ethics. More and more it is being thought that certainly the removal of original layers and patina is problematic. Equally, the addition of pieces to create the illusion of a complete, original object is now less frequently carried out – with inserted pieces usually being visually identified as such. This varies from country to country and from collection to collection, but certainly major museums in the UK have moved away from restoring objects in favour of showing them as they are now, thereby keeping that part of the biography of the object that postdates burial. However, old collections the world over have many thousands of objects that were conserved and restored in the early twentieth century. Many of these are very important pieces, and work on these pieces is often ongoing.

4.6.2   *Identifying Early Conservation*

There are several issues of concern for the scientist working on pieces that came into major collections in the early twentieth century and earlier. The first is identifying early conservation work. It is often unclear what work has been carried out on an object. Even in museums that keep good records of conservation work, it is often not entirely clear what was done in this early period. Whilst done with the best intentions, early conservation work often creates problems of its own which have to be addressed. It is perhaps helpful to give some examples. It was well understood from an early period that organic materials taken out of the ground and put into museum collections were subject to decay. It was clear that at least part of this decay was biological – the pieces were being consumed and/ or inhabited by mould, fungus and other biota. To deal with this ongoing problem, organics in museums were often treated wholesale with very poisonous chemicals, often containing significant amounts of mercury and arsenic (Goldberg 1996). The presence of such compounds is an ongoing health and safety risk to those handling the collections, and such treatments are very difficult to remove. Identifying which objects have been treated, with which compounds, and the degree of severity of such contamination is the work of the analyst and involves many of the techniques discussed above.

A second example of the role in identifying past treatments involves metals. Most metal objects, even in the very dry climate of Egypt, are badly corroded when recovered from excavations. Early work in museums sought to combat this corrosion and return an object to its pre-corroded state. One of the major methods by which this was done was electrolysis. Gänsicke et al. (2003b) provide a very good account of the use of electrolysis in the Egyptian collection of the Museum of Fine Art in Boston. As discussed above, the collection contains many of the features that are typical of much of early restoration work (Gänsicke et al. 2003b: 196–197). They note that it is unclear which pieces of the collection have been worked upon; the precise methodology is often missing, some records merely state 'cleaned by electrolysis' or note that some parts were 'so badly oxidised that they could not be preserved' and appear to have been discarded. It is therefore difficult to know with some bronze pieces whether the appearance of the object now bears any relationship to its appearance when it came out of the ground. Their appearance can also be seriously misleading. Some scholars believe that at least some of the Egyptian bronze statues would have been not bronze in colour, but would instead have been black. This would allow bright inlays of silver and gold to clearly stand out and the colour would have had very important symbolic and ritual meanings. It is probable that most traces of such a black patina would have been lost in electrolysis, thereby potentially seriously affecting the interpretation of the object.

Perhaps one of the most widespread problems with early conservation is the use of glues, fillers and coatings. These are often mainly organic compounds that have been applied to an object in an attempt to reassemble, consolidate and preserve it. The problem is created by the interaction of these compounds with the object over a long period of time. Where inappropriate compounds have been used, cracking and flaking can occur and coatings can darken with age and cause other damage. As often with the questions posed to an analyst, the first necessity is to know what compound has been used on an object. This is therefore one of the significant roles of a scientist in a major museum collection – identifying compounds used in old restoration, along with devising and advising on systems of their removal and repair.

4.6.3 Modern Analysis in Collections

The removal of old conservation work naturally leads on to new conservation work on objects. Once again the role of the analyst revolves around the experimental production and development of conservation techniques and their application. The role, however, also extends to advice on objects not in need of conservation. In this sort of work the methodology employed is very similar to that discussed in the preceding chapters. The questions asked, however, tend not to directly concern how and where an object was made, but more commonly question whether an object is actually what it appears to be. For an excellent modern account of the work of a museum scientist in the identification of problematic pieces, Craddock (2009) cannot be beaten for the breadth and detail of the work described. All major museum collections have objects that have question marks against their provenance and authenticity. This is certainly true in those Egyptian collections that do not consist entirely of excavated material, and even in these, later intrusive pieces derived from older historical excavations and not identified as such can be a real issue. Analysts tend therefore to be approached by museum curators who have an object that appears to them to be stylistically problematic for its period or object type. Analysis can then be carried out to determine whether the composition of the object is consistent either with other objects of the same period in the collection or with published reference pieces. This can be very successful with materials such as pigments, glass, copper alloys and ceramics. It can also have success with precious metals, but these tend to be more difficult because of the continual re-use and recycling of these metals.

The real problem comes with stone objects (see Whitehouse (1989) for an example and Kruglov (2010) for more general comments). Stone was frequently used in Egyptian objects of all periods and they form an important part of collections. Often the size of sample required to properly identify a stone is prohibitive, especially for small, relatively complete objects, and so only visual or non-destructive surface analysis is possible. At times, this does allow for a stone type to be recognised. If it is a type that is not found in Egypt or surrounding states and has never previously been seen in Egyptian objects, this can be sufficient to rule out the possibility that the item is a genuine ancient object. However, it is much more complicated if either the stone is of a common type found throughout Egypt or can be positively identified as an Egyptian stone type that was used in antiquity. Just because the stone is of the right composition does not mean, of course, that it was carved thousands of years ago. Most of the ancient Egyptian quarries have been readily accessible and exploitable in modern times, and industrious locals have been fashioning pieces from these stones both as tourist souvenirs and as deliberate copies. In this case, the composition of the stone is therefore not really of any assistance. Direct absolute dating of the stone is usually impossible, so this leaves only an assessment of how the stone was carved – ancient technology or modern equipment? This is often more subjective, and leaves an object in 'limbo', favoured by some, castigated by others. Over time, some objects are reviewed by a series of curators and scientists and the object can go on and off display, and the doubts as to its authenticity can be expressed or not expressed.

Box 4.9   Documenting Archaeological Surfaces with Reflectance Transformation Imaging (RTI)
Kathryn E. Piquette

What is Reflectance Transformation Imaging (RTI)?

Reflectance Transformation Imaging (RTI) refers to a type of digital photography where image capture, processing and manipulation are augmented or extended through computational techniques. RTI entails taking multiple high-resolution photographs of an artefact or other material surface with light systematically applied from a different position for each exposure. These multiple captures are amalgamated using mathematical algorithms such as polynomial texture mapping (Malzbender et al. 2001a; Schroer et al. 2009) or spherical and hemispherical harmonics (Mudge et al. 2010; cf. Havemann et al. 2008; Willems et al. 2005).

The image file format produced, e.g. *.ptm (Malzbender et al. 2001b) or *.rti (Mudge et al. 2010), can be opened with freeware viewers (e.g. RTIViewer 2010) which enable the user to virtually relight the target surface, giving the 2D image a 3D appearance and thus throwing into relief fine surface details. Visualisation is further improved by increasing the level of zoom and applying rendering modes that mathematically enhance surface shape and colour attributes (e.g. specular enhancement, diffuse gain). These visualisation capabilities can reveal surface information not readily apparent under the conditions of first-hand observation. Such capabilities make RTI a powerful documentation and analytical tool for material culture while facilitating accessibility and dissemination.

Types of RTI

The two main RTI methods are distinguished primarily by the way illumination is applied during photography:

- A rig equipped with a series of fixed lights, such as an arc or hemisphere (e.g. Earl et al. 2011: 2, figures 1-2);
- A hand-held light source used to create a virtual lighting dome (Schroer 2012: figure 3)

For both types of RTI the target surface and camera must remain fixed during the capture process; only the light source should move.

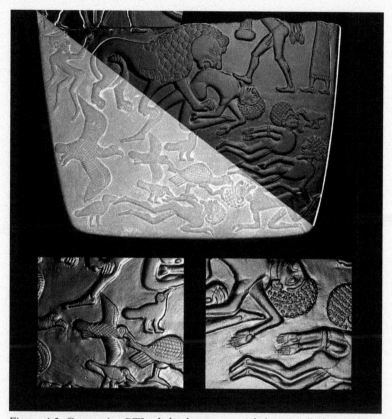

Figure 4.2 Composite RTI of the lower part of the so-called Battlefield Palette, lit under normal light or default illumination (left) and using the diffuse gain rendering mode (right).

The lower left callout shows surface marks indicative of adjustment to birds' beak area, tail length and wing width. The lower right callout shows tool marks, slips and other marks around the figures' hands. Both are shown using the specular enhancement rending mode. Mudstone, 28 cm x 20 cm, perhaps from Abydos, c.3150 BCE. British Museum, EA 20791. (Photograph Kathryn E. Piquette, courtesy of the Trustees of the British Museum.)

RTI lighting rigs are ideal for imaging quantities of small objects relatively quickly, e.g. impressed sealings, amulets, tablets, lithics, coins (Earl et al. 2011; Mudge et al. 2005). Depending on the extent of automation RTI capture and processing can be completed in about 5 minutes. The design of most fixed-lighting rigs constrains the position and size of target surfaces, however, and can be cumbersome to transport (but see below). Fixed-lighting rigs are thus more amenable to use in museums and photographic laboratory settings.

RTI using hand-held illumination, referred to as Highlight RTI (H-RTI), offers a more flexible method with respect to target size, location and orientation (Mudge et al. 2006). Reflective spheres, such as ball bearings or billiard balls, are placed in the shot to record the light position for each exposure. Fitting software (e.g. RTIBuilder) relies on the highlight recorded in the sphere to calculate the coordinates required for amalgamating the multiple exposures into a single image file. While H-RTI is more time-consuming with regard to shot set up, capture and data processing, ranging from 45 minutes to an hour depending on surface size and capture environment, its portability and flexibility make it ideal for field as well as museums and collections use. Either type of RTI can also be combined with microscopy (e.g. Artal-Isbrand et al. 2011; Schroer 2012: 41–42).

Suitability for Archaeological Materials

RTI relies on the behaviour of light and shadow as it plays across surface topography, such as the marks of inscription or other morphological features (e.g. Kleinitz 2012; Padfield et al. 2005; Piquette et al. 2011). Enhancing the visibility of faint ink, staining or other very low-lying features may require a different technique, such as RTI combined with multi-spectral imaging or the latter on its own.

Most archaeological materials are suitable for RTI, from fired/ un-fired clay, metal, stone, shell, bone and ivory to faience, leather, wood, papyrus and other botanical remains, parchment, paper and textile (e.g. Earl et al. 2010a; Piquette in press). Materials with high specularity, such as glass, translucent or reflective stone, polished metal, etc., can present challenges. Techniques such as shooting against a light box (A Serotta pers comm July 2012) or applying different processing methods can improve results for such surfaces attributes.

Museum and Field Applications

Although a relatively new technology, adoption of RTI in the US and European museum and cultural heritage sectors is well underway (Earl et al. 2010b). In contrast, adoption within Egyptology to date has been limited to a handful of museum- and field-based projects (below). Nevertheless, the success of RTI in both museum and field contexts has been demonstrated by numerous other projects (e.g. Earl et al. 2010a; 2010b).

New designs are making fixed-light RTI more portable and thus field-deployable (e.g. RTI on jar stopper sealings at Deir el-Barsha, Willems et al. 2009: 21, n. 60, pls 4A-B, D). As part of a recent AHRC-funded Reflectance Transformation Imaging Systems for Ancient Documentary Artefacts (RTISAD) project, a collaboration between the University of Oxford and the University of Southampton, a more portable, affordable and faster fixed-light dome system was designed (Earl et al. 2011). This included refining the capture and processing workflow for small, relatively flat objects in large quantities, including Egyptian objects. The mudstone relief-carved Battlefield and Hunters Palettes were imaged as part of this project with compelling results (Piquette 2015). Particular lighting angles reveal tool marks and surface transformations indicative of stages in the carving process, such as the cutting of channels to block out animal and human figures, carving and sculpting, the addition of interior details, and polishing (Fig. 4.2).

The suitability of H-RTI for museum objects that cannot be moved easily, such as monumental sculpture or reliefs installed in gallery spaces was recently demonstrated by capture of the decorated side panels of the colossal Min statues in the Ashmolean Museum. Superb results obtained with H-RTI in field conditions for rock art (see Earl et al. 2010a: 219-220; Piquette et al. forthcoming) and graffiti (Kleinitz 2012) attest to the potential for such work on ancient Egyptian imagery. Indeed, a case study on an Old Kingdom tomb inscription at Qubbet el-Hawa during the Universidad de *Jaén 2012 season demonstrated the feasibility of H-RTI on larger fixed surfaces in hard-to-reach locations.* The application of H-RTI to selected small finds, i.e. ceramics, basketry, beadwork, faience, glass, stone inscriptions, human remains, with equipment set up in a tomb next to the excavation area proved equally successful. Archival quality records of the artefacts not only aided archaeological drawing, epigraphy, conservation, and subsequent study and dissemination, but also provided a preservation solution for delicate finds under threat of losing colour, form or configuration before detailed post-excavation documentation and conservation could take place. A potentially critical use of RTI is for cultural heritage under threat, e.g. inscriptions, rock art or other material culture confronting loss, whether through development projects or cultural vandalism.

Natural and artificial light source control has long presented challenges to archaeologists, finds specialists, epigraphers, conservators, curators, photographers and others dealing with cultural heritage. RTI overcomes such problems through computational augmentations of digital photography, especially with regard to illumination control. In addition to the suite of conventional and digital technologies archaeologists already employ, RTI promises to greatly enhance research, as well as increase access to cultural heritage for teaching, learning and public engagement.

References

Artal-Isbrand P., Klausmeyer P. & Murray W. (2011) An Evaluation of Decorative Techniques on a Red-Figure Attic Vase from the Worcester Art Museum using Reflectance Transformation Imaging (RTI) and Confocal Microscopy with a Special Focus on the "Relief Line". Worcester: Worcester Art Museum. http://culturalheritageimaging.files.wordpress.com/2011/09/artal-isbrand_mrs2010_publicationmanuscript.pdf

Earl G., Basford PJ., Bischoff AS., Bowman A., Crowther C., Hodgson M., Martinez K., Isaksen L., Pagi H., Piquette KE. & Kotoula E. (2011) Reflectance Transformation Imaging systems for ancient documentary artefacts. In: Bowen JP., Dunn S. & Ng K. (eds.) *EVA London 2011: Electronic Visualisation and the Arts.* BCS. pp. 147-154.

Earl G., Beale G., Martinez K. & Pagi H. (2010a) Polynomial texture mapping and related imaging technologies for the recording, analysis and presentation of archaeological materials. *International Archives of Photogrammetry, Remote Sensing and Spatial Information Sciences* 38: 218-223.

Earl GP., Martinez K. & Malzbender T. (2010b) Archaeological applications of Polynomial Texture Mapping: analysis, conservation and representation. *Journal of Archaeological Science* 37: 2040-2050.

Havemann S., Settgast V., Fellner D., Willems G., Van Gool L., Müller G., Schneider M., & Klein R. (2008) The presentation of cultural heritage models in Epoch. In: Arnold D., Van Gool L., Niccolucci F. & Pletinckx D. (eds.) *Open Digital Cultural Heritage Systems Conference.* Budapest: EPOCH/3D-COFORM Publication. pp. 32-46.

Kleinitz C. (2012) Reflectance Transformation Imaging (RTI) in der Bestandsdokumentation der Sekundärbilder und -inschriften von Musawwarat es Sufra im Rahmen des Musawwarat Graffiti Project. *Der Antike Sudan: Mitteilungen der Sudanarchäologischen Gesellschaft zu Berlin* 23: 7-20.

Malzbender T., Gelb D. & Wolters H. (2001a) Polynomial Texture Maps. In: *SIGGRAPH '01 Proceedings of the 28th annual conference on Computer graphics and interactive techniques.* New York: ACM Press. pp. 519-528. http://www.hpl.hp.com/research/ptm/papers/ptm.pdf

Malzbender T., Gelb D. & Wolters H. (2001b) Polynomial Texture Map (.ptm) File Format, *HP Labs Technical Report HPL-2000-38R1.* Palo Alto: Hewlett-Packard Laboratories. http://www.hpl.hp.com/techreports/2001/HPL-2001-104.pdf

Mudge M., Malzbender T., Shroer C. & Lum M. (2006) New transformation imaging methods for rock art and multiple-viewpoint display. In: Ioannides M., Arnold D., Niccolucci F. & Mania K. (eds) *The 7th International Symposium in Virtual Reality, Archaeology and Cultural Heritage (VAST 2006)*. Aire-la-Ville: Eurographics Association. pp. 195-202.

Mudge M., Schroer C., Earl G., Martinez K., Pagi H., Toler-Franklin C., Rusinkiewicz S., Palma G., Wachowiak M., Ashley M., Matthews N., Noble T. & Dellepiane M. (2010) Principles and practices of robust, photography-based digital imaging techniques for museums. In: Artusi A., Joly-Parvex M., Lucet G., Ribes A. & Pitzalis D. (eds.) *Proceedings of the 11th VAST International Symposium on Virtual Reality, Archaeology and Cultural Heritage (VAST 2010)*. Aire-la-Ville: Eurographics Association. pp. 111-137.

Mudge M., Voutaz J.-P., Schroer C. & Lum M. (2005) Reflection Transformation Imaging and virtual representations of coins from the Hospice of the Grand St. Bernard. In: Mudge M., Ryan N. & Scopigno R. (eds.) *Proceedings of 6th International Symposium on Virtual Reality, Archaeology and Cultural Heritage (VAST 2005)*. Aire-la-Ville: Eurographics Association. pp. 29–39.

Padfield J., Saunders D. & Malzbender T. (2005) Polynomial Texture Mapping: A new tool for examining the surface of paintings. Preprints of the 14th Triennial Meeting. *ICOM Committee for Conservation* 1: 504-510.

Piquette KE. (2015) Documenting early Egyptian imagery: analysing past technologies and materialities with the aid of Reflectance Transformation Imaging (RTI). In: Graff G., Jimenez-Serrano A. & Bailly M. (eds) *Préhistoires de l'écriture: Iconographie, pratiques graphiques et émergence de l'écrit dans l'Égypte prédynastique*. Aix-en-Provence: Presses Universitaires de Provence (Préhistoires de la Méditerranée). pp. 81-105.

Piquette KE., Graff G. & Bailly M., with contributions by Kelany A. & El-Bialy M. (forthcoming) Documenting a new hunting scene from Wadi Abu Subeira with Reflectance Transformation Imaging. In Midant-Reynes B. & Tristant Y. (eds.) *Egypt at its Origins 5*. Leuven: Peeters.

Piquette KE., Dahl JL. & Green J. (2011) *Exploring Ancient Writings at the Ashmolean Museum with Advanced Digital Technologies*. Oxford: Ashmolean Museum Department of Antiquities. http://www.ashmolean.org/departments/antiquities/research/research/rtisad/

Schroer C. (2012) Advanced imaging tools for museum and library conservation and research. *Bulletin of the American Society for Information Science* 38: 38-42.

Schroer C., Bogart J., Mudge B. & Lum M. (2009) *Guide to Highlight Image Capture*. Cultural Heritage Imaging. http://culturalheritageimaging.org/What_We_Offer/

Willems G., Verbiest F., Moreau W., Hameeuw H., Van Lerberghe K. & Van Gool L. (2005) Easy and cost-effective cuneiform digitizing. In: Mudge M., Ryan N. & Scopigno R. (eds.) *Proceedings of 6th International Symposium on Virtual Reality, Archaeology and Cultural Heritage (VAST 2005)*. Aire-la-Ville: Eurographics Association. pp. 73-80.

Willems H., Vereecken S., Kuijper L., Vanthuyne B., Marinova E., Linseele V., Verstraeten G., Hendrickx S., Eyckerman M., Van den Broeck A., Van Neer W., Bourriau J., French P., Peeters C., De Laet V., Mortier S. & De Kooning Z. (2009) An industrial site at al Shaykh Sa'id/Wadi Zabayda. *Ägypten und Levante* 19: 293-331.

Various RTI resources are available on the Cultural Heritage Imaging (CHI) website: http://culturalheritageimaging.org

4.6.4   *Storage and Display*

Finally, the end point of the work on conservation will be the final storage and/or display of an object. Modern display cabinets can be far removed from the simple glass boxes of fifty years ago. In the modern museum, every aspect of the storage and display of an object has been considered (Lee & Thicket 1996). Once again, the guidelines put together for how an object should be stored will be based on experimental work and analysis, as well as the long experience of the conservators involved. The immediate packaging of an object will be selected so that the object does not react with the packing material. It will be stored with other objects that also will not react together. The store room should be both temperature and humidity controlled and suitable for the object types they contain. Even the cases and shelving should be designed so that vibration from people moving past or traffic outside does not cause the object damage or to move around on the shelf. All objects on display need to be carefully monitored through the day and through the year to check that the temperature and humidity in the display cases do not vary greatly and do not fall outside the bracket determined as safe for the object. A dedicated team of conservators, backed by scientific training, should also be on hand to advise on new displays and review and maintain the safety of existing displays. Science plays an important and indeed growing part in all of this work.

5 Ankh, Wedja, Seneb at Tell el-Amarna

In this book, we have demonstrated the potential of undertaking an integrated approach to the use of scientific techniques and methodologies within the study of ancient Egypt. We hope that we have clearly presented the scientific methods of analysis available, and their potential applications to Egyptologists, in association with the explicit integration of such methods into Egyptian research through the best-practice exemplars provided as boxed case studies. In this chapter, we aim to synthesise these ideas further using the site of Tell el-Amarna as a model of this approach. This chapter thus acts as an example of how the methods previously discussed can be and have been applied to develop this more nuanced approach to Egyptian archaeology. The ancient Egyptian phrase 'ankh, wedja, seneb' means 'life, prosperity, health' and was suffixed to names of Egyptian Pharaohs or their households, and from the New Kingdom onwards was used as a parting benediction at the end of letters. We feel that the ideas encompassed by the epithet are those covered in Egyptian scientific studies and so the phrase can be viewed as a metaphor for the benefits of using scientific approaches within Egyptology and Egyptian archaeology.

Amarna, the city of Akhenaten and Nefertiti, provides an admirable example for many of the scientific methods that we have described. Many of the techniques described earlier have been used at the site of Amarna by large numbers of specialists, including archaeological scientists, under the careful directorship of Professor Barry Kemp with the Egypt Exploration Society (EES) and the Amarna Trust. The synthetic approach to Egyptian studies is well exemplified by the excellent monograph discussing the site, entitled *The City of Akhenaten and Nefertiti*, by Professor Barry Kemp (2012a). This chapter draws heavily on this volume, as well as the wide and varied range of publications on data from the site, due to its integration of the scientific results into an anthropological, historical and Egyptological framework.

Tell el-Amarna, in Middle Egypt, was the chosen location for the short-lived capital of Egypt during the early New Kingdom; it was constructed on the orders of the 18th Dynasty ruler Akhenaten, and abandoned following his death. In tandem with changes in New Kingdom theology and religious beliefs, the city, located as part of a sacred territory called "The Horizon of the Sun's Disc" (Kemp 2012a: 17), was constructed according to the Pharaoh's vision and placed the purified cult of the sun (the Aten) at its centre. This transition, both in creed and aspects of sociality, involved the relocation of the capital of Egypt away from Thebes, in Upper Egypt. Such a move is unusual but not unknown, for example the 19th Dynasty development of a new capital city at Per-Ramesses in a strategic eastern Delta location (van de Mieroop 2010: 73), but the formation of Amarna, and its associated spiritual and religious 'hinterland', involved rapid social upheaval. The reign of Akhenaten lasted only 17 years (*c.* 1355–1338 BC; Kemp 2012a: 304), and by his Year 4 the decision had been made to found a city to act both as home of the Aten cult and as the main royal residence (Dodson 2009: 8). This new city was called Akhetaten and its limits were set by a series of boundary stelae dating from Years 4 and 5. The city became the capital by Year 9, and was completed by Year 12. Akhetaten ceased to function as a capital by Year 2 of Tutankhamun's reign (Manning et al. 2013: 122), and was abandoned soon thereafter. Due to the site having been occupied for such a short time, it has great potential as a case study to demonstrate the potential for scientific studies. Finally, and potentially most importantly, the integrated use of data obtained has been an important characteristic of the overall Amarna research project throughout its design and implementation.

Amarna was never a 'lost' city and so never had to be 'found' using the modern methods described earlier in this volume. It existed as a visible ruin close to the Nile and near to inhabited villages. The first known map of Amarna was made by Edmé Jomard during the Napoleonic expedition in 1798–9 (published in *Déscription de l'Égypte. Antiquités*, Planches IV (Paris 1817), Pl. 63.6). Sir John Gardner Wilkinson visited Amarna in 1824 and 1826 and mapped many of the important individual buildings on the site (first published in *Manners and Customs of the Ancient Egyptians* (1837), Pl. VI). A Prussian expedition led by Karl Richard Lepsius visited Amarna in the 1840s and developed the most thorough map of the site, including the residential area to the south of the central city. The plans were undertaken by Erbkam (and published as *Denkmaeler aus Aegypten und Aethiopien* (1849–1859) I: Abth. 64). Petrie also excavated at the site over the winter and spring of 1891–2, and included a limited study of the desert hinterland. The first areas of Amarna to be excavated by the Borchardt expedition, of the Deutsche Orient-Gesellschaft between 1911 and 1914, appear to coincide with areas shown as lacking standing walls in these early maps, implying that they may have been selected for excavation due to their relatively undisturbed nature (Kemp & Garfi 1993: 17, 19). This was a large-scale and methodological excavation, and led to the development of a 200m grid system which is still of use for current research (Kemp & Garfi 1993). Between 1914 and 1921 the Egypt Exploration Society took over the Amarna concession, with work being undertaken by Eric Peet and Leonard Woolley. Their work eventually linked back with the preceding Borchardt excavations, and integrated the Borchardt 200m grid into their map of the site (published as Griffith 1924: 304, Pl. XXXVI). Deriving partially from older maps and plans, Waddington, working under the direction of John Pendlebury, created a master plan of Amarna (partially published as Pendlebury 1933a: 117; Pendlebury 1933b: 630). Between 1979 and 1988 the city was remapped using modern scientific methods (Kemp & Garfi 1993, with early progress reports published as Kemp 1983: 22, Figure 7; Kemp 1984–5: map 1).

Although the basic site was well known from the early plans and the topographic mapping by Kemp & Garfi (1993), the internal detail and structure of the site was further delineated using modern scientific methods. Topographic survey of the site and hinterland, undertaken by Helen Fenwick (2004), was used with correlated grids and data employed as a base map for further work, such as at the Stone Village (Stevens 2012a). Amarna was placed within its greater hinterland context through topographic survey, with certain areas exposed to more detailed treatment, such as the survey of the alabaster quarries at Hatnub (Shaw 1986). Furthermore, the intensive use of geophysical survey and resistivity in Egyptian contexts was pioneered at Amarna, with magnetometry indicating the presence of possible ovens or hearths, as well as kilns used for production of vitreous materials. These were subsequently ground-truthed (Mathieson 1995). Using a protonmagnetometer, Mathieson examined the area where Petrie (1894, pl. xxxv) had found moulds, in order to see whether they correlated with an area where pieces of slag had been found, which would be consistent with high-temperature firing. Mathieson (2007: 161–162) was successful in locating the area and excavation revealed kilns which had produced the slag found on the surface.

As noted above, the small time frame for Akhetaten lends itself well to intensive study and scrutiny and provides an excellent framework for radiocarbon measurements. Very early in the work of the Amarna project, in 1982, a series of 48 samples were obtained for radiocarbon dating (Switsur 1984). Some of these samples, although too small for use at that time, have now been dated using AMS, and present a good example of how scientific methods continually develop and improve, and hence demonstrate the importance of collecting and curating well-contextualised material for future scientific developments. The original dates obtained from five samples and the more recent samples are all in relatively close agreement, focusing on the 14th century BC (Manning et al. 2013: 124–125; Switsur 1984). Issues associated with wiggles in the calibration curves, offsets resulting from the Egyptian growing season (Dee et al. 2010) and the choice of resolution curve employed mean that some uncertainty remains in the dating, but the short-lived samples from the Amarna period date to 1349–1319 BC at 68.2 per cent probability (using a 10-year resolution curve) (Manning et al. 2013: 138).

In a similar vein, Amarna has been used as a location within which to test and develop new approaches to landscape archaeology. Examples include an aerial photographic survey of Amarna and the surrounding areas, undertaken in 2002 and 2007 by Gwil Owen using a helium balloon (Kemp 2010; Stevens 2012a: 14) and the photographs geo-referenced (Kemp 2010: 28); a 3D laser scan made of the Stone Village in 2008 by the Center for Advanced Spatial Technologies at the University of Arkansas (CAST) (Stevens 2012a: 12); and in 2008, Malcolm Williamson (University of Arkansas) used the Main Site as a test for the actual recording of landscapes using 3D laser scanning (Stevens 2012a: 12). At Amarna, the ancient regional environment is studied holistically. The importance attached to this integrated approach is notable in publications: in 2005, Kemp stated that "These are early days. Research into the past relationship between human population and local environment in Middle Egypt has barely begun" (Kemp 2005: 31). Within environmental archaeology, the movements of the Nile River are of particular importance as we try to establish answers to the following questions: where was the Nile located during Akhenaten's reign? and how did this impact upon the development of Akhetaten? The river levels were lower at the time, and so it is likely that the desert ran to the water's edge all along where the city of Amarna was built (Kemp 2012a).

Similar to the work undertaken at Luxor to investigate movements of the Nile over time (Box 2.6), the Nile in the Amarna region also has branched and changed its location; these changes have been studied by examining the locations of ceramic sherd findspots and drill cores. The course of the river in the 1720s has been suggested as possibly being one arm of a branching river, due to insufficient room between the then existing villages for a major waterway to run (Kemp 2005: 29). Using such approaches, the waterfront at Amarna has been recreated by synthesising depictions from the tomb of *May* (no. 14) with the archaeological evidence derived at least partially from science.

The local environment impacted upon both the people living at Akhetaten and their social organisation. During the New Kingdom, there was limited ability to raise water from the Nile; indeed the first depiction of a shaduf derives from one of the rock tombs at Amarna (Kemp 2012a: 50). Much of the water used at Amarna came from wells, wells that it has been possible to locate through geophysical survey, many of which are located in public areas. These then supplied small groups of houses which were huddled around them – although larger houses had their own wells within the walled compound that surrounded the home (Kemp 2012a: 51). Furthermore, following Egyptian norms, the gardens were laid out in the form of a grid of low mud ridges. Each grid square was one cubit (52cm) across, and each square contained a bed of fertile soil. Some gardens had water runnels, but most relied on watering by hand (Kemp 2012a: 52–55). This would have been physically hard work, as evidenced by the macroscopic analysis of skeletal lesions related to repeated activity causing bone deformation and modification (Box 3.3). The selection of Amarna as the chosen home of the Aten was therefore rather problematic.

Given the difficulty of cultivating the desert with gardens, Kemp (2012a) suggests that the stone architecture formed a link to the environment. At least three types of architectural column are recognised and found at Amarna. The vegetation represented on column capitals provides artistic evidence of the plant species being cultivated around the city. "A regular find of excavations across the city is a small bunch of grapes modelled in faience ... and with a hole for suspension" (Kemp 2012a: 56). These architectural ornaments sometimes appear to have been made to fit to beams, and further form a link to the greening of the environment. Although not scientific in nature, this architecture forms a link with studies of the ancient environment and scientifically testable hypotheses regarding the available resources.

The important parts of Akhetaten, such as the temples, were built from stone, and these stones were re-used as a cheap source of building material after the abandonment of the site. During the Amarna period, the building blocks were standardised and set at one cubit in length, by (ideally) half that in width, and a height of approximately just under half a cubit. Most of the building stone used at Amarna was limestone, although harder stones, such as red granite, black speckled granodiorite and quartzite were also used (Kemp 2012a). Travertine, in early publications sometimes referred to as alabaster, was also used in the Great Palace, such as for carved balustrades bordering ramps to doorways (Kemp 2012a). Experimental archaeology has shown that gypsum can be made to set into a plaster by heating to a relatively low temperature to reduce to a powder that can be mixed with water to form a setting plaster and visual analysis has demonstrated that this was used to create foundations for stone buildings. The thick concrete base to buildings appears to have been made from gypsum paste mixed with sand grains, reinforced with chippings of sandstone and limestone, and gypsum was also used to make mortar (Kemp 2012a). The source of the Amarna gypsum is unknown, although cones of gypsum have been found. Analysis of the composition of mud-brick and the local sediments was undertaken using sieving and hydrometry (French 1984: 192), and these studies suggest that the bricks were made using locally available raw materials, with some wind-blowing input resulting in high medium-sized sand content (French 1984: 194, 201). Petrology has enabled the identification of quartz, calcite crystals, olivine, plagioclase feldspar and opaque oxides of iron within the silt bricks (Nicholson 1985: 149). The marl bricks were found to be easier to thin-section, and this revealed that they included quartz, mica, feldspar, augite, hornblende and iron oxide (Nicholson 1985: 150). Some of the royal buildings used bricks made from Nile mud and plant fragments, but most used desert gravels, sand and crumbly marl (Kemp 2012a: 69–70).

Transport of items to Amarna and distribution within the city were complex problems. Such issues are usually considered in terms of moving large stone blocks for construction, e.g. the talatat, including the quarry structure and the use of ramps (Klemm & Klemm 2008). For example, harder stones moved some distance, such as red granite, black speckled granodiorite and quartzite, were used for prestige items such as statues, but also for querns (Kemp 2012a), but even the movement of marls and silts for ceramic vessel production requires some organisation, and a distinction appears to be made by potters regarding marl and silt from the Middle Kingdom (Bourriau et al. 2000: 122). Such trading systems provide regional lynchpins and provide nodes for dating (both in terms of absolute and relative dates). Even items as ubiquitous as mud-bricks (and their associated soils) have the potential to inform in terms of their composition and manufacture. The bricks across Amarna differ in terms of their composition, with different silt/sand ratios being found even between three bricks from the Small Aten temple (Kemp 2000: 81). The distinctive reddish-yellow colour of the desert marl at Amarna is due to its high oxidised iron content (Kemp 2000: 82) and such patterning in colouration aids in visual macroscopic differentiation on the basis of composition.

Many of the buildings were plastered and/or painted, and diverse scientific analyses have been undertaken. For example, large areas of wall paintings are still found *in situ* at the North Palace. Common decorative themes, for both interior and exterior walls, include papyrus plants with birds and butterflies, and birds feeding under the supervision of men clad in white clothing (Kemp 2012a: 57). Painted floors survive at both the Great Palace and the Maru-Aten sun-temple: a thick hard gypsum backing brought out the intensity of the floor colours (Kemp 2012a: 59). Faience tiles were also used for decoration; it is thought they were used for floor areas (Kemp 2012a: 59). The plaster under the decoration itself varies, with some being made from mud, and other sections of more complex compositions. SEM-EDX analysis has demonstrated that the top layer of the pavement plaster is almost pure calcium carbonate ($CaCO_3$) with very little gypsum (Weatherhead 2007: 362, 371). In addition, wet chemical analysis has shown the existence of both calcium sulphate ($CaSO_4$) and $CaCO_3$ in the plasters (Weatherhead 2007: 371). Some mud plaster on walls appears to have been given a 'primer' coat, and at times, using 10´ magnification with a hand lens, a white undercoat below a creamy-yellow ground colour can be seen, such as on sections of the Princesses panel (curated in the Ashmolean Museum, Oxford; Weatherhead 2007: 140–142). Analysis of the composition of the plaster using XRD has enabled the methods of painting to be ascertained. In many areas, the set substance was gypsum with limestone existing as grains within the structure. In these situations, the top surface was prepared with lime plaster so that paint could be applied in true fresco manner (i.e. paint is applied to wet plaster), although the fresco secco technique was also used, especially for green and blue colours (Weatherhead 2007: 365–367, with EDX microanalysis of pigments pp. 372–375). Early SEM, XRD and EPMA studies of the pigments, such as of these fantasy tableaux, already demonstrated the presence of Egyptian blue and the compositions of the green and red colours to be chrysocolla and haematite respectively (Weatherhead & Buckley 1989: 205–209). In addition to the famous painted limestone plaster head of Nefertiti (Berlin, ÄM 21300), a series of portrait sculptural heads, made in plaster, were recovered in the early twentieth century. CT scanning of these heads has shown that they were cast in several stages. Several different layers of plaster are clearly distinguishable in the CT images due to differences in the stages of drying when the plaster was poured into the mould (Seyfried 2012a).

The use of the fresco secco technique for the plant-like colours demonstrates the importance of the environment at Amarna. But the city needed timber for the construction of roofs, doors, windows etc. Wood for this usually differs from the wood required for burning, such as for cooking and manufacturing. Hand specimen identification of the charcoal derived from the burning of these latter uses of wood indicates that the population were primarily burning acacia wood (Gerisch 2007; Kemp 2012a: 73). This tree grows along the Nile Valley, but some of the other woods identified by microscopic and macroscopic examination, such as cedars, must have been transported some distance. Apart from cedar imported from Lebanon, all the sources of charcoal were indigenous to Egypt. Moving the wood from across the Nile or from a distance would have required administrative and physical effort. When attempting to reconstruct the Amarna landscape, it is worth noting that there would have been groves of trees between the cultivated fields that were remnants of earlier gallery woodland. Furthermore, these groves are noted in the second set of Boundary Stelae (see Murnane & Van Siclen 1993: 102, 106).

Charcoal remains are abundant across Amarna. Almost 7000 pieces of charcoal have been recovered by dry sieving and/or hand selection from the Amarna vitreous materials manufacturing site, O45.1, with over 6000 of these pieces being macro- or microscopically identifiable pieces of wood charcoal, with the rest as non-identifiable wood and bark charcoal (Gerisch 2007: 169). Charcoal was found not only in the bases and rake-outs of kilns, but also as inclusions in larger-sized slag lumps. Using microscopic analysis, nine wood taxa were identified; 95 per cent are Nile acacia (*Acacia nilotica*), with the other 5 per cent including tamarisk (*Tamarix* sp.), acacia (*Acacia* sp.), sycamore fig (*Ficus sycamorus*) and white acacia (*Faidherbia albida*). In addition, there were also some mimusops (*Mimusops* sp.), balanos (*Balanites aegyptiaca*), Lebanon cedar

(*Cedrus libani*) and Palmae. The slag lumps only contained Nile acacia charcoal. The charcoal in the vitreous materials manufacturing area was much more limited in species diversity than in other areas of Amarna. Whereas in most samples from the vitreous materials manufacturing area, only one or two taxa were identified, mostly Nile acacia with some tamarisk, by contrast, there were 22 taxa found in the Workmen's Village, 21 taxa in the Main City and workshop area Q58.4 (Gerisch 2007: 170). In the House of Ranefer there were 13 taxa, and in the Central City 14 taxa. The high frequency of acacia is likely to result from its frequency in the landscape, as well as its preferential use due to the fact that it is a wood that burns slowly and steadily, without too much smoke, and has a high calorific value (Gerisch 2007: 170). The other species are of lower quality than acacia as they burn more rapidly and produce more smoke. Furthermore, there are texts from Deir el-Medina which point to the supply of firewood being managed (Gerisch 2010: 407), so it is very possible that this was also the case for the Workmen's Village and the Stone Village, and potentially the Main City too. Although counts of wood or charcoal fragments are commonly given in anthracological reports, we suggest that, due to fragmentation, these might be meaningless unless considered and compared in terms of weight, volume or other quantifiable measure. The development of a more formalised volumetric method by which proportions of woods, charcoals and other such items can be evaluated is required, rather than simple counts of numbers of fragments and/or pieces.

It is unclear whether raw wood or prefabricated charcoal was used as the principal fuel for manufacturing, but there are remains of uncharred and incompletely charred wood and bark, thereby suggesting the use of at least some wood as the burning fuel. In principle, either wood or charcoal may be used as fuel for glass, faience and pottery production. With no artificial air supply, wood fires can get up to 700°C, charcoal to 1000°C (Gerisch 2007: 172). Where artificial air supplies are available, wood and charcoal fires can reach the temperature at which kilns normally fail, approximately 1200°C. A temperature of around 1700°C, far higher than accessible using ancient technologies, is required to liquefy silicious sand. If an alkaline flux, such as plant ash, is added, the temperature required can be lowered to around 900°C, although a replica kiln at Amarna (Kiln 3) achieved and maintained a temperature of 1100°C (Nicholson & Jackson 2007).

As part of the 'Amarna Glass Project' (Nicholson 2007: v), experimental work was undertaken in association with the scientific analysis of the faience, glass, frits etc. Since frits, faience and glass are found together in the same workshops, such as at site O45.1, their production is considered to have been linked (Nicholson & Peltenburg 2000: 187). The Amarna Glass Project aimed to investigate i) whether a sufficiently high temperature could be reached in one of these types of kilns to produce glass, ii) to find out whether a forced draught was necessary to achieve temperatures high enough to produce such glass, and iii) to ascertain whether glass could actually be made from the available (suitable) raw materials. The replica kiln furnace reached 1150°C after just over 6 hours of firing. An ingot of deep cobalt blue glass was also produced, suggesting a single-stage process for glass manufacture (Nicholson 2007: 97). Future experimental archaeological research might include the use of an authentic plant ash alkali, a natron alkali source, use of mixed alkali sources, use of authentic Nile silt crucibles, with acacia and/or sycamore wood for fuel etc. These experimental works must be studied in association with the laboratory-determined results of analyses, such as exploration of efflorescence glazing of faience (Nicholson & Peltenburg 2000: 190). X-ray spectra of Amarna glasses have identified their major components, such as soda, lime and silica, but also other elements including antimony (Nicholson & Henderson 2000: 208–210), with the latter not only giving the yellow and white Amarna glasses their colour, but also rendering them opaque.

With such a large excavation assemblage, specific recovery and sampling strategies for all items and classes of artefact are vital. For example, the retrieval of wood and charcoal during excavation of the Stone Village was undertaken manually during excavation or by dry sieving, usually with a 4mm mesh (Gerisch 2012). The wood debris and charcoal from Amarna were then analysed for anatomical structures using a binocular microscope with magnifications of 40× and 80× for the sorting part of species identifications, with a high power reflected light microscope (60–500×) employed to note greater detail. This involved using existing breaks in the charcoal fragments or by breaking parts off in the three characteristic sections of wood (transversal, tangential longitudinal and radial longitudinal). Furthermore, as noted earlier, it is important to develop reliable mechanisms for geographic and temporal comparisons and contrasts. Counts of fragments are common in many bioarchaeological or anthracological reports, but these may be meaningless due to fragmentation and repeat counting. In the same way as NISP (number of identified specimens), MNI (minimum number of individuals) and MNE (minimum number of elements) are commonly reported zooarchaeologically and provide contrasting data, weights or other methods are required for archaeobotany. Significant intra-Amarna differences were noted in the wood and charcoal, and were argued by Gerisch (2012: 54) to demonstrate aspects of the Amarna supply system. Taxa missing in the Stone Village relative to the Workmen's Village include a number of fruit trees and palms. More types of imported wood and timber are found at the Workmen's Village, including Lebanese cedar. It is further possible that fruit and vegetables were more prominent in the diet at the Workmen's Village than in the Stone Village (Gerisch 2012: 55).

Other botanical material is similarly well preserved at Amarna, and varying methods of sampling and study have been used in order to answer different questions regarding different parts of the overall site (Stevens 2012b). These have the advantage of demonstrating differences, but make some syntheses and evaluation more problematic. Two methods for large-scale recovery of plant remains are available: dry sieving or water flotation. Traditionally in Egyptian archaeology, dry sieving is undertaken to recover desiccated material such as seeds and fruit stones, but this method can damage some fragile plant remains (Smith 2003: 27). In the Main City, Stevens & Clapham (2010: 427) sieved material through a stack of sieves at 4mm, 2mm and 1mm, and the finer silt/sand (the <1mm fraction) was floated to separate the remaining organics using a 300μm mesh. At the Late Antique monastery at Kom el-Nana, by contrast, after the identification and removal of large stoned fruits such as date and peach, archaeobotanical recovery was undertaken entirely using flotation of 15–20 litre soil samples, leading to a 5 litre volume of flots (Smith 2003: 26–27). A pair of 1.7mm and 0.5mm meshes was used as it was found that quartzite crystals in the sediment blocked the 0.3mm sieve typically used to collect flots in Near Eastern archaeology (Nesbitt 1993). Identification was typically made using a lower-power binocular microscope, with magnification of between 10× and 45×. Comparison was made with a variety of reference materials, comprising economic plants brought from England, Egyptian markets, samples collected from local farmers' fields and the local wadis, and modern specimens at the Herbarium of Cairo University. Where suitable reference material for identification was not available, use was also made of the various Egyptian floras, including Täckholm's *Student's Flora of Egypt* (1974), *Flora of Egypt* by Loutfy Boulos (1999; 2000; 2002; 2005) and *Domestication of Plants in the Old World* by Zohary & Hopf (1988; 1994; 2000; Zohary et al. 2012).

Botanical material has been used in palaeoenvironmental and dietary reconstruction. The latter has also been reconstructed through the use of specialists, from other interlocking aspects of bioarchaeology, such as human skeletal studies and zooarchaeology. These methods provide broad-sweeping views of diet but lack the detail and clarity of isotopic methods. Unfortunately, given the inability to export samples for testing, isotopic reconstruction of human diet at Amarna is impossible. Nutritional stress at Amarna is suggested by the relatively short heights of the adults calculated from the osteology (Kemp et al. 2013) and the palaeopathologically determined delay in skeletal maturation relative to dental development in the juveniles (Dabbs & Zabecki 2012). In addition, cribra orbitalia (i.e. additional porosity in the eye sockets) is macroscopically identifiable in both children and adults, suggesting a diet deficient in iron and other nutrients (Kemp et al. 2013).

Residues on the surfaces of potsherds from the Workmen's Village, studied using high magnification light microscopy, demonstrated that cereal chaff or bran was embedded within them (Samuel 1994: 156). Microscopic study of the cellular patterns indicates that these cereal remains were emmer, and desiccated emmer chaff has also been recovered from around a limestone mortar (Samuel 1999: 131). SEM analysis of the residues on the sherds has demonstrated that the cereal starches found had been exposed to a number of different processes and that the residues are likely the remains of beer (Samuel 1994: 156). This is important as iron absorption in the body is inhibited by phytates, such as found in bran (Hallberg 1987). The diet at Amarna appears to have been weighted toward cereals, consumed as bread and beer, thereby potentially resulting in iron deficiency anaemia. SEM of the desiccated beer residues from the Workmen's Village has shown that the yeast was still alive at the point of the drying out of the sample (Samuel 2000: 547–549). Dates were previously widely considered to have been a key ingredient in Egyptian beer, but SEM of modern and ancient date tissue does not match any of the Amarna beer residues (Samuel 2000: 548–549).

Querns and quern fragments have been found throughout Amarna, with a large concentration within the official quarter of the city. A further concentration was found near the Great Aten Temple bakeries, supporting the suggestion that they were the official 'grinderies' supplying flour to the temple bakers and potentially the nearby palace (Samuel 1994: 157). Households had their own facilities for flour grinding, but there were also the official 'grinderies'. Large houses at Amarna frequently have granaries, whereas small houses do not (Samuel 1999: 134). This means that the state had some mechanism for controlling and producing flour on industrial scales (Samuel 1994: 157), with wealthy households being "conduits of grain to the poorer members of the population" (Samuel 1999: 134). Emmer is not free-threshing, and thus requires pounding in a limestone mortar to remove the husks. Not all houses possess a mortar, but those houses without commonly had access to a mortar placed in a nearby street (Samuel 1999: 134). By contrast, most houses have querns. Visual and petrographic examination has shown that the querns employed for grinding were usually made from quartzite or granite (Kemp 2012a; Samuel 1994), and thus must have been supplied into the area. Production of bread and beer thus required both social and economic interactions, and can be considered in terms of a *chaîne opératoire* (see Section 4.3.1: *Chaîne opératoire?*). Samuel (1999: 135–140) argued that this *chaîne opératoire* impacted upon the structural organisation of the houses and streets within the city.

Zooarchaeological analysis has shown that meat was obtained from cattle, sheep, goats and pigs (Kemp 2012a: 219–220), with pig husbandry being particularly noted from carefully laid out pig-pens in the Workmen's Village (Kemp 2012a: 297–298). Examination and magnification of fibres found around these animal pens indicated that the fibres were hog bristles and so support the hypothesis of pig-rearing in the area (Appleyard 1986). In terms of meat supply, due to their large body size, cattle are dominant, but in terms of the minimum number of individual animals recovered from Amarna, pig and sheep/goat are more common (Kemp 2012a). From macroscopic osteological study of the butchery patterns on the cattle bones, the range of joints used has been noted to be much greater than the variety depicted in offering scenes (Kemp 2012a: 111). Over 2000 animal bone fragments were recovered, with over 600 visually identified, from the industrial area of the main city alone (Payne 2007: 165). A 5mm sieve was used to collect bones. Except for possible gazelle, the other identified bones from this portion of the site derive from domestic animals. The large number of fish remains suggests their importance within the diet (Payne 2007: 166). In this part of Amarna, the zooarchaeological assemblage is weathered, and Payne (2007: 166) notes that this may account for the very low incidence of macroscopically visible butchery marks. By contrast, the pig bones from the Workmen's Village showed much evidence of dismemberment, with some filleting butchery marks (Luff 1994: 159). The pig carcasses are argued, on the basis of the zooarchaeological study, to have been stiff at the point of secondary butchery, suggesting that joints might have been preserved through drying, smoking and/or salting. The organisation of the cuts upon the bones suggests that a few individuals, potentially just one person, were involved in this pig butchery at the Workmen's Village (Luff 1994: 162). The lack of hind limbs in the animal bone assemblage from the Workmen's Village implies trade in these cuts of meat. All across Amarna, wild animal bones are rarely found, but Kemp (2012a: 150–151) has suggested that animals and birds depicted on walls of the North Palace might have formed a fantasy or "tableau of the gifts of life".

The Amarna cattle are relatively large animals as the cattle bones from grid 12 in the Main City overlap in zoological size with osteological measurements obtained from regional *Bos primigenius*. From the direct measurement of the breadth of the distal portion of the tibia, the Amarna cattle have the skeletal size of the smaller wild males of *Bos primigenius* (Legge 2010: 450–451). Legge (2010: 451) argues that this suggests that there was much investment in the appropriate cattle feed, including carbohydrate-rich foodstuffs – probably grain – and possibly also fodder crops as are given to modern water buffalo. This matches tomb and temple depictions of cattle being hand-fed large pieces of items that could be bread. Unfortunately, bovine laminitis is caused by carbohydrates comprising a large portion of the cattle's diet, with the effects being inflammation of the feet and continued growth of upturned hoof. In modern cattle this condition is associated with over-nutrition, usually by a high-carbohydrate diet. Both inflammation of bovine feet and upturned hooves are depicted in ancient scenes, such as in the Tomb of Meryra II at Amarna where "several servants bring forward, for sacrifice or feasting, bouquets, fowl, and three stalled oxen, whose misshapen hoofs show their fat condition" (Davies 1905: 42, Pl XXXVII, Pl XL). This is also recognised archaeologically at Amarna, where "grossly overgrown hoof sheaths" have been found in the preserved remains of cattle (Legge 2010: 451). This demonstrates the importance of an integrated and synthetic approach to the study of archaeological and artistic data in order to better understand both human and animal conditions.

The zooarchaeological study at Amarna has not only aided the reconstruction of the diet of the human population, but also the study of the processing marks on the bones aid in reconstruction of the production, trade and consumption mechanisms. These cut marks provide information regarding the slaughtering methods, carcass dressing, food processing methods for initial consumption or storage, de-horning, skinning, tanning, glue-making, grease production and bone working etc. In this sense, animals should be regarded as commodities and studied similarly to other 'objects', such as in terms of a *chaîne opératoire* or study of provenance. Strontium or oxygen isotope sampling (such as using LA-ICP-MS) along the length of an animal tooth would enable the mobility of the animal to be reconstructed, as has been undertaken for cattle (Montgomery et al. 2010; Viner et al. 2010) or sheep (Balasse et al. 2009; 2012; Henton 2012; Henton et al. 2010). Sampling many points along one animal's teeth may potentially enable transhumance to be identified. Furthermore, this method may allow seasonality in kill-off times to be recognised, and hence the animal husbandry organisation and structure at Amarna could be better modelled. The importance of seasonality has been noted at Dakhleh oasis in terms of human reproduction (Williams et al. 2012) and death (Williams 2008), and thus deserves further attention at Amarna.

How were the foodstuffs stored? The pottery corpus from Amarna is vast, with several published corpora resulting, such as Hope (1991), specialising in the blue-painted pottery, and Rose (2007) demonstrating the breadth of 18th Dynasty material from the site. Examination in the field of ceramic sherds with fresh breaks was made with a 10× hand lens (French 1986: 152; Rose 2007: 11). Thin sections of ceramic sherds were also studied using a polarising microscope to assess for the size classes of the inclusions (Nicholson & Rose 1985: 152–164). This was a relatively pioneering study for its integrated use of statistical analyses in association with ceramic petrology. Thin-section analysis of Amarna ceramic fabrics has also been undertaken in order to compare and contrast with other Egyptian sites (Rose 2007: 11). The blue pigment on Amarna blue-painted ware ceramics has been identified as cobalt aluminate, most likely applied post-firing (Loschwitz 2012), but, using EPMA, cobalt has not been identified in any samples of pigment applied to walls (Rose 2007: 18; Weatherhead & Buckley 1989: 237–238).

What were the implications of storing foods in such ceramic vessels? Scientific analyses of both the grain remains and associated detritus can aid in reconstruction of such aspects of daily life. As noted earlier, SEM and high-resolution light microscopy have been used to identify cereal chaffs and bran embedded in sherds, enabling their identification as deriving from emmer (Samuel 1994: 156; 1999: 131). Additionally *Sitophilus granarius* (the grain weevil) was identified microscopically (for details, see below) in both desiccated and charred cereal debris (Panagiotakopulu & Buckland 2010: 456): this insect can destroy a large proportion of grain produced and stored, thereby having economic and potentially health impacts upon the Amarna population.

And how were the foodstuffs disposed of? After dry sieving of soil samples deriving from sealed contexts in the Main City, using 5mm and 300μm stacked sieves, insect remains residues were examined under a stereo-microscope for identification purposes. Puparia of Sarcophagidae flies (flesh flies) were identified from above the floors of the House of Ranefer (Panagiotakopulu & Buckland 2010: 454). These remains, associated with the lack of zooarchaeological bone, demonstrated that meat waste or offal was thrown or dumped into the building after the useable meat had been removed. Similarly *Simex lectularius* (the common bedbug) was identified from beneath the floor of Ranefer's house (Panagiotakopulu & Buckland 2010: 455) and had already been identified in the Workmen's Village, thereby implying that ectoparasites were found in both richer and poorer parts of Amarna. The most frequently identified insect from beneath floors was the *Musca domestica* (common house fly) (Panagiotakopulu & Buckland 2010: 456).

Amarna has provided an excellent venue for the study and reconstruction of population sizes and demography. From estimates of the area of cultivable land, the fields of Akhetaten could have supported a population as large as 45,000, although Kemp (1989: 269) estimates the actual city population to have varied between 20,000 and 50,000. Similarly, from estimates of the number of actual houses in the city, and the number of people per household, the estimated population of Akhetaten becomes around 30,000 (Kemp 2012a: 271–272). In addition, the actual social composition of the population needs to be considered, such as the number of slaves and the size of the average family. A city of this size would require complex provisioning and industrial organisation, thereby impacting upon individual lives and their activities and workloads. Stevens (2012a: 422) thinks, despite the Stone Village generally having a mixed population, it is quite possible that there was a higher proportion of males than females in the community because of the supposed connection of the village with tomb building. She notes that we do have to be careful in considering 'gender-specific' occupations (Stevens 2012a: 422). Palaeodemography has an important role to play in such reconstructions.

Our understanding of the lives of the people from Amarna has developed tremendously as a result of the excavation of one of the cemeteries for commoners. Geophysical survey suggests that one cemetery, the South Tombs Cemetery, contained as many as 3000 graves, with some graves having contained more than one body (Kemp 2012a: 256). The graves are characterised by overall simplicity and a consistency in burial approach

(Kemp et al. 2013: 67). Most graves appear to have been marked by low cairns of limestone boulders, sometimes topped with memorial stelae, although a few were topped by miniature limestone pyramids (Kemp 2012a; Kemp et al. 2013). The best preserved stela depicts a couple seated side by side, and emulates the informality of royal Amarna-period artistic representation (Kemp 2012a). Although the burials lack macroscopically identifiable signs of deliberate mummification, the bodies are entirely skeletonised and well preserved, with some hair preserved (Kemp et al. 2013). There are also clusters suggestive of family grouping (Kemp 2012a: 260); these would benefit from further scientific investigation, such as ancient DNA kinship analysis of the human remains. Despite past grave-robbing activity, there are some grave goods found, such as pottery vessels. There are, however, also undisturbed burials that, upon excavation, contained not a single funerary offering. Some burials were placed in wooden coffins, but most were placed in tightly rolled and lashed mats of rushes (Kemp 2012a: 260; Kemp et al. 2013: 67). Scientific analysis would permit the identification of the source of the wood and the type of rushes used. By contrast, the rich were buried in rock tombs, with their names added to the list of recipients of offerings in the House of the Aten (Kemp 2012a: 262). There is thus variation in the degree of resourcing of burials, with a minority able to afford painted coffins or carved grave markers. This burial variation is, however, small relative to diversity in the size of houses within the city (Kemp 2012a: 262), where the houses of the rich and poor are distinguished more by size than by design (Kemp 2006: 311). Such variation in size might be studied using spatial technologies. This patterning might also be linked with the relative social fluidity of New Kingdom Egypt (Kemp 1989: 307; 2006: 313; Spence 2010).

As noted earlier, workloads at Amarna were likely to have been heavy. Gross skeletal analysis has shown that degenerative joint disease of the upper limb affected over 65 per cent of the adults buried in the South Tombs Cemetery (Kemp et al. 2013: 71). Visual analysis of this upper limb degenerative joint disease has shown that it was often not too severe, whereas that in the lower limb, affecting nearly 50 per cent of adults, was often more severe. The people buried in the South Tombs Cemetery also display more pronounced musculoskeletal stress markers than other Egyptian skeletons, with the lower legs similarly exhibiting heavier workload relative to the upper limbs (Kemp et al. 2013: 74; Zabecki 2009). Ascertained from macroscopic visual study, the frequency of adult trauma, such as the fracturing of bones, was apparently also relatively high, and thus is consistent with people undertaking heavy, hard and potentially dangerous manual labour. Furthermore, although the subadults exhibited little visually identifiable skeletal trauma, both the two individuals with bony fractures were teenagers and the fractures (spondylolysis and a fractured foot bone) were likely work related. This potentially indicates that the young were working and labouring early in adolescence (Kemp et al. 2013: 72–73).

As assessed using standard osteological methods, the age at death was relatively young in the South Tombs assemblage, with the modal age at death being approximately 35 years of age (Kemp 2012a: 228). This pattern is consistent with epidemics of disease and mortality (Chamberlain 2006). Macroscopically and microscopically identifiable episodes of enamel hypoplasia record periods of physiological stress in childhood. These episodes might be linked to (small) waves of disease, leading to stress in subadults but mortality in some young adults, but sparing the older or elderly individuals (similar to the morbidity patterning of the 2009 swine 'flu pandemic; Writing Committee of the WHO Consultation 2010). Later Hittite records claim that Egyptian prisoners of war from the Amarna period spread an epidemic within Hittite society (Kemp 2012a: 228). Given the multiple inhumations within some graves, it is possible that these were either members of families or individuals who died at the same time and were thus buried together. If, in the future, export of samples for scientific analysis was permitted, or suitable equipment was available in laboratories within Egypt, then it might be possible to study the ancient DNA to ascertain whether the people buried in the South Tombs Cemetery, amongst others, suffered from epidemic or pandemic infectious diseases. The people lived in close proximity to their animals, and coprolite samples taken from around the animal pens from the Workmen's Village included *Ascaris* and *Taenia* eggs, identified by parasitological study (Donald 1984: 56–58). Taenia is a tapeworm, and commonly found in humans, dogs and cats. Ascaris is a parasitic nematode, the roundworm. Although it is unclear from which form of Ascaris these eggs derive, the pig form or the human form, infestation with roundworms can cause morbidity by compromising the individual's nutritional status (Hall et al. 2008), and thereby reducing disease resistance.

The fate of the royal family from Amarna is potentially less clear. Although the body of Tutankhamun from tomb KV62 is well known, and infected with malaria (Hawass et al. 2010: 643–646), the body of Akhenaten is unidentified. The body found in KV55, which has been purported to be Akhenaten (Hawass et al. 2010), may firstly be too biologically young to be Akhenaten (Baker 2010) and the STR (short tandem repeat) genetic data supports the proposal that body KV55 might be that of Smenkhare (Phizackerley 2010). The ethnicity and biological affinities of members of the royal family are also unknown, with Nefertiti having been suggested to have obscure origins, potentially from the Syrian region (Dodson 2009: 16). Such aspects could be further elucidated by further genetic studies, developing from those so far undertaken (Hawass et al. 2010), in association with isotopic studies of strontium and/or oxygen. Comparisons of diet between Amarna commoners and the royal family could be undertaken were carbon and nitrogen isotope signatures available from all such samples.

What did the people wear at Amarna? Over 4000 fragments of textile have been studied, mostly through lower power optical microscopy (Kemp & Vogelsang-Eastwood 2001). Flax (*Linum usitatissimum*) is the dominant raw material, making up 84 per cent of the textile assemblage (Kemp & Vogelsang-Eastwood 2001: 25). Some wool derives from sheep, but a little derives from goat hair. The sheep wool is relatively coarse wool, with both the warp and the weft being approximately 20 microns in diameter, although there are some pieces that are over 50 microns in diameter (Kemp & Vogelsang-Eastwood 2001: 35–53). In their study, Kemp & Vogelsang-Eastwood also note that the textiles are usually made with yarn that has a diameter of 0.2–0.3mm. The weaves used, both a tabby weave and a basket weave, usually involve two fringes; the selvedge and the warp fringe. Some pieces, such as linen cloths <224> and <226>, appear to have been sharply cut, such as with a knife. These fragments might benefit from SEM analysis in order to see the detail of the implement used for their cutting.

Production of such fabrics requires a manufacturing *chaîne opératoire*, which results in associated debris and equipment. At Amarna, for fabric weaving, these include whorls for spinning and loom equipment. Wooden spinning whorls were mostly sycamore fig and lebanon cedar (identified by visual analysis), although they were also made from limestone, and clay etc. Wear patterns are noted on the spinning bowls and provide evidence of the directionality of the yarn (Kemp & Vogelsang-Eastwood 2001: 291–295). This could be further elucidated, including potentially the identification of different wear traces from different threads, using SEM. Similarly friction grooves are found on loom items such as wooden shed-sticks, warp spacers etc. (Kemp & Vogelsang-Eastwood 2001). The bone points used in weaving (pointed bone spatula) also exhibit abrasion scratches and polishing. SEM study of these has indicated that the linear grooves are scratches from their manufacture and that the wear gloss derives from mechanical polish resulting from use as the striations are all in one direction (Kemp & Vogelsang-Eastwood 2001: 370–372). Fabrics, however, were not only employed for clothing or funerary purposes. Cloths were used for wicks, and GCMS analysis of such wicks showed presence of both moronic acid and oleanonic acid (Serpico 1996: 183–186). These are constituents of pistacia resin, and *Pistacia* resins have been identitified by GCMS from the contents of the Canaanite amphorae at Amarna (Serpico & White 2000; Stern et al. 2003). Other wicks contained traces of palmitic and stearic acids, and likely derived from animal fats and vegetable oils.

Of course, the city of Amarna cannot be considered in isolation; its links to other parts of Egypt and the rest of the world are as important and interesting as the city itself. Scientific analysis can and has been used to investigate these links. A useful way to characterise this is to think of what might be coming into the city from outside and what might be leaving the city to go elsewhere. This includes raw materials, finished objects, food and drink, animals and people. It is easier to pick out those things that are imported into Amarna that those that it exports. The reason for this is fairly obvious – it is relatively easy to see what resources the city might have in its immediate surroundings, and to identify those things that do not appear to fit in with them. Identifying exports is much more difficult as will be explained below.

Imports into Amarna have been identified by a whole range of techniques. There is a host of raw materials that are not found, or not found in sufficient quantities, within the city area that were obviously used within the city. Some of these have been discussed above, for example wood used as fuel and building material. Others include ceramics (see Box 4.1) and all metals, both base and precious. Thin section analysis is the main way of determining foreign ceramic wares when their form or decoration is not specific to a particular region. Even when it is, further analysis can pinpoint exactly from where the ceramics came. An example of this is the neutron activation analysis carried out by Mommsen and colleagues, who analysed Mycenaean sherds found at Amarna and were able to show they most probably originated from the northern Argolid, around Mycenae-Berbati (Mommsen et al. 1992). This sort of analysis has not only been carried out on ceramic vessels; many of the clay cuneiform tablets known as the Amarna letters, found in the Office of the Correspondence of Pharaoh, building Q42.21, have been analysed in an attempt to determine from where they might have come. A significant number of these letters are from places that cannot be 'placed' or definitely located on the map. As a result, chemical and petrographic analysis has narrowed down the possibilities for them. One of the more interesting results relates to those letters originating from 'Alashiya' which Goren and colleagues have traced, by the very specific mineralogy of the clays of these tablets, to the Troodos Mountains in Cyprus (Goren et al. 2003a; 2003b). Sourcing of metals at Amarna initially relied on trace element techniques, but increasingly isotopic analysis has been employed. Copper has been subjected to lead isotopic analysis (Stos-Gale et al. 1995; see Box 4.5 for an analogous study) and shown to derive most probably from mines in Cyprus. Lead itself, in the form of fishing weights and lead sulphide-based kohls, has had the same analysis and probably originates from lead mines on the Red Sea Coast (Shortland et al. 2000). Hard stone from the Palace and Temples has been identified as coming from the granite quarries around Aswan and elsewhere.

Identifying people who might have been born elsewhere and come to Amarna is of great potential interest. As noted earlier, strontium isotopic analysis of tooth enamel has been widely employed over the last few years to populations outside of Egypt in an attempt to identify foreign born individuals in cemeteries. Similarly, DNA analysis has the potential to do the same thing. It is clear that there were foreigners at Amarna. They are mentioned in texts and are shown in tomb scenes. Foreigners are depicted as mercenaries, as part of Akhenaten's bodyguard, as musicians and in other roles (Kemp 2012a: 269). They were almost certainly resident in the city at least for periods of time. Despite the potential, isotopic analysis has not been carried out on skeletal material from Amarna.

As noted above, tracing material produced in Amarna and found elsewhere is problematic. To do this requires confident identification of something unique about the raw materials or the way that they are processed at Amarna. Unfortunately, there is very little that is geologically or geographically unique about the area surrounding Amarna. It is very similar to considerable stretches of the Nile Valley from the First Cataract to the Delta. This would make it almost impossible to identify a person who was born and lived at Amarna and subsequently lived and died elsewhere. It might be possible to identify them as 'Egyptian' but beyond that seems very unlikely. In contrast, objects that were definitely made in Amarna have been found elsewhere. A good candidate is the Golden Throne of Tutankhamen, now in the Cairo Museum (JE62028). Found by Carter in the Tomb of Tutankhamen in the Valley of the Kings, the iconography of the Throne shows the King sitting under the Aten's disc in a classic Amarna Period pose. This identifies the throne as having been made at the start of Tutankhamen's reign since he relatively quickly moved away from the Aten cult and restored the older worship patterns. It therefore seems highly likely that the throne was made at Amarna; scientific analysis, however, plays no part in this attribution.

Perhaps the best examples of items that scientific analysis has played a part in identifying as coming from Amarna are those that were found on the wreck of the Ulu Burun ship (see Box 4.6 for background). The many glass ingots found on the ship can be broadly divided into two categories: light-blue, copper-coloured ingots which are relatively tall and narrow, and dark-blue, cobalt-coloured ingots which are relatively broad and flat. Analysis of the glass of the cobalt-blue ingots by microprobe and LA-ICP-MS shows it to be identical to the cobalt-blue glass found at Amarna. This is thought to have been made from a cobalt source from the Western Oases. This makes the dark-blue glass almost certainly Egyptian in source, but does not pin it uniquely to Amarna. However, the work of Turner and Paul Nicholson takes this further. It was noticed that the glass ingots were the right size and shape to fit into the cylindrical vessels found at Amarna associated with vitreous materials production, that Petrie had thought to be stands (Jackson & Nicholson 2013). These vessels occasionally have glass adhering to them and appear to have been used either in glassworking or melting (Nicholson et al. 1997). Thus the dark-blue ingots from the Ulu Burun wreck are the right composition and the right size to have been made at Amarna. The balance of probabilities has to favour that this was where they were made – a rare example of an identified Amarna export.

"The creation of Amarna led to the uprooting and migration of thousands of people. Did they go by choice?" (Kemp 2012a: 41). The creation of the city, or as Kemp (2012a: 20) argues the 'urban village' of Amarna, enables a pristine palimpsest of New Kingdom life to be studied. "The scale and variety of Amarna's archaeology allows for the development of broad research themes. One is the location and institutional context of manufacture" (Kemp 2012b: 53). The integrated approach, synthesising scientific methods with more traditional archaeology and Egyptology, has the potential to greatly elucidate features of the site and its organisation. In addition, this approach enables the people to be highlighted together within their role in the development of the cult of the Aten and the formation of the city complex. Amarna was not wholly destroyed after the death of Akhenaten, and, as demonstrated in this chapter, scientific approaches to Egyptian archaeology are aiding in its further reconstruction and understanding. Amarna has also provided an opportunity for archaeological science to 'break out' of the appendix or scientific report in that the results are actively integrated into synthetic critique and evaluation. The scientific aspects of research at Amarna have been sufficiently integrated into all the recent research that aspects of these scientific studies comprise portions of recent exhibitions without attracting comment for their scientific nature. Scientific study is an *integral part* of Amarna research and the framing of research questions. For example, in the recent Berlin exhibition to mark the centenary of the discovery of the famous Nefertiti bust, aspects of the scientific studies are thoroughly integrated within the exhibition and companion catalogue (Seyfried 2012b). It is this approach including the active integration of archaeological science within Egyptian archaeology and Egyptology that we advocate, and hope that this volume has demonstrated the immense potential and importance for the study of ancient Egypt.

Appendix I

King List and Summary of Egyptian Chronology

Palaeolithic

Lower Palaeolithic	*c.* 700/500,000–200,000 BP
Middle Palaeolithic	*c.* 200,000–35,000 BP
Upper Palaeolithic	*c.* 35,000–24,000 BP
Late Palaeolithic	*c.* 24,000–10,000 BP
Epipalaeolithic	*c.* 10,000–7000 BP

Predynastic   c. 5300–3000 BC

Lower Egypt

Neolithic	*c.* 5300–4000 BC
	(or *c.* 6400–5200 BP)
Maadi Cultural Complex	*c.* 4000–3200 BC

Upper Egypt

Badarian period	*c.* 4500–3800 BC
Naqada I period	*c.* 4000–3600 BC
Naqada II period	*c.* 3600–3300 BC

After *c.* 3300 BC the same chronological sequence is used throughout Egypt

Naqada III/'Dynasty 0'	*c.* 3300–3100 BC

Early Dynastic period   3100–2686 BC

1st Dynasty   *3100–2890 BC*

Narmer
Aha
Djer
Djet

Queen Merneith
Den
Anedjib
Semerkhet
Qa^ca

2nd Dynasty	**2890–2686** BC

Hetepsekhemwy
Raneb
Nynetjer
Weneg
Sened
Peribsen
Khasekhemwy

Old Kingdom	**2686–2125** BC

3rd Dynasty	**2686–2613**

Djoser (Netjerikhet)	2686–2667
Sekhemkhet	2667–2659
Khaba	2659–2656
Nebka	2656–2637
Sanakht?	
Huni	2637–2613

4th Dynasty	**2613–2494**

Sneferu	2613–2589
Khufu (Cheops)	2589–2566
Djedefre (Redjedef)	2566–2558
Khafre (Chephren)	2558–2532
Menkaure (Mycerinus)	2532–2503
Shepseskaf	2503–2498

5th Dynasty	**2494–2345**

Userkaf	2494–2487
Sahure	2487–2475
Neferirkare	2475–2455
Shepseskare	2455–2448
Raneferef	2448–2445
Nyuserre	2445–2421
Menkauhor	2421–2414

Djedkare	2414–2375
Unas	2375–2345

6th Dynasty	*2345–2181*
Teti	2345–2323
Userkare (a usurper)	2323–2321
Pepy I (Meryre)	2321–2287
Merenre	2287–2278
Pepy II (Neferkare)	2278–2184
Neitiqerty Siptah	2184–2181

7th and 8th Dynasties	*2181–2125*
Numerous kings, called Neferkare	

First Intermediate Period   2160–2055 BC

9th & 10th Dynasties (Herakleopolitan)	*2160–2025*
Khety (Meryibre)	
Khety (Nebkaure)	
Khety (Wahkare)	
Merykare	

11th Dynasty (Thebes only)	*2125–2055*
(Mentuhotep I (Tepy-a: 'the ancestor'))	
Intef I (Sehertawy)	2125–2112
Intef II (Wahankh)	2112–2063
Intef III (Nakhtnebtepnefer)	2063–2055

Middle Kingdom   2055–1650 BC

11th Dynasty (all Egypt)	*2055–1985*
Mentuhotep II (Nebhepetre)	2055–2004
Mentuhotep III (Sankhkare)	2004–1992
Mentuhotep IV (Nebtawyre)	1992–1985

12th Dynasty	*1985–1795*
Amenemhat I (Sehetepibre)	1985–1965
Senusret I (Kheperkare)	1965–1920
Amenemhat II (Nubkaure)	1922–1878

Senusret II (Khakheperre)	1880–1874
Senusret III (Khakaure)	1874–1855
Amenemhat III (Nimaatre)	1855–1808
Amenemhat IV (Maakherure)	1808–1799
Queen Sobekneferu (Sobekkare)	1799–1795

13th Dynasty	*1795–after 1650*

Hor (Awibre)
Khendjer (Userkare)
Sobekhotep III (Sekhemrasewadjtawy)
Neferhotep I (Khasekhemre)

Sobekhotep IV (Khaneferre)	*c.* 1725

14th Dynasty	*1750–1650*

Series of minor rulers who may be contemporary with the 13th or the 15th Dynasty.

Second Intermediate period	**1650–1550** BC

15th Dynasty (Hyksos)	*1650–1550*

Salitis/Sheshi

Khyan (Seuserenre)	*c.* 1600
Apepi (Aauserre)	*c.* 1555

Khamudi

16th Dynasty	*1650–1550*

Abydos Dynasty contemporary with 16th Dynasty
Minor Hyksos rulers who are contemporary with the 15th Dynasty

17th Dynasty	*1650–1550*

Intef (Nubkheperre)

Taa (Senakhtenra/Seqenenre)	*c.* 1560
Kamose (Wadjkheperre)	1555–1550

New Kingdom	**1550–1069** BC

18th Dynasty	*1550–1295*

Ahmose (Nebpehtyre)	1550–1525

Amenhotep I (Djeserkare)	1525–1504
Thutmose I (Aakheperkare)	1504–1492
Thutmose II (Aakheperenre)	1492–1479
Thutmose III (Menkheperre)	1479–1425
Hatshepsut (Maatkare)	1473–1458
Amenhotep II (Aakheperure)	1427–1400
Thutmose IV (Menkheperure)	1400–1390
Amenhotep III (Nebmaatre)	1390–1352
Amenhotep IV/Akhenaten (Neferkheperurawaenre)	1352–1336
Nefernefruaten (Smenkhkare)	1338–1336
Tutankhamen (Nebkheperure)	1336–1327
Ay (Kheperkheperure)	1327–1323
Horemheb (Djeserkheperure)	1323–1295

19th Dynasty	*1295–1186*
Ramesses I (Menpehtyre)	1295–1294
Sety I (Menmaatre)	1294–1279
Ramesses II (Usermaatre Setepenre)	1279–1213
Merenptah (Baenre)	1213–1203
Amenmessu (Menmire)	1203–1200
Sety II (Userkheperure Setepenre)	1200–1194
Saptah (Akehnrasetepenre)	1194–1188
Tausret (Sitrameritamun)	1188–1186

20th Dynasty	*1186–1069*
Sethnakhte (Userkhaure Meryamun)	1186–1184
Ramesses III (Usermaatre Meryamun)	1184–1153
Ramesses IV (Heqamaatre Setepenamun)	1153–1147
Ramesses V (Usermaatre Sckheperenre)	1147–1143
Ramesses VI (Nebmaatre Meryamun)	1143–1136
Ramesses VII (Usermaatre Setepenre Meryamun)	1136–1129
Ramesses VIII (Usermaatre Akhenamun)	1129–1126
Ramesses IX (Neferkare Setepenre)	1126–1108
Ramesses X (Khepermaatre Setepenre)	1108–1099
Ramesses XI (Menmaatre Setepenptah)	1099–1069

Third Intermediate Period   1069–664 BC

21st Dynasty	*1069–945*
Smendes (Hedjkheperre Setepenre)	1069–1043

Amenemnisu (Neferkare)	1043–1039
Psusennes I (Pasebakhaenniut)	
(Akheperre Setepenamun)	1039–991
Amenemope (Usermaatre Setepenamun)	993–984
Osorkon the elder (Akheperre Setepenre)	984–978
Siamun (Netjerkheperre Setepenamun)	978–959
Psusennes II (Pasebakhaenniut)	
(Titkheperure Setepenre)	959–945

22nd Dynasty	*943–736*
Sheshonq I (Hedjkheperre)	943–922
Osorkon I (Sekhemkheperre)	922–888
Takelot I (Hedjkheperre)	888–872
Osorkon II (Usermaatre)	872–831
Sheshonq III (Usermaatre)	831–791
Sheshonq IV (Hedjkheperre Setepenre)	791–779
Pimay (Usermaatre)	779–773
Sheshonq V (Aakheperre)	773–736

23rd Dynasty	*736–666*
Osorkon IV (Usermaatre)	736–716+
Gemenefkhonsubak (Shepseskare Irenre)	
Pedubast III (Sehetepibenre)	–666

Thebes	*–753*
Horsieset I (Hedjkheperre Setepenamun)	–840
Takelot II (Hedjkheperre Setepenre)	834–810
Pedubast I (Usermaatre)	824–800
Shoshenq VI (Usermaatre Meramun)	800–794
Osorkon III (Usermaatre)	791–762
Takelot III (Usermaatre Setepenamun)	768–753

24th Dynasty (Lower Egypt)	*734–721*
Tefnakhte	734–726
Bakenrenef	726–721

25th Dynasty	*–656+*
Kashta (Maare)	–754
Piye (Menkheperre)	754–722
Shabaqo (Neferkare)	722–707

Shabitqo (Djedkaure)	707–690
Taharqo (Khunefertemre)	690–664
Tanutamani (Bakare)	664–656+

Late Period   664–332 BC

26th Dynasty (Saite Period)   664–525

(Nekau I	672–664)
Psamtek I (Wahibre)	664–610
Nekau II (Wehemibre)	610–595
Psamtek II (Neferibre)	595–589
Apries (Haaibre)	589–570
Ahmose II (Amasis) (Khnemibre)	570–526
Psamtek III (Ankhkaenre)	526–525

27th Dynasty (1st Persian period)   525–404

Cambyses	525–522
Darius I	522–486
Xerxes I	486–465
Artaxerxes I	465–424
Darius II	424–405
Artaxerxes II	405–359

28th Dynasty   404–399

Amyrtaios	404–399

29th Dynasty   399–380

Nepherites I (Nefaarud)	399–393
Hakor (Achoris) (Khnemmaatre)	393–380
Nepherites II	*c.* 380

30th Dynasty   380–343

Nectanebo I (Kheperkare)	380–362
Teos (Irma atenre)	362–360
Nectanebo II (Senedjemibre Setepenanhur)	360–343

2nd Persian period   343–332

Artaxerxes III Ochus	343–338
Arses	338–336
Darius III Codoman	336–332

Ptolemaic Period   332–32 BC

Macedonian Dynasty   *332–305*

Alexander the Great	332–323
Philip Arrhidaeus	323–317
Alexander IV	317–310

Ptolemaic Dynasty

Ptolemy I Soter I	305–285
Ptolemy II Philadelphus	285–246
Ptolemy III Euergetes I	246–221
Ptolemy IV Philopator	221–205
Ptolemy V Epiphanes	205–180
Ptolemy VI Philometor	180–145
Ptolemy VII Neos Philopator	145
Ptolemy VIII Euergetes II	170–116
Ptolemy IX Soter II	116–107
Ptolemy X Alexander I	107–88
Ptolemy IX Soter II (restored)	88–80
Ptolemy XI Alexander II	80
Ptolemy XII Neos Dionysos (Auletes)	80–51
Cleopatra VII Philopator	51–30
Ptolemy XIII	51–47
Ptolemy XIV	47–44
Ptolemy XV Caesarion	44–30

Roman Period   30 BC–AD 311

Augustus	30 BC–AD 14
Tiberius	AD 14–37
Gaius (Caligula)	37–41
Claudius	41–54
Nero	54–68
Galba	68–69
Otho	69
Vespasian	69–79
Titus	79–81
Domitian	81–96
Nerva	96–98
Trajan	98–117
Hadrian	117–138
Antoninus Pius	138–161
Marcus Aurelius	161–180

Lucius Verus	161–169
Commodus	180–192
Septimius Severus	193–211
Caracalla	198–217
Geta	209–212
Macrinus	217–218
Didumenianus	218
Severus Alexander	222–235
Gordian III	238–242
Philip	244–249
Decius	249–251
Gallus and Volusianus	251–253
Valerian	253–260
Gallienus	253–268
Macrianus and Quietus	260–261
Aurelian	270–275
Probus	276–282
Diocletian	284–305
Maximian	286–305
Galerius	293–311
Constantius	293–306
Constantine I	306–337
Maxentius	306–312
Maximinus Daia	307–324
Licinius	308–324
Constantine II	337–340
Constans (co-ruler)	337–350
Constantius II (co-ruler)	337–361
Magnetius (co-ruler)	350–353
Julian the Apostate	361–363
Jovian	363–364
Valentinian I (west)	364–375
Valens (co-ruler, east)	364–378
Gratian (co-ruler, west)	375–383
Theodosius (co-ruler)	379–395
Valentinian II (co-ruler, west)	383–392
Eugenius (co-ruler)	392–394

All dates given here are approximate, and based on or derived from Dodson (2009; 2012; 2014), Hornung et al. (2006c), Marée (2010), and Shortland & Bronk Ramsey (2013).

Appendix II

Summary of Some Major Analytical Techniques and Methods

Although by no means exhaustive, this appendix aims to give the Egyptologist or archaeologist working in Egypt some appreciation for how analysis is carried out, using some of the various different scientific techniques considered in the text. This appendix provides a brief outline of the basic aspects of analysis that are of importance. While some appreciation of the physical principles behind each of the techniques is given, this is done on quite a superficial level. The idea of this section is to allow non-scientists to appreciate and interpret the data produced by the techniques, their strengths and weaknesses, and why some techniques are preferred to others. In addition, brief considerations for applicability or sampling procedures are also given. Further information on all techniques can be found in standard works, such as Brown & Brown (2011), Brothwell & Pollard (2001), Malainey (2011) or Pollard & Heron (2008).

Before summarising the methods and techniques used in this text, brief mention will be made of accuracy and precision. These are distinct and different. Accuracy refers to how certain the data are to *being* the *real* state. A method is accurate if it includes the correct (true) result. Precision refers to how close the data are to the *real* state. The closer the data to the actual state, the more precise is the technique. For example, suppose a mummy contains the remains of an individual who was 29 years of age at death. An accurate age would include this number, so an ageing methods that indicates that the mummy is that of an adult is accurate, but it is only precise if that age range is narrowed down – such as to between 25 and 35 or similar. However an ageing method that estimates the age of the individual as between 40 and 50 would be precise but not accurate! Precision also refers to the ability to deliver reproducible results, such as by repeating analyses or measurements undertaken in the same way. The mean and standard deviation thus provide a confidence interval for the data. Assuming a normal distribution of data, the standard deviation (1σ) above and below the mean value cover 68 per cent of the range of the possible data, and 2σ cover 95 per cent of the distribution. It is therefore imperative to remember that when considering the mean and 1 standard deviation (i.e. mean $\pm$ 1σ), the true value can fall outside of this range 32 per cent of the time.

It is also important to remember the sensitivity of the methods being used. These are the detection limits (usually of the equipment or methods being employed). There are further issues which relate to the reliability and replicability, and hence impinge on sensitivity. For example, digital sliding callipers might have the ability to measure to 0.001mm, but it is not possible for a human to repeat an osteological measurement of bone to that degree of precision (as dust might be present in the callipers or the cortical surface might erode etc.). It is therefore imperative to be aware of the sensitivity, precision and accuracy being claimed and to judge these in the context of the method employed and its norms.

Scientific analyses usually produce a series of results. The form of these results varies, but the archaeologist or Egyptologist normally interprets the results in the context of the greater archaeological or Egyptological milieu. The significance of the results can only be understood from their archaeological context. It is therefore imperative to synthesise the science with the archaeology throughout the research process. The analytical scientist might require a series of previous data or assumptions in order to undertake the analyses. Having the archaeological scientist within the research project means that these assumptions or previous data (e.g. archaeological contexts or degree of potential contamination) can be appropriately considered or selected, and thus suitable analyses undertaken. The archaeology or Egyptology and the science must be thoroughly integrated and synthesised.

Dating Methods

Although various different dating methods have been discussed in this text, the primary method of use in Egypt is radiocarbon-based. We have provided very brief details also for dendrochronology and trapped charge dating methods. For details on the precautions that should be taken when sampling for dating, we recommend Malainey (2011) or Pollard & Heron (2008).

Radiocarbon Dating (including Accelerator Mass Spectrometry AMS)

Introduction

Radiometric dating relies on the emission of radiation and the half-life of the radioactive isotopes being studied. The most commonly used method in archaeology today is AMS ^{14}C although, at the time of writing, a facility for such analyses does not exist in Egypt.

Principles

Rather than using emission rates to estimate the amount of radioactive carbon present in a sample, AMS determines the relative amount of ^{14}C atoms in the sample. Using this method a mass spectrometer connected to a particle accelerator sorts the carbon ions on the basis of mass and measures their relative concentrations.

The sample is combusted to produce CO_2 which is then converted to graphite. This solid carbon is placed in a vacuum and is bombarded with caesium ions. The carbon is ionised and passes magnets operating like those in normal mass spectrometers. The analyser magnets separate out the carbon ions by mass as heavier ions are less deflected by the magnet than the lighter ions, meaning that the latter have tighter curves to their pathway.

Sample preparation

Samples can be obtained from a wide variety of materials, with the quality depending mostly on the integrity of the sample. This is considered in much greater detail in Section 2.1.2.2: Radiocarbon Measurements. Samples are usually subjected to pre-treatment to eliminate extraneous carbon. Most samples are crushed and washed in deionised water. Following this, further pre-treatment might occur, such as, for charcoal or wood, use of hydrochloric acid to dissolve carbonates and/or sodium hydroxide to remove secondary organic acids, followed by a final acid wash to remove the carbon dioxide. Samples are then converted into CO_2, usually by combustion, although for carbon in shell or some other samples, acid hydrolysis might be used. For liquid scintillation counting (LSC, as is available at the IFAO) the CO_2 is converted into acetylene or benzene for analysis.

Advantages

The samples required are very small (can be < 1mg). Analysis is very rapid and can be precise. Additional use of calibration methods and Bayesian modelling of dates can increase precision and accuracy still further. Outside of Egypt, the method is easily and cheaply available.

Disadvantages

The only real disadvantage is that the method is destructive as a (micro) sample has to be taken. Note must be taken of temporal fluctuations in the carbon curves, but these are considered by the actual dating analyst during analysis processing. Care must also be taken when converting radiocarbon ages into calendar dates.

Dendrochronology

Introduction

Tree-ring dating is deceptively simple as (almost) everyone is aware of the concentric rings within wood.

Principles

Some species of tree add their annual growth increments in two parts – spring wood (or early wood) and summer wood (late wood). When seen on the end-grain of a section of tree, these increments appear like rings. Trees in a given climatic region are affected in similar ways and so tree rings can be matched (crossdated) and so each increment can be assigned to a particular year

Sample preparation

Full cross-sections are preferred as these provide the greatest quantity of information. Where this is impossible, coring may be undertaken. Occasionally, high resolution photography of end-grains will permit dating. In order to avoid spurious apparent tree-ring matches, samples usually require at least 100 rings. In addition, multiple sections of variation in ring-width are preferred. It is important to include as much of the sapwood as possible and avoid areas with knots, cracks or other blemishes that might interfere with the patterning. Presence of the bark enables the year of felling to be identified. The surface is prepared for study with fine sandpaper or use of a razor blade. Binocular dissecting microscopes are used to identify and measure the rings.

Advantages

The method can be applied to dry wood, wet wood (although rarely found in Egypt) or burned wood (charcoal). A felling time can sometimes be identified when a fragment of bark (and thus the final ring) is also visible.

Disadvantages

The method only works on species that have clear annual growth rings. Reuse of wood from earlier felling can cause confusion and difficulty. This is important in Egypt when certain woods might have been an important resource. Cross-dating can only be done with areas of similar climatic regimes, and thereby can limit comparator samples.

*Trapped Charge Dating (including Electron Spin Resonance
ESR, Thermoluminescence TL and Optically Stimulated
Luminescence OSL)*

Introduction

Trapped charged dating methods comprise a series of methods that
consider the effect of time on the accumulation of electrons and holes in
the crystal lattice of certain minerals. In archaeology, ESR is mostly used
for dating tooth enamel, TL for sediments and burnt flint whereas OSL is
used primarily for dating of sediments.

Principles

When a mineral is formed, all the electrons are in the ground state. This
also occurs when the mineral is "reset". Naturally occurring radioisotopes
emit rays which ionise atoms by exciting the electrons and move them
to higher energy shells. This leaves a positively charged hole near the
original shell and a negative charge in the higher energy area. After a
period of time, the electrons return back to their original states. Natural
minerals, however, contain defects in which either these excited electrons
or the holes can be trapped.

For a measurement of luminescence, the trapped electrons have to be
activated by heating or by light exposure. A "reset" occurs when there is
heating or exposure to light and thus restores the trapped electrons to
their original (natural) states.

In ESR spectrometry, the sample is placed into a microwave cavity
located in a magnetic field. The trapped electron or hole has a magnetic
moment and can be flipped to the opposite direction by the absorption of
microwave energy. ESR measures the microwave absorption with respect
to the magnetic field.

TL activates the trapped electrons by heating, leading to light emission
when the electrons recombine with the luminescence centres. At high
temperatures, the electrons are excited and thus light is emitted. The light
photons are converted to electric pulses via a photomultiplier and this
light emission is plotted as a glow curve relative to heating temperature.

By contrast, in OSL light from a narrow frequency range is shone onto
the sample. Again this activates the electrons which combine with the
luminescence centres, thereby emitting light. The light emitted is plotted
against the time elapsed after the light was turned on.

Sample preparation

The main issue with these methods is the normal requirement to take
samples in the dark and ensure that they are not exposed to light or heat.

Advantages

The methods are suitable for dating samples that cannot easily be dated by other means. They can be reasonably accurate and precise. The methods are applicable for ages beyond the limit of radiocarbon, such as potentially several million years old and hence are applicable to fossil material and palaeoenvironmental reconstruction.

Disadvantages

The requirement to maintain samples in dark and cool conditions is problematic. Furthermore, neither the equipment nor the analyst are easily available and so the methods can be quite costly.

Surface Study Methods

The following methods are applicable when wishing to study the structure, morphology or composition of the surface of an object or item. These methods focus on surfaces and so might not provide information on the internal structures or compositions.

Scanning Electron Microscope (SEM) – usually with Energy Dispersive Spectrometry

Introduction

The SEM is one of the most common pieces of analytical equipment in scientific laboratories and is an almost standard laboratory "workhorse". It is an imaging tool, designed to take pictures at very high magnification. As such it is extremely useful if looking at surfaces or sections of three-dimensional objects. In addition to the imaging capability, modern SEMs are frequently fitted with one or more spectrometers, such as an Energy Dispersive Spectrometer (EDS), thereby making an SEM-EDS. The EDS allows the elemental composition of the item to be determined. Furthermore, recent advances in SEM technology have included the development of Controlled Pressure or Variable Pressure SEM.

Principles

The SEM consists of an electron gun which is usually mounted pointing down into a sample chamber. The electron gun fires electrons through collimators and lenses into the sample chamber where the electrons hit the sample. The electrons fired at the sample excite the electrons orbiting the atoms at the surface of the sample, thereby making them more energetic. Over a short period of time, these latter electrons lose this energy by giving off an X-ray photon. X-rays given off by the sample or backscattered from the sample are used to create an image of the surface. The energy of the X-rays emitted is characteristic of the element involved and the number of X-rays given off is proportional to the amount of that element in the sample (at least at the surface). The different X-rays given off by the sample are picked up by the detector, which distinguishes between them on the basis of their different energies, hence Energy Dispersive Spectrometer (EDS). The EDS measures all the different energies, and the elements are revealed by more counts at their particular characteristic energies. The compositional aspect is usually depicted as a series of (compositional) peaks on a graph where keV is plotted on the x-axis and the number of photons detected is plotted on the y-axis. Most elements, especially high atomic number elements, give off multiple peaks. Complex algorithms are used to turn these peaks into concentrations in the sample. Broadly speaking, the higher the peaks, the more of the element in the sample, however the height of the peaks is affected also by the presence of other elements in the sample.

Sample Preparation

As a sample almost always has to be taken, this method is micro-destructive. The sample must be dry, so the method is highly suitable in most Egyptian contexts. It is especially useful for metals, ceramics, stone, glasses and glazes.

Three possibilities for sample preparation exist. The first and most commonly used in the past was to take a fragment of the object needing analysis, mount it in a resin block and polish it flat. The whole block was then coated in a conducting material, such as carbon or gold (to avoid a build-up of charge on the sample affecting both the image and the analysis). The second possibility is to mount the fragment on a carbon pad on a small stub and again apply a conductive coat to it. This is much quicker and is safer with very small or delicate samples which might be damaged or lost in polishing. The third approach is to put the whole object into the vacuum chamber. This has the advantage that a sample does not have to be taken from the object. Unless the object itself is metal and therefore is conductive, for a conventional SEM the object will still have to be coated with a conducting material. Furthermore, there is also a chance of the object shattering under the vacuum, and so, for example, this means that in practice, whole glass objects are rarely put in conventional SEM. New controlled pressure SEMs (CP-SEM) employ a new technology to conduct away the charge build up and permit whole objects to be put in and analysed without a conductive coating. They still operate under a vacuum and so there is still the possibility of damage to or shattering of the object.

Advantages

The SEM is a relatively cheap and quick technique for analysis and exists in many university science departments. They are comparatively easy to run and maintain. SEM is one of the best ways to obtain microscopic images of samples, usually at higher magnifications than possible using optical microscopes. SEM-EDS is very flexible as it is capable of analysing all elements from carbon (atomic number 6) to uranium (atomic number 92). The results produced, if run carefully with the right standards, are reasonably accurate.

Disadvantages

Apart from when using CP-SEM, a sample has to be taken from the object. This is intrinsically (micro)destructive. Furthermore, the sample normally requires coating with a conductive substance, thereby rendering it less useful afterwards in other analyses. To obtain the best compositional results, the sample should be as flat as possible. As a result, best results tend to be derived from polished blocks, whereas stub mounting small fragments tends to produce less reliable results. The major disadvantage, however, of the SEM-EDS is the resolution of the EDS spectrometer as most spectrometers have relatively low energy resolution. This means that instead of obtaining tall and sharp peaks with background on either side, peaks tend to overlap and merge into one another. This can often be resolved, but some co-occurring elements have peaks that are very close together. For glass, some of the worst examples are Fe Kb (7.057 keV) and Co Ka (6.929 keV), and Pb La (10.548 keV) and As Ka (10.543 keV). Cobalt can therefore be difficult to measure, especially in small quantities when iron is present, and similarly, arsenic is hard to measure when lead is also present.

Limits

Detection limits for the SEM-EDS are normally in the order of 0.1–0.5 per cent. Precision tends not to be not as good as some other methods, such as the microprobe (SEM-WDS).

Microprobe (SEM-WDS)

The microprobe is very closely related to the SEM and works using wavelength dispersive spectrometry (WDS). In modern microprobes, SEM imaging is included so that phases can be quickly identified and targeted. The basic system configuration usually has to be set up in favour of either imaging (SEM-EDS) or analysis (microprobe, SEM-WDS).

Principles

The microprobe is very similar to the SEM as it also has both an electron gun and a sample chamber. Furthermore, the physical principle of characteristic X-rays being given off after excitation by an electron stream is the same. The difference is that the wavelength of the X-rays are measured rather than their energy. The wavelength of the X-ray is element specific. The X-rays are diffracted through different crystals within the WDS detector and the diffraction angle produced is characteristic of the X-ray wavelength. The detector swings round various different positions in order to measure the intensity at the various characteristic angles and the associated background intensity. These intensity data provide the compositional information. As the detector has to be moved into different positions, each element has to be measured sequentially rather than simultaneously. As a result, analysis times can be long, although many microprobes now have multiple spectrometers (up to five), with each working on a different element, in order to shorten the analysis time. Furthermore prior knowledge or information is required. Unlike SEM-EDS, the elements being investigated must be imputed in advance, and you cannot later go back and look for another element without completely re-running the analysis.

Sample Preparation

Samples for analysis must be polished perfectly flat. Any slope or irregularity on the sample affects the geometry of the whole system and directly impacts on the results. Microprobe analyses are therefore normally carried out on samples mounted on specially polished and coated resin blocks. Sample size is restricted by the need to be able to mount and polish effectively.

Advantages

Microprobe analyses tend to be more precise and accurate than SEM-EDS. This is because measuring wavelengths gives the system a higher peak resolving power and because there are fewer problematic overlapping peaks. A microprobe can therefore run on most elements with an average detection level down to 0.05–0.02 per cent, and hence has been the equipment of choice for giving good quality basic compositional analyses.

Disadvantages

Microprobes are not as common as SEM-EDS and access can be difficult. They need constant care and calibration, and can require a lengthy set up time in advance of study. Even a well-run microprobe can take a day merely to calibrate and be ready to run on the 20–25 elements. Analysis time also tends to be slow. In addition, the analysis area of a microprobe is very small, typically around a few tens of microns across. This means that obtaining a reasonable average for a sample can require many points.

X-Ray Fluorescence (XRF)

Introduction

This is a non-destructive technique that is ideal for initial studies of small surface features. There are two main forms of XRF: benchtop and handheld (HH-). XRF permits analysis of surface area composition.

Principles

A high energy X-ray beam is aimed at the surface of a sample, and, like in SEM, this excites the electrons orbiting the atoms at the surface of the sample, thereby moving them into higher energy shells. These latter electrons quickly lose this energy in the form of emitted X-rays. These X-rays are characteristic of the transitions between the shells and therefore indicate which element is present at the surface of the sample. Light elements have few shells whereas heavy elements have many shells. The quantity of X-rays emitted is proportional to the amount of that element in the sample (at least at the surface).

Sample Preparation

No sample preparation is required and whole objects can be analysed as long as they can fit inside the sample chamber (when using a bench-top XRF). Handheld XRF is portable and the equipment can fit into a small suitcase, but the sample must be placed in direct contact with the tip of the instrument. This highlights the fact that the area being studied can only be on the absolute surface and, when using the benchtop XRF, must be uppermost on the object being studied. This makes internal surfaces of vessels almost impossible to analyse unless broken or fragmented.

Advantages

The process is entirely non-destructive and no sample preparation is required. These are huge advantages of this method as an initial mechanism for compositional analysis. The method is very quick and can also cover and identify a large range of elements. The detection limits are usually reasonably good, and normally are of the order of 1000ppm or better. The handheld method is extremely useful for large immobile objects and permits analysis in situ. The benchtop version is more suitable to studying and mapping variation in pigment or surface over larger areas.

Disadvantages

Both methods require the use of X-rays. The handheld form includes an open source of X-rays and hence strict operating procedures must be followed to minimise risks to health and safety. The area being studied is small and may not be representative of other surface features (such as those that are not easily accessible or are on undersides). Quantification precision and accuracy are not always as high as might be hoped with 10–20 per cent errors often occurring. In addition, interpretation of the spectra produced is not simple.

X-Ray Diffraction (XRD)

Introduction

This method bears some similarity to XRF in that X-rays are fired towards a sample. The difference is that in XRD it is the diffraction of the X-rays by the lattice spacing of the sample that is important.

Principles

Crystalline substances consist of regular and repeating lattices of atoms. These lattices usually contain many repeated planes of atoms. When a crystal lattice is placed into an X-ray beam, the lattice structure diffracts the X-rays along the planes of the crystal lattice. The diffraction is related to the spacing of the planes within the lattice. This spacing is characteristic of the mineral phase present. The X-rays detector sweeps through the angles of possible diffraction and measures the intensity of the X-rays at the different angles.

Sample Preparation

Most XRD is undertaken on powdered samples of material placed onto special slides. It is also possible to carry out XRD on single crystals and some complete objects. The quantity of sample to be powdered is usually quite large.

Advantages

The method can give good identifications of crystalline compounds and can identify both organic and some inorganic materials. In certain circumstances, quantitative results can also be produced.

Disadvantages

The method is normally destructive as a sample usually has to be taken and powdered. Where only traces of compounds are present, these cannot usually be identified due to the detection limits of the method. The results spectra can be confusing and interpretation and understanding relies on good libraries of similar comparator spectra.

Fourier Transform Infra-Red Spectroscopy (FTIR)

Introduction

This is a technique that measures the absorption of light to characterise the structure of materials.

Principles

This method uses a modified microscope in order to measure the absorption of light. A broadband light source shines through an interferometer so that a beam of light of multiple wavelengths is formed. These are then shone at the sample. The wavelengths of the light change through the analysis. The light interacts with the sample and the sample absorbs light of specific wavelengths depending on its internal structure. The absorption is caused by chemical bonds being excited at specific light wavelengths, thereby absorbing the energy. An absorbed light spectrum is thus formed.

Sample preparation

Very small samples are needed and can be mounted directly into the FTIR equipment. The sample should ideally comprise only a few compounds; with larger numbers of constituent compounds it becomes hard to distinguish each one from the results.

Advantages

The method can provide very good identification of both organic and some inorganic compounds, but the former are generally better, such as identifying pigments, waxes or resins.

Disadvantages

There are two main disadvantages. The first is the requirement to take a sample, so the method is (micro)destructive. The second is that the resulting spectra are often confusing and so useless unless there is a similar sample in the comparator libraries of spectra for identification. In addition, detection is limited and so trace quantities cannot often be identified.

Raman Spectroscopy

Introduction

Like FTIR, this method involves the study of changes in vibrational energy. Raman spectroscopy involves the analysis of monochromatic light scattered by a molecule.

Principles

After excitation by irradiation, a molecule usually returns to its initial energy level after releasing the additional energy received. A few molecules in an irradiated sample may not return to their original state, but instead move to a lower vibrational energy level which may be at either a higher or lower energy level than the original initial state. The energy scattered by these molecules (Raman scattering) may be at either a higher or lower frequency than the incident beam. Unlike in FTIR, covalent bonds (which do not absorb infrared energy) can be studied by Raman spectroscopy. Raman spectroscopy requires extremely intense incident light, so lasers are commonly used, with the scattered radiation detected at 90° to the incident beam.

Sample preparation

Very small samples are needed and can be mounted directly into the Raman spectroscopy equipment.

Advantages

Raman spectroscopy can be untaken on aqueous solutions, crystals and films (unlike FTIR). Raman spectra are also easier to interpret than FTIR spectra.

Disadvantages

Like FTIR, there are two main disadvantages. The first is the requirement to take a sample (albeit very small), so the method is (micro)destructive. The second is that the equipment is not commonly available and so cost can be an issue.

Compositional Analyses

The following series of methods enable aspects of the internal composition to be understood. Some are also applicable to samples taken from the surface and so may be integrated with the above methods.

Inductively Coupled Plasma Mass Spectrometry (ICPMS)

The technique uses a plasma (ICP) to ionise a sample and then measures composition using a spectrometer. The methods belongs to a group of techniques including inductively coupled plasma optical emission spectrometry (ICP-OES) and inductively coupled plasma atomic emission spectrometry (ICP-AES).

Principles

For conventional ICPMS, the sample must be dissolved in a solution. The solution is then sprayed into a plasma torch where it is sequentially vapourized, atomized, and ionized at a temperature of several thousand degrees Celsius. The ions enter a quadrupole mass spectrometer, comprising four metal bars arranged parallel to each other. The bars have an electric charge applied across them meaning that only certain ions of specific mass-to-charge ratios will reach the detector at the end. The equipment can be set up to measure either specific masses by very carefully choosing the voltage applied, or can measure a whole range of masses by sweeping through the various voltages and measuring all the ions that hit the detector. The various intensities of the ions hitting the detectors for each mass can be used to calculate the concentration of the ion in the plasma and hence the concentration of the element in the original sample. It is normally necessary to know the concentration of a certain element in the sample in order to undertake a calibration comparison with that known element. This is usually done by specifically adding a rare element to the solution in a known dose (a "spike") or by measuring an element using another method and using this to calibrate all the others.

Sample Preparation

Samples may be prepared for ICPMS by either dissolving in a solution or by using laser ablation (LA-ICPMS). Dissolving some materials in acid is not easy and can require concentrated acids. It is difficult to ensure that the solution represents the composition of the full sample rather than only a portion of the sample dissolving and hence potentially being unrepresentative. Laser ablation instead uses a laser to remove a very small sample: this new sample is then carried in an argon or helium stream into the plasma flame. This removes the difficulties of dissolving a sample and the process of direct ablation leaves a sampling scar of less than 0.1mm in diameter. For calibration to produce absolute values, one element requires measurement using another technique. The most common method is normally to measure either calcium or silicon by microprobe in advance of LA-ICPMS.

Advantages

(LA-) ICPMS is theoretically capable of measuring almost all elements from lithium (atomic number 3) to uranium (92), with the exception of the noble gases and carbon, nitrogen, oxygen, fluorine and chlorine. The method has very low detection limits, being able to identify parts per billion (or, in comes cases, parts per trillion). This means that far more compositional elements can be identified and quantified. The sample size used in LA-ICPMS is very small, and the technique is fairly rapid, so, after the equipment is set up, each analysis takes a couple of minutes depending on the number of elements and the degree of precision. As the chamber is not under vacuum, it is also possible to put items or objects directly into the sample chamber for analysis.

Disadvantages

LA-ICPMS is not common laboratory equipment and so access can be quite difficult and requires a relatively complex set-up. Detection limits of different elements vary greatly. Just as SEM has overlapping energy peaks, LA-ICPMS has a similar problem caused by two types of isobaric and polyatomic inferences. LA-ICPMS determines elemental compositions by mass-to-charge ratio, but some elements have isotopes that have equal weight. For example, both iron and nickel have isotopes of weight 58, hence the intensity measured at mass-to-charge ratio 58 will contain both iron and nickel. This isotope can be ignored, or a different isotope for these elements could be used. Polyatomic interferences occur when the plasma does not completely breakdown the sample into monatomic ions. This is particularly significant in the very high abundance ions. Argon is the carrier gas bringing the sample into the plasma and has atomic mass 40. If combined with oxygen from the air, argon oxide has a mass of 56 which is exactly the same as the most abundant isotope of iron.

Gas Chromatography (GC) – usually with Mass Spectrometry (-MS)

Introduction

Gas chromatography is a separation technique, used to separate out the individual constituents of mixtures of organic compounds. A sample is dissolved in a fluid, and passed through or over a stationary phase which uses the different physical and chemical properties of the different constituents to separate them out. Gas chromatography is commonly used in association with mass spectrometry, but there are other forms of chromatography sued in archaeology, such as liquid chromatography.

Principles

GC is used to separate out organic compounds, or derivatives of such compounds. The sample is dissolved in a carrier gas, usually helium or hydrogen, and then passed over or through liquid coated on the interior of a silica column. The column is in an oven and the temperature of the oven slowly rises. Initially the vector gas evaporates and leaves the sample as a liquid residue. As the temperature rises, the constituents of the sample with progressively higher boiling points evaporate and move through the column. Separation of the components is thus based on the different volatilities of the components and their interactions with the stationary phase.

Sample preparation

Samples are usually injected into the GC column in solution. Dissolving the samples in solution varies in difficulty, and non-volatile solid samples can be prepared using pyrolysis.

Advantages

GC-MS is commonly available and so has a relatively low cost, although use of autosamplers, which automatically syringe samples into the column, are variably available and therefore may make analysis expensive.

Disadvantages

By its very nature, the method is intrinsically destructive. The size of the sample required depends upon the material in which the sample is found and the nature of the sample itself.

Isotopic Analysis

Introduction

Archaeology uses both light (carbon, nitrogen, oxygen, sulphur, etc.) and heavy (strontium, lead) isotopes. The former are commonly used to study diet in both humans and animals, whereas the latter are used for provenancing both inanimate objects and once living humans or animals. Some isotopes, such as some of strontium (e.g. ^{90}Sr and ^{85}Sr), are radioactive and decay. Others, such as ^{12}C, are stable.

Principles

Isotopes are alternative forms of the same element, and thus have different elemental masses depending on the number of neutrons in the element's nucleus. When studying diet, stable isotopes of carbon and nitrogen in bone collagen and/or bioapatite are quantified by isotope ratio mass spectrometry. This works because the different isotopes, such as ^{14}N and ^{15}N, have slightly different mass-to-charge ratios.

For carbon and nitrogen studies, a magnetic mass spectrometer is used (rather than the quadrupole system mentioned earlier for ICPMS). The collagen or other material is combusted to convert it to carbon dioxide and nitrogen, before electron ionisation and passage through a mass spectrometer. The ions of the different isotopes are deflected to different extents by the magnet in the mass spectrometer, where they are then counted by different Faraday collectors.

Heavy isotopes, such as strontium, are normally studied using either Thermal Ionisation Mass Spectrometry (TIMS), Plasma (exposure) Ionisation Mass Spectrometry (PIMS) or ICPMS.

TIMS vaporises a sample by heating it on a wire in a vacuum where it is ionised and accelerated down a thin flight tube. The ions are then made to pass round a curve in the tube by subjecting them to a magnetic field. The heavier ions have a higher momentum so will deviate from their path less than the lighter ions. Thus the different isotopes are separated by mass. In a normal TIMS, the magnetic field is frequently changed to allow ions of the different isotopes to hit the single faraday cup detector and be measured. In the more modern multi-collector TIMS (MC-TIMS), a number of Faraday collectors are fitted, permitting simultaneous measurement of a number of isotopes. As it is usually the ratio of the various isotopes being studied that is important, the absolute count of each isotope in the sample is not as relevant. As above, laser ablation may also be undertaken to obtain and vaporise the sample.

Sample Preparation

Sample preparation depends upon both the elements and the nature of the object or item being studied. Lead isotopes are obtained from glass by dissolving in hot strong acids followed by redeposition and purification by electrolysis, before further evaporation. Similarly, oxygen isotopes are released from carbonates by acid digestion. In contrast, oxygen isotopes are released from phosphates by pyrolysis. Although they need chemical purification, strontium isotopes are similarly ionised by heating (TIMS) or by plasma exposure (PIMS), although laser ablation (LA-) of the sample is preferred where possible as this removes the purification requirement. At times, strontium isotopic analysis is undertaken after acid digestion, such as for glass. Laser ablation, however, is significantly faster than the chemical processing methods and hence is preferable.

Advantages

Isotopic analysis is one of the few techniques that can be used to track provenance. For example, for metals, the nature, trace element characteristics and bulk analysis can change radically between the raw ore and the finished object. Strontium isotope analysis can be used to identify lifetime or seasonal migration of an individual by sampling repeatedly along the enamel of a tooth using laser ablation. Strontium isotope ratios that differ from those in the surrounding rock indicate migration from an area with a different rock Sr isotope signal, expressed as $^{87}Sr/^{86}Sr$ (such as in Box 3.9). Similarly, limestone has the strontium isotopic ratios that were present in the ocean when it was formed. This means that this ratio can be passed on to a glass if the limestone (or other item, such as a shell) is used either directly in the glass as the lime component or as a glassmaking sand or as lime derived from limestone in a plant ash. As a result, glasses made in different geological areas will thus have different strontium isotope ratios. Other isotopic analyses provide compositional information, such as human or animal diet (see Box 3.2).

Disadvantages

Isotopic studies are, by definition, destructive. The actual size of sample, and therefore the level of destruction, depends upon the material being studied and the isotopes being studied. For example, strontium isotope analysis of glass uses larger samples than Lead Isotope Analysis (LIA). Laser ablation, such as for strontium analysis of teeth, requires only micro-destruction. The methods are also relatively costly, but again the actual costs depend upon the isotopes and samples being studied. Strontium analyses are significantly more expensive than carbon or nitrogen analyses.

A further complication arises if the isotope ratio changes (fractionates) through the chain of transitions from source to final vessel. For example, nitrogen levels change with each trophic level in a food web, so carnivores have higher nitrogen values than herbivores. As long as this fractionation is understood, and (where possible) comparator samples studied, these are not problematic and can easily be resolved. Similarly, reservations have been expressed as to whether the lead isotope ratios change as a result of the mixing and recycling of metals, and regarding how to cope with multiple ore sources existing with the same isotopic ratios and how to define the isotopic field of an ore source.

Ancient DNA Analysis (aDNA)

Introduction

The set of DNA molecules that are needed to construct and maintain an organism form a store of biological information called the genome. Multicellular animals have two distinct parts to their genome: the nuclear genome (found in the cell nucleus) and the mitochondrial genome (found in the energy-generating organelles of the cell). The composition of the genome can be studied in full detail (sequencing) or sections of it can be studied for patterns. Most aDNA studies rely on extraction of the DNA from the archaeological items, and include sequencing.

Principles

DNA in archaeological samples is usually extracted and amplified using the polymerase chain reaction (PCR). This acts like a DNA photocopy machine through denaturing the DNA strands by heating, and then using these two separated strands as a template for the synthesis of its complementary nucleotide by adding primers and thereby enabling these new DNA strands to form. This process is repeated multiple times. The results of the PCR are usually analysed by agarose gel electrophoresis. DNA molecules have negative electric charges. The agarose gel is under electric charge and therefore separates out the DNA fragments on the basis of their length, as smaller lengths of DNA migrate faster towards the positive pole than do longer fragments. If aDNA is found to have been successfully extracted (due to bands being made on the agarose gel), this amplified sample can then be sequenced. Sequencing of aDNA is usually preceded by cloning. Recently next generation sequencing (NGS) has started to be used for aDNA studies as this enables entire genomes to be studied.

Sample Preparation

Depending upon the DNA being studied, the nature of precautions being taken in sample collection vary. Generally samples are extracted by drilling (often under ultraviolet light, in a positive airflow clean environment). Obtaining the small sample from which DNA is to be obtained thus needs to take place in a laboratory environment. The DNA itself can then be extracted from this sample by suspension in water or a similar weak buffer. Bone samples are sometimes soaked in a chelating agent in order to weaken the inorganic matrix and thereby increase the yield of aDNA. Centrifuging is used to separate out the dissolved DNA from the solids. Purification of aDNA is undertaken, often using silica binding. The small sample of DNA extracted is then amplified using PCR. Given that PCR is so sensitive, it is imperative to avoid contamination (as modern DNA, being larger than aDNA, will be replicated more in the PCR process).

Brown and Brown (2011: 141) provide detailed precautions that should be taken by field archaeologists when handling specimens destined for aDNA analysis. The samples required are small – usually a few grams. They recommend wearing clean gloves when excavating or handling material. Ideally these should be disposable medical latex gloves, but other gloves are suitable as long as care is taken to avoid contamination through touch (e.g. of your hair or skin), and that new gloves are used for each new sample. The tools used to actually excavate or hold the samples, such as tweezers, should be sterilised repeatedly and certainly between samples. This can be done by heating with a cigarette lighter flame. Mouths and noses should be covered, such as with surgical masks or scarves, and hair tied back or covered to avoid contaminating the sample. Samples must be kept dry and kept in a clean and (preferably sterile) airtight container, such as a screw top bottle. Samples should be kept out of direct sunlight and kept cool, such as being stored in a refrigerator. Longer term storage should be in a freezer. Note must also be made of soil conditions around each sample, such as pH, presence of metal objects etc. Finally, if studying human aDNA, all archaeologists involved in the sampling or excavation should have their own DNA sampled by saliva swab (or other non-invasive method) to eliminate contamination.

Neutron Activation Analysis (NAA)

Introduction

This method is used to detect and quantify elements in inorganic compounds, and hence is used extensively in provenancing studies of ceramics, lithics and glass. Samples are placed in vials and irradiated in the core of nuclear reactor. This converts them to radioactive isotopes. After cooling, their emissions of radiation (usually gamma (γ) radiation) are measured.

Principles

Samples and standards are usually carried inside the core of a nuclear reactor for irradiation using pneumatic tubes. The irradiation converts the sample into radioactive isotopes, with the length of irradiation normally being about 3 to 5 times the half-life or the targeted elements (and can thus vary from a few seconds to several hours). After irradiation, the samples are highly radioactive. The cooling period of the sample varies depending on the irradiation time, but can also last several days. Analysis comprises detection of the γ radiation emitted by the sample. Depending on the strength of the emitted radiation signal is a measure of the concentration in the sample. Samples with complex compositions may require several irradiations and counts.

Sample preparation

Samples are usually required to be powdered prior to analysis, although some small samples may be suitable for analysis whole without preparation. The sample size required varies tremendously, from a few mg to several grams. Any contaminants must be removed from the samples in advance of irradiation (especially if the sample is taken with a metal drill bit).

Advantages

The sensitivity is very high, with detection limits normally being a few ppm. In addition, very little sample preparation is required prior to analysis. Furthermore, although a sample is taken, the sample is not actually destroyed by the analysis.

Disadvantages

Not all elements can be analysed using NAA. For analysis to work, neutron capture by the atom must result in the formation of a radioactive isotope. In addition, the radioactive decay must have a suitable half-life for monitoring and the emission must be free of interference. The main disadvantage, however, is the need for a neutron generator (usually a nuclear reactor). This is obviously not easily or commonly available and hence the cost of analysis can be high.

Site Survey

To conclude this appendix, we also want to draw attention to the different considerations that should be made when undertaking site survey and the types of questions that might be answered using the different methods described in Section 2.2: Finding Sites.

A series of potential questions are presented as a flow chart to enable decisions as to which methods can be applied and in what order they should be undertaken.

Is the area under investigation expansive or quite limited?
If expansive → then use magnetometry and ground penetrating radar (GPR)
Is the area being investigated close to modern buildings?
Yes → GPR or electrical resistance tomography (ERT)

Is the area close to the water table?
    Yes → magnetometry or ERT
    No → magnetometry or ERT or GPR
Is it anticipated that the features under investigation lie very deep below the surface?
    Yes → ERT or GPR or seismic survey
    No → ERT or GPR or magnetometry
Is the terrain very uneven or very steep?
    Yes → magnetometry or ERT
    No → magnetometry or ERT or GPR
Is the survey essentially being undertaken to assess the scope of sub-surface features prior to invasive investigations?
    Yes → magnetometry or GPR
    No → magnetometry or GPR or ERT
Is the objective of the survey to produce a horizontal plan or is a vertical slice more important, or both?
    Horizontal plan → magnetometry and GPR
    Vertical slice → ERT
    Both → GPR
Is a production site (e.g. one using pyrotechnology) anticipated/known?
    Yes → magnetometry
    No → magnetometry or GPR or ERT
Is a cemetery site anticipated/known?
    Yes → magnetometry or GPR
    No → magnetometry
Is it important to be able to detect which features are overlying which?
    Yes → GPR or ERT
    No → magnetometry or GPR or ERT
Will the data be correlated with environmental core samples?
    Yes → ERT
    No → magnetometry or GPR or ERT

References

Abbas AM, MA Atya, EA Al-Sayed & H Kamei. (2004) Assessment of groundwater resources of the Nuweiba area at Sinai Peninsula, Egypt by using geoelectric data corrected for the influence of near surface inhomogeneities. *Journal of Applied Geophysics* 56: 107–122.

Abdallatif TF, SE Mousa & A Elbassiony. (2003) Geophysical Investigation for Mapping the Archaeological Features at Qantir, Sharqyia, Egypt. *Archaeological Prospection* 10: 27–42.

Abd El-Ghani MM & AG Fahmy. (2001) Analysis of Aquatic Vegetation in Islands of the Nile Valley (Egypt). *International Journal of Ecology and Environmental Sciences* 27: 1–11.

Adams AE, WS Mackenzie & C Guilford. (1984) *Atlas of the sedimentary rocks under the microscope.* New York: John Wiley.

Agarwal SC & BA Glencros.s (eds.) (2011) *Social Bioarchaeology.* Oxford: Blackwell.

Agarwal SC & SD Stout. (eds.) (2003) *Bone loss and osteoporosis: an anthropological perspective.* New York: Kluwer Academic/Plenum Publishers.

Aitken MJ. (1985) *Thermoluminescence Dating.* Orlando: Academic Press.

Aitken MJ. (1998) *Introduction to Optical Dating.* Oxford: Clarendon Press.

Allsop JM. (1992) The British Geological Survey: Geoprospection Techniques Applied to the Archaeological Landscape. In: Spoerry P. (ed.) *Geoprospection in the Archaeological Landscape.* Oxford: Oxbow. pp. 121–140.

AlQahtani SJ, HM Liversidge & MP Hector. (2010) Atlas of tooth development and eruption. *American Journal of Physical Anthropology* 142: 481–490.

Ambrose SH & MJ DeNiro. (1986) The isotopic ecology of East African mammals. *Oecologia* 69: 395–406.

Andres W & J Wunderlich. (1992) Environmental Conditions for Early Settlement at Minshat Abu Omar, Eastern Nile Delta, Egypt. In: van den Brink ECM. (ed.) *The Nile Delta in Transition; 4th–3rd Millennium BC.* Tel Aviv: Netherlands Institute of Archaeology and Arabic Studies in Cairo. pp. 157–166.

Andrews C. (1984) *Egyptian Mummies.* London: British Museum Press.

Andrews K & RCP Doonan. (2003) *Test Tubes and Trowels: Using science in archaeology.* Oxford: Tempus.

Anthony DW. (1990) Migration in archeology: The baby and the bathwater. *American Anthropologist* 92: 895–914.

Appleby J. (2010) Why We Need an Archaeology of Old Age, and a Suggested Approach. *Norwegian Archaeological Review* 43(2): 145–168.

Appleby J. (2011) Bodies, burials and ageing: the temporality of old age in prehistoric societies. *Oxford Journal of Archaeology* 30: 231–246.

Appleyard HM. (1986) Appendix: Report on two fibre diameter samples from el-Amarna. In: Kemp BJ. (ed.) *Amarna Reports III*. London: Egypt Exploration Society. pp. 75–79.

Arensburg B & I Hershkovitz. (1988) Cranial deformation and trephination in the Middle East. *Bulletins et Mémoires de la Société d'Anthropologie de Paris* 5: 139–150.

Armelagos GJ, KH Jacobs & DL Martin. (1981) Death and Demography in Prehistoric Sudanese Nubia. In: Humphreys SC & H King. (eds.) *Mortality and Immortality: The Anthropology and Archaeology of Death*. London: Academic Press. pp. 33–57.

Armelagos GJ, DP Van Gerven, DL Martin & R Huss-Ashmore. (1984) Effects of Nutritional Change on the Skeletal Biology of Northeast African (Sudanese Nubian) Populations. In Clark JD & SA Brandt. (eds.) *From Hunters to Farmers: The Causes and Consequences of Food Production in Africa*. London: University of California Press. pp. 132–146.

Arnold DE. (1971) Ethnomineralogy of Ticul, Yucatan Potters: Etics and Emics. *American Antiquity* 36: 20–40.

Arnold D. (1988) Pottery. In: Arnold D. (ed.) *The Pyramid of Senwosret I. The Metropolitan Museum of Art Expedition*. New York: MMA. pp. 106–147.

Arnold D. (1991) *Building in Egypt: Pharaonic Stone Masonry*. Oxford: Oxford University Press.

Arnold D. (1993) Techniques and traditions of manufacture in the pottery of ancient Egypt. In: Arnold D & JD Bourriau. (eds.) *An Introduction to Ancient Egyptian Pottery*. Mainz am Rhein: Philipp von Zabern. pp. 9–141.

Arnold D. (1997) *The Royal Women of Amarna*. New York: Metropolitan Museum of Art.

Arnold D & JD Bourriau. (eds.) (1993) *An Introduction to Ancient Egyptian Pottery*. Mainz am Rhein: Philipp von Zabern.

Arnold JR & WF Libby. (1949) Age determinations by radiocarbon content; checks with samples of known age. *Science* 110: 678–680.

Arnold D, Bourriau J & H-Å Nordström. (1993) *An Introduction to Ancient Egyptian Pottery*. Mainz am Rhein: P. von Zabern.

Aspinall A, Gaffney C & A Schmidt. (2008) *Magnetometry for Archaeologists (Geophysical Methods for Archaeology)*. Plymouth: Altamira Press.

Aston BG, Harrell JA & I Shaw. (2000) Stone. In: Nicholson PT & I Shaw (eds.) *Ancient Egyptian Materials and Technology*. Cambridge: Cambridge University Press. pp. 5–77.

Atya MA, H Kamei, AM Abbas, FA Shaaban, AGh Hassaneen, MA Abd Alla, MN Soliman, Y Marukawa, T Ako & Y Kobayash. (2005) Complementary Integrated Geophysical Investigation around Al-Zayyan Temple, Kharga Oasis, Al-Wadi Al-Jadeed (New Valley), Egypt. *Archaeological Prospection* 12: 177–189.

Aufderheide AC. (2003) *The Scientific Study of Mummified Human Remains*. Cambridge: Cambridge University Press.

Aufderheide AC & C Rodríguez-Martín. (1998) *The Cambridge Encyclopedia of Human Paleopathology*. Cambridge, UK: Cambridge University Press.

Aufrere S. (2001) The Egyptian Temple, substitute for the mineral universe. In: Davies WV. (ed.) *Colour and Painting in Ancient Egypt*. London: British Museum Press. pp. 158–163.

Bailey DM. (1999) Sebakh, sherds and survey. *Journal of Egyptian Archaeology* 85: 211–218.

Baillie MGL. (1982) *Tree-Ring Dating and Archaeology*. London: Croom-Helm.

Baillie MGL. (1995) *A slice through time: Dendrochronology and precision dating*. London: Batsford.

Baines J & J Málek. (1980) *Atlas of Ancient Egypt*. Oxford: Phaidon.

Baker BJ. (2010) King Tutankhamun's Family and Demise. Letter to the Editor in response to Hawass et al. 'Ancestry and pathology in King Tutankhamun's family.' *Journal of the American Medical Association* 303: 2471–2472.

Baker BJ, TL Dupras & MW Tocheri. (2005) *The Osteology of Infants and Children*. College Station, TX: Texas A&M University Press.

Balasse M, L Boury, J Ughetto-Monfrin & Tresset A. (2012) Stable isotope insights (δ18O, δ13C) into cattle and sheep husbandry at Bercy (Paris, France, 4th millennium BC): birth seasonality and winter leaf foddering. *Environmental Archaeology* 17: 29–44.

Balasse, M, I Mainland & Richards, M. (2009) Stable isotope evidence for seasonal consumption of marine seaweed by modern and archaeological sheep in the Orkney archipelago (Scotland). *Environmental Archaeology* 14: 1–14.

Balasse M, B Smith, SH Ambrose & S Leigh. (2003) Determining sheep birth seasonality by analysis of tooth enamel oxygen isotope ratios: the Late Stone Age site of Kasteelberg (South Africa). *Journal of Archaeological Science* 30: 205–215.

Bannister B. (1963) Dendrochronology. In: Brothwell DR & E Higgs. (eds.) *Science in Archaeology*. New York: Thames & Hudson. pp. 161–176.

Bard KA. (2008) *An Introduction to the Archaeology of Ancient Egypt*. London: Wiley-Blackwell.

Barker P. (1993) *Techniques of Archaeological Excavation*. London: Routledge.

Barkoudah J & J Henderson. (2006) Plant ashes from Syria and the manufacture of glass: ethnographic and scientific aspects. *Journal of Glass Studies* 48: 297–321.

Bárta M & V Brůna. (2005) Satellite imaging in the pyramid fields. *Egyptian Archaeology* 26: 3–7.

Beck A. (2006) Google Earth and World Wind: remote sensing for the masses? *Antiquity* 80 (308). http://www.antiquity.ac.uk/projgall/beck308/

Beck LA. (1995) Regional cults and ethnic boundaries in 'Southern Hopewell'. In: Beck LA. (ed.) *Regional Approaches to Mortuary Analysis*. New York: Plenum Press. pp. 167–187.

Bergmann M & M Heinzelmann. (2003) Schedia (Kom el-Gizah and Kom el-Hamam, Department of Beheira) *Report on the Documentation and Excavation Season 18th March – 18th April 2003. Archive Report.*

Bergmann M. & M Heinzelmann. (2004) *Schedia, Alexandria's Harbour on the Canopic Nile (Kom el Giza/ Department of Beheira). Preliminary Report on the Second Season 2004.* Available at: http://www.schedia.de/.

Bietak M. (1975) *Tell el-Dab^ca: Der Fundort im Rahmen einer archäologisch-geographischen Untersuchung über das ägyptische Ostdelta (Vol. 2).* Vienna: Verlag der Österreichischen Akademie der Wissenschaften.

Bietak M & F Höflmayer. (2007) Introduction: High and Low Chronology. In: Bietak E & E Czerny. (eds.) *The Synchronisation of Civilisations in the Eastern Mediterranean in the Second Millennium B.C.* Contributions to the Chronology of the Eastern Mediterranean 9. Wien: Verlag der Österreichischen Akademie der Wissenschaften. pp. 13–23.

Bietak M, I Forstner-Müller & T Herbich. (2007) Geophysical Survey and its Archaeological Verification. Discovery of a new palatial complex in Tell el-Dabᶜa in the Delta, In: Hawass Z & J Richards. (eds.) *The Archaeology and Art of Ancient Egypt: Essays in Honor of David B. O'Connor.* Cairo: American University in Cairo Press. pp. 141–147.

Bimson M, S La Neice & M Leese. (1982) The characterisation of mounted garnets. *Archaeometry* 24: 51–58.

Binford LR. (1971) Mortuary Practices: Their Study and Their Potential. In: Brown JA. (ed.) *Approaches to the Social Dimensions of Mortuary Practices.* Washington DC: The Society for American Archaeology. pp. 6–29.

Bitelli G, MA Tini & L Vittuari. (2003) Low-height Aerial Photogrammetry for Archaeological Orthoimaging Production. In: *The International Archives of the Photogrammetry, Remote Sensing and Spatial Information Sciences* 34: 55–59.

Blondiaux J, A Alduc-Le Bagousse, C Niel, N Gabard & E Tyler. (2006) Relevance of Cement Annulations to Paleopathology. *Paleopathology Newsletter* 135: 4–13.

Bluszcz A. (2005) OSL Dating in Archaeology. In: EM Scott, AY Alekseev & G Zaitseva. (eds.) *Impact of the Environment on Human Migration in Eurasia.* Doetinchem: Springer. pp. 137–149.

Boccone S, M Cremasco, S Bortoluzzi, J Moggi-Cecchi & E Rabino Massa. (2010) Age estimation in subadult Egyptian remains. *Homo* 61: 337–358.

Bocherens H, M Mashkour, D Billiou, E Pellé & A Mariotti. (2001) A new approach for studying prehistoric managements in arid areas: intra-tooth isotopic analyses of archaeological caprine from Iran. *Compte Rendu de l'Académie des Sciences Paris Série IIA* 332: 67–74.

Boessneck J. (1988) *Die Tierwelt des Alten Ägypten.* Munich: Beck.

Bohnert M, Rost T & S Pollak. (1998) The degree of destruction of human bodies in relation to the duration of the fire. *Forensic Science International* 95: 11–21.

Boulos L. (1999) *Flora of Egypt. Vol.1: Azollaceae – Oxalidaceae.* Cairo: Al Hadara Publishing.

Boulos L. (2000) *Flora of Egypt. Vol. 2: Geraniaceae – Boraginaceae.* Cairo: Al Hadara Publishing.

Boulos L. (2002) *Flora of Egypt. Vol. 3: Verbenaceae – Compositae.* Cairo: Al Hadara Publishing.

Boulos L. (2005) *Flora of Egypt. Vol. 4: Monocotyledons (Alismataceae – Orchidaceae).* Cairo: Al Hadara Publishing.

Bourriau J. (2001) Change of body position in Egyptian burials from the mid XIIth Dynasty until the early XVIIIth Dynasty. In: Willems H. (ed.) *Social Aspects of Funerary Culture in the Egyptian Old and Middle Kingdoms.* Leuven: Peeters. *Orientalia Lovaniensa Analecta* 103: 1–20.

Bourriau J & P French. (2007) Imported amphorae from Buto dating from *c.*750 BC to the early Sixth Century AD. *Cahiers de la Céramique Égyptienne* 8: 115–134.

Bourriau JD, P Nicholson & PJ Rose. (2000) Pottery. In: Nicholson P & I Shaw. (eds.) *Ancient Egyptian materials and technology*. Cambridge: Cambridge University Press. pp. 121–147.

Bouwman AS, KA Brown, JNW Prag & TA Brown. (2008) Kinship between burials from Grave Circle B at Mycenae revealed by ancient DNA typing. *Journal of Archaeological Science* 35: 2580–2584.

Bradford JSP. (1957) *Ancient Landscapes: Studies in Field Archaeology*. London: Bell.

Bradley SM & SB Hanna. (1986) The effect of soluble salt movements on the conservation of an Egyptian limestone standing figure. *Studies in Conservation* 31(Suppl1): 57–61.

Bramanti B. (2013) The use of DNA analysis in the archaeology of death and burial. In: Tarlow S & L Nilsson-Stutz (eds.) *The Oxford Handbook of the Archaeology of Death & Burial*. Oxford: Oxford University Press. pp. 99–122.

Brand JP. (2000) *The Monuments of Seti I and their Historical Significance*. Leiden: Brill.

Brewer DJ & RF Friedman. (1989) *Fish and Fishing in Ancient Egypt*. Warminster: Aris & Phillips.

Brewer DJ, DB Redford & S Redford. (1994) *Domestic Plants and Animals: The Egyptian Origins*. Warminster: Aris & Phillips.

Brickley M & R Ives. (2008) *The Bioarchaeology of Metabolic Bone Disease*. San Diego: Academic Press.

Brickley M & JI McKinley. (eds.) (2004) Guidelines to the Standards for Recording Human Remains. IFA Paper No. 7. BABAO & IFA publication. http://babao.org.uk/HumanremainsFINAL.pdf

Bridges P. (1992) Prehistoric arthritis in the Americas. *Annual Review of Anthropology* 21: 67–91.

Brill RH. (1970) Lead and oxygen isotopes in ancient objects. *Philosophical Transactions of the Royal Society*, London. 269: 143–164.

Brill RH. (1999) *Chemical analyses of early glasses: Volume 1 (tables) and 2 (catalogue)*. Corning, NY: Corning Museum of Glass.

Brill RH, Barnes IL & B Adams. (1974) Lead isotopes in some ancient Egyptian objects. In: Bishay A. (ed.) *Recent advances in the science and technology of materials*. New York: Plenum. pp. 9–27.

Bronk Ramsey C. (2006) New approaches to constructing age models: OxCal4. *PAGES News* 14(3): 14–15.

Bronk Ramsey C. (2013) Using Radiocarbon Evidence in Egyptian Chronological Research. In: Shortland AJ & C Bronk Ramsey. (eds.) *Radiocarbon and the Chronologies of Ancient Egypt*. Oxford: Oxbow. pp. 29–39.

Bronk Ramsey C, M Dee, J Rowland, T Higham, S Harris, F Brock, A Quiles, E Wild, E Marcus & A Shortland. (2010) Radiocarbon-Based Chronology for Dynastic Egypt. *Science* 328: 1554–1557.

Bronk Ramsey C, SW Manning & M Galimberti. (2004) Dating the volcanic eruption at Thera. *Radiocarbon* 46(1): 325–344.

Brooks S & JM Suchey. (1990) Skeletal age determination based on the os pubis: comparison of the Ascadi-Nemeskeri and Suchey-Brooks methods. *Human Evolution* 5: 227–238.

Brothwell DR. (1989) The relationship of tooth wear to aging. In: Işcan MY. (ed.) *Age Markers in the Human Skeleton*. Springfield: Charles E Thomas. pp. 303–316.

Brothwell D. (2012) Tumors: Problems of differential diagnosis in paleopathology. In: Grauer AL. (ed.) *A Companion to Paleopathology*. Chichester: Wiley-Blackwell. pp. 420–433.

Brothwell DR & AM Pollard. (eds.) (2001) *Handbook of Archaeological Sciences*. Chichester: Wiley.

Brown JA. (ed.) (1971) *Approaches to the Social Dimensions of Mortuary Practices*. Washington DC: Memoir of the Society for American Archaeology.

Brown TA & KA Brown. (2011) *Biomolecular Archaeology: An Introduction*. Chichester: Wiley-Blackwell.

Brown TA, RG Allaby, KA Brown & MK Jones. (1993) Biomolecular archaeology of wheat: past, present and future. *World Archaeology* 25: 64–73.

Brown TA, MK Jones, W Powell & RG Allaby. (2009) The complex origins of domesticated crops in the Fertile Crescent. *Trends in Ecology & Evolution* 24: 103–109.

Brown VM & JA Harrell. (1995) Topographical and petrological study of ancient Roman quarries in the Eastern Desert of Egypt. In: Maniatis Y, N Herz & Y Bassiakis. (eds.) The study of Marble and other Stones used in Antiquity – ASMOSIA III, Athens. *Transactions of the Third International Symposium of the Associations for the Study of Marble and Other Stones used in Antiquity*. London: Archetype. pp. 221–234.

Brunton G. (1927) *Qau and Badari I*. London: BSAE and Bernard Quaritch.

Brunton G. (1937) *Mostagedda and the Tasian Culture*. London: Bernard Quaritch.

Brunton G & G Caton-Thompson. (1928) *The Badarian Civilisation and Predynastic Remains near Badari*. London: Bernard Quaritch.

Bubenzer O & H Riemer. (2007) Holocene Climatic Change and Human Settlement between the Central Sahara and the Nile Valley: Archaeological and Geomorphological Results. *Geoarchaeology: An International Journal* 22: 607–620.

Buckberry JL & AT Chamberlain. (2002) Age estimation from the auricular surface of the ilium: a revised method. *American Journal of Physical Anthropology* 119: 231–239.

Buckley SA & RP Evershed. (2001) Organic chemistry of embalming agents in Pharaonic and Graeco-Roman mummies. *Nature* 413: 837–841.

Buckley SA, KA Clark & RP Evershed. (2004) Complex organic chemical balms of Pharaonic animal mummies. *Nature* 431: 294–299.

Buikstra J & D Ubelaker. (eds.) (1994) *Standards for Data Collection from Human Skeletal Remains*. Fayetteville, AR: Arkansas Archaeological Survey.

Bunbury JM, Graham A & MA Hunter. (2008) Stratigraphic Landscape Analysis: Charting the Holocene Movements of the Nile at Karnak through Ancient Egyptian Time. *Geoarchaeology: An International Journal* 23: 351–373.

Bunbury JM, A Graham & KD Strutt. (2009) Kom el-Farahy: a New Kingdom island in an evolving Edfu floodplain. *British Museum Studies in Ancient Egypt and Sudan* 14: 1–23.

Burt BA & AI Ismail. (1986) Diet, nutrition, and food cariogenicity. *Journal of Dental Research* 65(Special Issue) 1475–1484.

Butzer KW. (1976) *Early Hydraulic Civilization in Egypt: A Study in Cultural Ecology*. Chicago: University of Chicago Press.

Butzer K. (2002) Geoarchaeological implications of recent research in the Nile Delta. In: van den Brink ECM & TE Levy. (eds.) *Egypt and the Levant: Interrelations from the 4th through the Early 3rd Millennium BCE*. Leciester: Leicester University Press. pp. 83–97.

Buzon MR. (2006) Biological and Ethnic Identity in New Kingdom Nubia: A Case Study from Tombos. *Current Anthropology* 47: 683–695.

Buzon MR. (2012) The Bioarchaeological Approach to Paleopathology. In: Grauer AL. (ed.) *A Companion to Paleopathology*. Chichester: Wiley-Blackwell. pp. 58–75.

Cabo LL, CP Brewster & JL Azpiazu. (2012) Sexual Dimorphism: Interpreting sex markers. In: Dirkmaat DC. (ed.) *A Companion to Forensic Anthropology*. Chichester: Wiley-Blackwell. pp. 248–286.

Caley ER. (1962) Analysis of Ancient Glasses 1790–1957: a comprehensive and critical survey. In: *The Corning Museum of Glass Monographs, vol. I*. Corning, NY: Corning Glass Center.

Campagno M. (2009) Kinship and Family Relations. In: Frood E & W Wendrich. (eds.) *UCLA Encyclopedia of Egyptology*. Los Angeles, CA: http://digital2. library.ucla.edu/viewItem.do?ark=21198/zz001nf68f

Campbell JB. (1996) *Introduction to Remote Sensing (2nd Edition)*. London: Guildford Press.

Capasso LL. (2005) Antiquity of Cancer. *International Journal of Cancer* 113: 2–13.

Caple C. (2006) *Objects: Reluctant Witnesses to the Past*. London: Routledge.

Carlson DS. (1976) Temporal Variation in Prehistoric Nubian Crania. *American Journal of Physical Anthropology* 45: 467–484.

Carlson DS & DP Van-Gerven. (1977) Masticatory Function and Post-Pleistocene Evolution in Nubia. *American Journal of Physical Anthropology* 46: 495–506.

Carlson DS & DP Van-Gerven. (1979) Diffusion, Biological Determinism, and Biocultural Adaptation in the Nubian Corridor. *American Anthropologist* 81: 561–580.

Carr C. (1995) Mortuary Practices: Their Social, Philosophical-Religious, Circumstantial, and Physical Determinants. *Journal of Archaeological Method and Theory* 2: 105–200.

Cartmell L, AC Aufderheide, A Springfield, C Weems & A Springfield. (1991) The frequency and antiquity of prehistoric coca-leaf-chewing practices in northern Chile: radioimmunoassay of a cocaine metabolite in human-mummy hair. *Latin American Antiquity* 2: 260–268.

Cashmore L. (2009) Can hominin 'handedness' be accurately assessed? *Annals of Human Biology* 36: 624–641.

Cashmore LA & SR Zakrzewski. (2009) The expression of asymmetry in hand bones from the medieval cemetery at Écija, Spain. In: Lewis ME & M Clegg. (eds). *Proceedings of the Ninth Annual Conference of the British Association of Biological Anthropology and Osteology*. BAR International Series 1918. Oxford: BAR. pp. 79–92.

Cashmore LA & SR Zakrzewski. (2013) Assessment of musculoskeletal stress marker (MSM) development in the hand. *International Journal of Osteoarchaeology*. 23: 334–347.

Cashmore L, Uomini N & A Chapelain. (2008) The evolution of handedness in humans and great apes: a review and current issues. *Journal of Anthropological Sciences* 86: 7–35.

Castillos JJ. (1982) *A Reappraisal of the Published Evidence on Egyptian Predynastic and Early Dynastic Cemeteries*. Toronto: private publication.

Castillos JJ. (1997) New Data on Egyptian Predynastic Cemeteries. *Révue d'Egyptologie*. 48: 251–256.

Caton-Thompson G. (1927) Exploration in the Northern Fayum. *Antiquity* 1: 326–348.

Caton-Thompson G & EW Gardner. (1934) *The Desert Fayum*. London: Royal Anthropological Institute of Great Britain and Ireland.

Chagula W. (1960) The age at eruption of third permanent molars in male East Africans. *American Journal of Physical Anthropology* 18: 77–82.

Chamberlain AT. (2006) *Demography in Archaeology*. Cambridge: Cambridge University Press.

Chapman R. (2003) Death, society and archaeology: the social dimensions of mortuary practices. *Mortality* 8: 305–312.

Chapman RW. (2005) Mortuary analysis. A matter of time? In: Rakita GFM, JE Buikstra, LA Beck & SR Williams. (eds.) *Interacting with the Dead: Perspectives on Mortuary Archaeology for the New Millennium*. University Press of Florida: Gainesville, FL. pp. 25–40.

Chapman R. (2007) Mortuary Rituals, Social Relations, and Identity in Southeast Spain in the Late Third to early Second Millennia B.C. In: Laneri N. (ed.) *Performing Death: Social Analyses of Funerary Traditions in the Ancient Near East and Mediterranean*. Chicago: Oriental Institute. pp. 69–79.

Childe VG. (1952) The birth of civilization. *Past and Present* 2: 1–10.

Christensen AM. (2002) Experiments in the combustibility of the human body. *Journal of Forensic Sciences* 47: 466–470.

Cichocki O. (2006) Dendrochronology. In: Hornung E, R Krauss & DA Warburton. (eds.) *Ancient Egyptian Chronology*. Leiden: Brill. pp. 361–368.

Clark A. (1996) *Seeing Beneath the Soil: Prospecting Methods in Archaeology*. Second Edition. London: Batsford.

Clutton-Brock J. (2012) *Animals as domesticates: a world view through history*. East Lansing: MSU Press.

Collier P. (2002) The Impact on Topographic Mapping of Developments in Land and Air Survey: 1900–1939. *Cartography and Geographic Information Science* 29: 155–174.

Conolly J & M Lake. (2006) *Geographical Information Systems in Archaeology*. Cambridge: Cambridge University Press.

Conyers LB. (2012) *Interpreting Ground-Penetrating Radar for Archaeology*. Walnut Creek: Left Coast Press.

Conyers LB & D Goodman. (1997) *Ground-Penetrating Radar: An Introduction for Archaeologists*. Oxford: Altamira Press.

Cope TA & HA Hosni. (1991) *A key to Egyptian grasses*. Kew: Royal Botanic Gardens.

Copley MS, PJ Rose, A Clapham, DN Edwards, MC Horton & RP Evershed. (2001) Processing palm fruits in the Nile Valley: Biomolecular evidence from Qasr Ibrim. *Antiquity* 75: 538–542.

Cowell RM. (1987) Scientific Appendix I: chemical analyses. In: Davies WV. (ed.) *Catalogue of Egyptian Antiquities in the British Museum*. London: British Museum Press.

Cowley D. (ed.) (2011) *Remote Sensing for Archaeological Heritage Management*. Budapest: Europae Archaeologiae Consilium.

Craddock PT. (1985) Three thousand years of copper alloys: From the Bronze Age to the Industrial revolution. In: England PA & L Van Zelst. (eds.) *Application of Science in Examination of Works of Art*. Proceedings of the Seminar: September 7–9, 1983. Boston: Museum of Fine Arts. pp. 59–67.

Craddock PT. (1995) *Early Metal Mining and Production*. Edinburgh: Edinburgh University Press.

Craddock PT. (1998) The black bronzes of Egypt. In: Esmael FA. (ed.) *Proceedings of the First International Conference on Ancient Mining and Metallurgy and Conservation of Metallic Artefacts*. Cairo: Supreme Council for Antiquities. pp. 89–119.

Craddock PT. (2009) *Scientific Investigation of Copies, Fakes and Forgeries*. London: Elsevier.

Craddock PT & J Lang. (2003) *Mining and Metal Production through the Ages*. London: British Museum Press.

Crawford OGS. (1923) Air Survey and Archaeology. *The Geographical Journal* 61: 342–360.

Crawford OGS. (1929) *Air-Photography for Archaeologists*. London: HMSO.

Creasman PP. (2014) Tree Rings and the Chronology of Ancient Egypt. *Radiocarbon* 56: 85–92.

Creasman PP. (2015) The Potential of Dendrochronology in Egypt: Understanding Ancient Human/Environment Interactions. In: Ikram S., Kaiser J. & R Walker. (eds.) *Egyptian Bioarchaeology: Humans, Animals, and the Environment*. Leiden: Sidestone Press. pp. 201–210.

Creasman PP & D Sassen. (2011) Remote sensing. In: Wilkinson RH (ed.) *The Temple of Tausret: The University of Arizona Egyptian Expedition Tausret Temple Project, 2004–2011*. Tucson: University of Arizona. pp. 150–159.

Crubézy É, Th Janin & B Midant-Reynes. (2002) *La nécropole prédynastique d'Adaïma*. FIFAO 47. Cairo: IFAO.

Dabbs GR & M Zabecki. (2012) Subadult age estimation at Tell-el Amarna: a systematic, site-specific approach. *American Journal of Physical Anthropology* 147(S54): 124.

Darby WJ, P Ghalioungui & L Grivetti. (1977) *Food: The Gift of Osiris*. London: Academic Press.

Dasen V. (1993) *Dwarfs in Ancient Egypt and Greece*. Oxford: Oxford University Press.

David AR. (2000) Mummification. In: Nicholson PT & I Shaw. (eds.) *Ancient Egyptian Materials and Technology*. Cambridge: Cambridge University Press. pp. 372–389.

David AR & MR Zimmerman. (2010) Cancer: an old disease, a new disease or something in between? *Nature Reviews Cancer* 10: 728–733.

Davies NdG. (1905) *The Rock Tombs of El Amarna. Part 2. The Tombs of Panehesy and Meryra*. London: Egypt Exploration Fund.

Davies NdGD. (1935) *Paintings from the Tomb of Rekhmire*. New York: Metropolitan Museum of Art.

Davies NdGD. (1943) *The Tomb of Rekhmire at Thebes.* New York: Metropolitan Museum of Art.

Davies WV. (2001) *Colour and Painting In Ancient Egypt.* London: British Museum Press.

Dean JS. (1996) Dendrochronology and the Study of Human Behavior. In: Dean JS, DM Meko & TW Swetnam (eds.) *Tree-rings, Environment, and Humanity.* Tucson: University of Arizona Press. pp. 461–471.

Dean JS. (2009) One hundred years of dendroarchaeology: Dating, human behavior, and past climate. In: Manning S & MJ Bruce. (eds.) *Tree-rings, Kings, and Old World Archaeology and Environment: Papers Presented in Honor of Peter Ian Kuniholm.* Oxford: Oxbow Books. pp. 25–32.

Debono F & B Mortensen. (1990) *El Omari: A Neolithic Settlement and Other Sites in the Vicinity of Wadi Hof.* Mainz am Rhein: Philipp von Zabern.

Dee MW. (2013a) A radiocarbon-based chronology for the Middle Kingdom. In: Shortland AJ & C Bronk Ramsey. (eds.) *Radiocarbon and the Chronologies of Ancient Egypt.* Oxford: Oxbow. pp. 174–181.

Dee MW. (2013b) A radiocarbon-based chronology for the New Kingdom. In: Shortland AJ & C Bronk Ramsey. (eds.) *Radiocarbon and the Chronologies of Ancient Egypt.* Oxford: Oxbow. pp. 65–75.

Dee MW, F Brock, SA Harris, C Bronk Ramsey, AJ Shortland, TFG Higham & JM Rowland. (2010) Investigating the likelihood of a reservoir offset in the radiocarbon record for ancient Egypt. *Journal of Archaeological Science* 37: 687–693.

Dee MW, JM Rowland, TFG Higham, AJ Shortland, F Brock, SA Harris & C Bronk Ramsey. (2012) Synchronising radiocarbon dating and the Egyptian historical chronology through improved sample selection. *Antiquity* 86: 868–883.

Degryse P & J Schneider. (2008) Pliny the Elder and Sr-Nd isotopes: tracing the provenance of raw materials for Roman glass production. *Journal of Archaeological Science* 35: 1993–2000.

Degryse P & AJ Shortland. (2009) Trace elements in provenancing raw materials for Roman glass production. *Geologica Belgica* 12: 135–143.

Degryse P, A Boyce, N Erb-Satullo, K Eremin, S Kirk, R Scott, AJ Shortland, J Schneider & M Walton. (2010) Isotopic discriminants between Late Bronze Age glasses from Egypt and the Near East. *Archaeometry* 52: 380–388.

de Jong T. (2006) The Heliacal Rising of Sirius. In: Hornung E, R Krauss & DA Warburton. (eds.) *Ancient Egyptian Chronology.* Leiden: Brill. pp. 432–438.

Demirjian A, H Goldstein & JM Tanner. (1973) A new system of dental age assessment. *Human Biology* 45: 211–227.

Denton J & AR David. (2005) Metabolic bone disease as a result of pregnancy and or famine in an ancient Egyptian mummy. *Journal of Bone and Mineral Research* 20: S421.

Depuydt L. (2006) The Foundations of Day-exact Chronology 690 BC–332 BC. In: Hornung E, R Krauss & DA Warburton. (eds.) *Ancient Egyptian Chronology.* Leiden: Brill. pp. 458–470.

Devereux BJ, GS Amable & P Crow. (2008) Visualisation of LiDAR Terrain Models for Archaeological Feature Detection. *Antiquity* 82: 470–479.

Déscription de l'Égypte. Volume IV Antiquités (1809–1828) Plan of Amarna is p. 128, Planche 63, Figure 4. Complete copy available at http://descegy.bibalex.org/

De Vito C & S Saunders. (1990) A discriminant function analysis of deciduous teeth to determine sex. *Journal of Forensic Sciences* 35: 845–858.

Di Iorio A, I Bridgewood, R Carlucci, F Bernardini, MS Rasmussen, MK Sørensen & A Osman. (2010) Automatic detection of archaeological sites using a hybrid process of Remote Sensing, GIS techniques and a shape detection algorithm. In: Reuter R. (ed.) *Proceedings of the 30th EARSeL Symposium, Remote Sensing for Science, Education, and Natural and Cultural Heritage. pp. 53–64. http:// www.earsel.org/?target=publications/proceedings/symposium-2010*

Dixon JE, JR Cann & C Renfrew. (1968) Obsidian and the Origins of Trade. *Scientific American* 218: 38–46.

Dodson A. (2009) *Amarna Sunset: Nefertiti, Tutankhamun, Ay, Horemheb, and the Egyptian Counter-reformation.* Cairo: AUC Press.

Dodson A. (2012) *Afterglow of Empire: Egypt from the fall of the New Kingdom to the Saite Renaissance.* Cairo: AUC Press.

Dodson A. (2013) The Ramesside Period: a case of overstretch? In: Shortland AJ. & C Bronk Ramsey (eds.) *Radiocarbon and the Chronologies of Ancient Egypt.* Oxford: Oxbow. pp. 146–152.

Dodson A. (2014) *Amarna Sunset: Nefertiti, Tutankhamun, Ay, Horemheb and the Egyptian Counter-Reformation.* Cairo: AUC Press.

Dodson A & S Ikram. (2008) *The Tomb in Ancient Egypt: Royal and Private Sepulchres from the Early Dynastic Period to the Romans.* London: Thames & Hudson.

Donald CR. (1984) Examination of the archaeological samples supplied by Barry J. Kemp from the Tell el- 'Amarna site. *Amarna Reports* I: 56–58.

Doneus M & C Briese (2011) Airborne laser scanning in forested areas – Potential and limitations of an archaeological prospection technique. In: Cowley D. (ed.) *Remote Sensing for Archaeological Heritage Management.* Budapest: Europae Archaeologiae Consilium. pp. 59–76.

Donoghue D. (2001) Remote sensing. In: Brothwell DR & AM Pollard. (eds.) *Handbook of Archaeological Sciences.* London: John Wiley and Sons. pp. 551–560.

Donoghue HD, OY-C Lee, DE Minnikin, GS Besra, JH Taylor & M Spigelman. (2010) Tuberculosis in Dr Granville's mummy: a molecular re-examination of the earliest known Egyptian mummy to be scientifically examined and given a medical diagnosis. *Proceedings of the Royal Society B* 277: 51–56.

Douglass AE. (1919) *Climatic Cycles and Tree Growth: Vol. I: A Study of the Annual Rings of Trees in Relation to Climate and Solar Activity.* Washington DC: Carnegie Institute of Washington.

Douglass AE. (1941) Crossdating in Dendrochronology. *Journal of Forestry* 39: 825–831.

Drennan RD. (2010) *Statistics for Archaeologists.* New York: Springer.

Duday H. (2009) *The Archaeology of the Dead: Lectures in Archaeothanatology.* Oxford: Oxbow Books.

Dupras TL & HP Schwarcz. (2001) Strangers in a strange land: stable isotope evidence for human migration in the Dakhleh Oasis. *Journal of Archaeological Science* 28: 1199–1208.

Dupras TL & MW Tocheri. (2007) Reconstructing infant weaning histories at Roman period Kellis, Egypt using stable isotope analysis of dentition. *American Journal of Physical Anthropology* 134: 63–74.

Edwards HGM. (2005) Overview: Biological materials and degradation. In: Edwards H & JM Chalmers. (eds.) *Raman Spectroscopy in Archaeology and Art History.* Cambridge: Royal Society of Chemistry. pp. 231–279.

Ehlers EG. (1987) *Optical Mineralogy.* Palo Alto: Blackwell Scientific Publications.

Ekengren F. (2013) Contextualizing grave goods: theoretical perspectives and methodological implications. In: Tarlow S & L Nilsson-Stutz. (eds.) *The Oxford Handbook of the Archaeology of Death & Burial.* Oxford: Oxford University Press. pp. 173–192.

Elliot Smith G & WR Dawson. (1924) *Egyptian Mummies.* London: Allen & Unwin.

Elliot Smith G & F Wood-Jones. (1910) *The archaeological survey of Nubia 1907–1908. Volume 2. Report on the human remains.* Cairo: National Printing Department.

Ellis CJ. (1992) A statistical analysis of Protodynastic burials in the 'Valley' Cemetery of Kafr Tarkhan. In: van den Brink ECM. (ed.) *The Nile Delta in Transition; 4th–3rd Millennium BC.* Tel Aviv: Netherlands Institute of Archaeology and Arabic Studies in Cairo. pp. 241–258.

El-Gamili MM, AS El-Mahmoudi, SSH Osman, AGH Hassaneen & MA Metwaly. (1999) Geoelectric resistance scanning on parts of Abydos cemetery region, Sohag Governorate, Upper Egypt. *Archaeological Prospection* 6: 225–239.

El-Qady G, C Sakamoto & K Ushijima. (1999) 2-D inversion of VES data in Saqqara archaeological area, Egypt. *Earth, Planets and Space* 51: 1091–1098.

Emery-Barbier A. (2008) Contribution of palynology and phytolithology to the study of Adaïma's graves. In: Midant-Reynes B & Y Tristant. (eds) *Egypt at its Origins 2.* Orientalia Lovaniensia Analecta. Leuven: Peeters. pp. 391–417.

Engelback R. (1922) *The Problem of Obelisks.* London: T Fisher Unwin.

English Heritage. (2010) *The Light Fantastic: Using airborne lidar in archaeological survey.* Swindon: English Heritage. http://www.english-heritage.org.uk/publications/light-fantastic/light-fantastic.pdf

Eriksson G. (2013) Stable isotope analysis of humans. In: Tarlow S & L Nilsson-Stutz. (eds.) *The Oxford Handbook of the Archaeology of Death & Burial.* Oxford: Oxford University Press. pp. 123–146.

Ewing GH. (1966) Functional Implications of the Morphology of Mesolithic Crania from Nubia. *American Journal of Physical Anthropology* 25: 214.

Ezzo JA. (1994) Putting the 'chemistry' back into archaeological bone chemistry analysis – modeling potential paleodietary indicators. *Journal of Anthropological Archaeology* 13: 1–34.

Faerman M, GK Bar-Gal, D Filon, CL Greenblatt, L Stager, A Oppenheim & P Smith. (1998) Determining the Sex of Infanticide Victims from the Late Roman Era through Ancient DNA Analysis. *Journal of Archaeological Science* 25: 861–865.

Fahmy AG. (2001) Plant remains in gut contents of ancient Egyptian Predynastic mummies (3750–3300 B.C.). *Online Journal of Biological Sciences* 1: 772–774.

Fahmy AG. (2003) A Fragrant Mixture: Botanicals from the Basket in B333. *Nekhen News* 15: 20.

Fahmy AG. (2004) Review Insights on Development of Archaeobotanical and Palaeo-ethnobotanical Studies in Egypt. In: Hendrickx S, RF Friedman, KM

Ciałowicz & M Chłodnicki. (eds.) *Egypt and its origins. Studies in memory of Barbara Adams.* Orientalia Lovaniensia Analecta 138. Leuven: Peeters. pp. 711–730.

Fahmy AG. (2005) Missing plant macro remains as indicators of plant exploitation in Predynastic Egypt. *Vegetation History and Archaeobotany* 14: 287–294.

Fairgrieve SI & JE Molto. (2000) Cribra orbitalia in two temporally disjunct population samples from the Dakhleh Oasis, Egypt. *American Journal of Physical Anthropology* 111: 319–331.

Falys CG, H Schutkowski & DA Weston. (2006) Auricular surface aging: worse than expected? A test of the revised method on a documented historic skeletal assemblage. *American Journal of Physical Anthropology* 130: 508–513.

Farmer JG, CL Sugden, AB Mackenzie, GH Moody & M Fulton. (1994) Isotopic ratios of lead in human teeth and sources of exposure in Edinburgh. *Environmental Technology* 15: 593–599.

Faulkner DK. (1986) The mass burial: an entomological perspective. In: Donnan CB & GA Cock. (eds.) *The Pacatnamu Papers. Fowler Museum of Cultural History, vol. 1.* Los Angeles, CA: University of California. pp. 145–150.

Fenwick H. (2004) Ancient roads and GPS survey: modelling the Amarna plain. *Antiquity* 78: 880–885.

Ferguson CW. (1970) Dendrochronology of bristlecone pine, *Pinus aristata*: establishment of a 7484-year chronology in the White Mountains of eastern-central California, USA. In: Olsson IU. (ed.) *Radiocarbon Variations and Absolute Chronology.* New York: John Wiley. pp. 237–259.

Ferguson JR. (ed.) (2010) *Designing Experimental Research in Archaeology: Examining Technology Through Production and Use.* Denver: University Press of Colorado.

Filon D, M Faerman, P Smith & A Oppenheim. (1995) Sequence analysis reveals a β–thalassaemia mutation in the DNA of skeletal remains from the archaeological site of Akhziv, Israel. *Nature Genetics* 9: 365–368.

Fischer HG. (1961) The Nubian Mercenaries of Gebelein during the First Intermediate Period. *Kush* 9: 44–80.

Fleming SJ. (1970) Thermoluminescent dating: refinement of the quartz inclusion method. *Archaeometry* 12: 133–45.

Fletcher M & G Lock. (2005) *Digging Numbers: Elementary Statistics for Archaeologists.* Oxford: Oxbow.

Flores DV. (2003) *Funerary Sacrifice of Animals in the Egyptian Predynastic Period.* BAR International Series 1153. Oxford: Archaeopress.

Flower RJ, K Keatings, M Hamdan & F Hassan. (2013) *Stephanodiscus* Ehr. species from Holocene sediments in the Faiyum Depression (Middle Egypt). *Phytotaxa* 127: 66–80.

Forstner-Müller I. (2008) *Tell el-Dabᶜa XVI, Die Gräber des Areals A/II von Tell el-Dabᶜa.* Vienna: Verlag der Österreichischen Akademie der Wissenschaften.

Forstner-Müller I, T Herbich, C Schweitzer & M Weissl. (2010) Preliminary Report on the Geophysical Survey at Tell el-Dabʿa/Qantir in Spring 2009 and 2010. *Jahreshefte des Österreichischen Archäologischen Institutes in Wien* 79: 67–85.

Fox CL, J Juan & RM Albert. (1996) Phytolith analysis on dental calculus, enamel surface, and burial soil: information about diet and paleoenvironment. *American Journal of Physical Anthropology* 101: 101–113.

France DL. (2009) *Human and Nonhuman Bone Identification: A Color Atlas.* Boca Raton: CRC Press.

Freestone IC & Y Gorin-Rosen. (1999) The great glass slab at Beth Shearim: an early Islamic glass-making experiment? *Journal of Glass Studies* 41: 105–116.

Freestone IC, Y Gorin-Rosen & MJ Hughes. (2000) Composition of primary glass from Israel. In: Nenna M-D. (ed.) *La Route du Verre: Ateliers primaires et secondaires du second millénaire av. J.-C. au Moyen Âge.* Paris: Travaux de la Maison de l'Orient Méditerranéen No.33, Jean Pouilloux. pp. 65–84.

French CAI. (1984) A sediments analysis of mud brick and natural features at el-Amarna. *Amarna Reports* I: 189–201.

French P. (1986) Late Dynastic pottery from the South Tombs. *Amarna Reports* III: 147–188.

Fried MH. (1967) *The evolution of political society: An essay in political anthropology.* New York: Random House.

Friedman FD. (ed.) (1998) *Gifts of the Nile – Ancient Egyptian faience.* London: Thames & Hudson.

Friedman R. (2003) A Basket of Delights: The 2003 Excavations at HK43. *Nekhen News* 15: 18–19.

Friedrich M, S Remmele, B Kromer, J Hofmann, M Spurk, KF Kaiser, C Orcel & M Küppers. (2004) The 12,460-year Hohenheim oak and pine tree-ring chronology from Central Europe—a unique annual record for radiocarbon calibration and paleoenvironment reconstructions. *Radiocarbon* 46(3): 1111–1122.

Fuller BT, JL Fuller, DA Harris & REM Hedges. (2006) Detection of breastfeeding and weaning in modern human infants with carbon and nitrogen stable isotope ratios. *American Journal of Physical Anthropology* 129: 279–293.

Fuller DQ, G Willcox & RG Allaby. (2012) Early agricultural pathways: moving outside the 'core area' hypothesis in Southwest Asia. *Journal of Experimental Botany* 63: 617–633.

Gaber S, AA El-Fiky, S Abou Shagar & M Mohamaden. (1999) Electrical Resistivity Exploration of the Royal Ptolemaic Necropolis in the Royal Quarter of Ancient Alexandria, Egypt. *Archaeological Prospection* 6: 1–10.

Gaffney C & J Gater. (1993) Development of Remote Sensing Part 2: Practice and method in the application of geophysical techniques in archaeology. In: Hunter J & I Ralston. (eds.) *Archaeological Resource Management in the UK.* Stroud: Alan Sutton. pp. 205–214.

Gaffney C & J Gater. (2003) *Revealing the Buried Past.* Stroud: Tempus.

Gaffney C, J Gater & S Ovendon. (1991) *The Use of Geophysical Survey Techniques in Archaeological Evaluations.* Institute of Field Archaeologists Technical Paper No. 9.

Gale NH. (1997) The isotopic composition of tin in some ancient metals and the recycling problem in metal provenancing. *Archaeometry* 39: 71–82.

Gale NH. (2009) A response to the paper of A M Pollard: What a long strange trip it's been: lead isotopes and archaeology. In: Shortland AJ, I Freestone & T Rehren. (eds.) *From Mine To Microscope: Advances in the Study of Ancient Technology.* Oxford: Oxbow Books.

Gale NH & ZA Stos-Gale. (1981) Ancient Egyptian silver. *Journal of Egyptian Archaeology* 67: 103–115.

Gale NH & ZA Stos-Gale. (1993) Comments on Budd P, Gale D, Pollard AM, Thomas RG & PA. Williams 'Evaluating lead isotope data: further observations', *Archaeometry* 35: 252–259.

Gale NH, AP Woodhead, ZA Stos-Gale, A Walder & I Bowen. (1999) Natural variations detected in the isotopic composition of copper: possible applications to archaeology and geochemistry. *International Journal of Mass Spectrometry* 184: 1–9.

Gale R, P Gasson, N Hepper & G Killen. (2000) Wood. In: Nicholson PT & I Shaw. (eds.) *Ancient Egyptian Materials and Technology*. Cambridge: Cambridge University Press. pp. 334–371.

Garvin HM. (2012) Adult Sex Determination: methods and application. In: Dirkmaat DC. (ed.) *A Companion to Forensic Anthropology*. Chichester: Wiley-Blackwell. pp. 239–247.

Garvin HM, NV Passalacqua, NM Uhl, DR Gipson, RS Overbury & LL Cabo. (2012) Developments in Forensic Anthropology: Age-at-Death estimation. In: Dirkmaat DC. (ed.) *A Companion to Forensic Anthropology*. Chichester: Wiley-Blackwell. pp. 202–223.

Gänsicke S, P Hatchfield, A Hykin, M Svoboda & CM-A Tsu. (2003a) The ancient Egyptian Collection at the Museum Of Fine Arts, Boston. Part 1: A Review of Treatments in the field and their consequences. *Journal of the American Institute for Conservation* 42: 167–192.

Gänsicke S, P Hatchfield, A Hykin, M Svoboda & CM-A Tsu. (2003b) The Ancient Egyptian Collection at the Museum of Fine Arts, Boston. Part 2: A review of former treatments at the MFA and their consequences. *Journal of the American Institute for Conservation* 42: 193–236.

Gerisch R. (2007) Appendix 3: Charcoal Remains. In: Nicholson P. (ed.) *Brilliant Things for Akhenaten*. London: Egypt Exploration Society. pp. 169–176.

Gerisch R. (2010) The wood fuel they burnt. In: Kemp K & A Stevens. (eds.) *Busy Lives at Amarna: Excavations in the Main City (Grid 12 and the House of Ranefer, N49.18). Volume I: The Excavations, Architecture and Environmental Remains*. London: Egypt Exploration Society. pp. 399–425.

Gerisch R. (2012) The Stone Village wood and charcoal. In: Stevens A. (ed.) *Akhenaten's Workers. The Amarna Stone Village Survey, 2005–2009. Volume I: The Survey, Excavations and Architecture*. London: Egypt Exploration Society. pp. 47–74.

Gerszten PC & E Gerszten. (1995) Intentional cranial deformation: a disappearing form of self-mutilation. *Neurosurgery* 37: 374–382.

Ghazala H, AS El-Mahmoudi & TF Abdallatif. (2003) Archaeogeophysical Study on the Site of Tell Toukh El-Qaramous, Sherkia Governate, east Nile Delta, Egypt. *Archaeological Prospection* 10: 43–55.

Ghazala H, A El-Shahat, O Abdel Raouf, P Wilson & Z Belal. (2005) Geoelectrical Investigations Around Sa el Hagar Archaeological Site, Gharbiya Governate, Nile Delta, Egypt. *Mansoura Journal of Geology and Geophysics* 32: 121–138.

Gibbon V, M Paximadis, G Štrkalj, P Ruff & C Penny. (2009) Novel methods of molecular sex identification from skeletal tissue using the amelogenin gene. *Forensic Science International* 3: 74–79.

Glencross B & SC Agarwal. (2011) An investigation of cortical bone loss and fracture patterns in the Neolithic community of Çatalhöyük, Turkey using metacarpal radiogrammetry. *Journal of Archaeological Science* 38: 513–521.

Godwin H. (1962) Half-life of Radiocarbon. *Nature* 195: 984.

Goedicke C. (2006) Luminescence Dating of Egyptian Artefacts. In: Hornung E, R Krauss & DA Warburton. (eds.) *Ancient Egyptian Chronology*. Leiden: Brill. pp. 356–360.

Goldberg L. (1996) A History Of Pest Control Measures in the Anthropology Collections, National Museum of Natural History, Smithsonian Institution. *Journal of the American Institute for Conservation* 35: 23–43.

Goldberg P & R Macphail. (2006) *Practical and Theoretical Geoarchaeology*. Oxford: Blackwell.

Goodman AH, GJ Armelagos, DP Van-Gerven & JM Calcagno. (1986) Diet and Post Mesolithic Craniofacial and Dental Evolution in Sudanese Nubia. In: David AR. (ed.) *Science in Egyptology*. Manchester: Manchester University Press. pp. 201–210.

Goren Y, S Bunimovitz, I Finkelstein & N Na'aman. (2003a) The location of Alashiya: new evidence from petrographic investigation of Alashiyan tablets from el-Amarna and Ugarit. *American Journal of Archaeology* 107: 233–255.

Goren Y, I Finkelstein & N Na›aman. (2003b) The Expansion of the Kingdom of Amurru according to the Petrographic Investigation of the Amarna Tablets. *Bulletin of the American Schools of Oriental Research* 329: 1–11.

Goren Y, I Finkelstein, N Na'aman & M Artzy. (2004) *Inscribed In Clay: Provenance Study of the Amarna Tablets and other Ancient Near Eastern Texts*. Tel Aviv: Emery and Claire Yass Publications in Archaeology.

Gosden C & Y Marshall. (eds.) (1999) The cultural biography of objects. *World Archaeology* 31(2): 169–178.

Gosselain OP & A Livingstone Smith. (1995) The ceramics and society project: an ethnographic and experimental approach to technological choices. In: Lindahl A & O Stilborg. (eds.) *The Aim Of Laboratory Analyses In Ceramics In Archaeology*. Stockholm: Kungl, Vitterhets Histoire och Antikvitets Akademien Konferenser. pp. 147–160.

Gowland R. (2006) Ageing the past: examining age identity from funerary evidence. In: Gowland R & C Knüsel. (eds.) *Social Archaeology of Funerary Remains*. Oxford: Oxbow. pp. 143–154.

Gowlett JAJ & REM Hedges. (eds.) (1986) *Archaeological Results from Accelerator Dating*. Oxford: Oxford University Committee for Archaeology.

Grabner M, D Salaberger & T Okochi. (2009) The need of high resolution mu-X-ray CT in dendrochronology and in wood identification. In: Zinterhof P, S Loncaric, A Uhl & A Carini. (eds.) *Proceedings of the 6th International Symposium on Image and Signal Processing and Analysis (ISPA)*. New York: IEEE. pp. 359–362.

Graham A, KD Strutt, VL Emery, S Jones & DS Barker. (2014) Theban harbours and waterscapes survey, 2013. *Journal of Egyptian Archaeology* 99: 35–52.

Graham A, KD Strutt, M Hunter, S Jones, A Masson, M Millet & BT Pennington. (2012a) Theban Harbours and Waterscapes Survey, 2012. *Journal of Egyptian Archaeology* 98: 27–42.

Graham A, KD Strutt, M Hunter, S Jones, A Masson, M Millet & BT Pennington. (2012b) Reconstructing Landscapes and Waterscapes in Thebes, Egypt. *eTopoi. Journal for Ancient Studies* 3: 135–142.

Grajetzki W. (2006) *The Middle Kingdom of Ancient Egypt*. London: Duckworth.

Gratuze B. (1999) Obsidian characterisation by laser ablation ICPMS and its application to prehistoric trade in the Mediterranean and the Near East. *Journal of Archaeological Science* 26: 869–891.

Grauer AL & P Stuart-Macadam. (1998) (eds.) *Sex and Gender in Paleopathological Perspective.* Cambridge: Cambridge University Press.

Graves R. (1955) (1992 edition.) *The Greek Myths.* London: Penguin.

Green S, S Greene & GJ Armelagos. (1974) Settlement and Mortality of the Christian Site (1050 AD–1300 AD) of Meinarti (Sudan). *Journal of Human Evolution* 3: 297316.

Griffith FLI. (1924) Excavations at El-'Amarnah, 1293–1924. *Journal of Egyptian Archaeology* 10: 299–305.

Grilleto RR. (1977) Carie et Usure Dentaire Chez les Égyptiens Prédynastiques et Dynastiques de la Collection de Turin (Italie). *L'Anthropologie* 81: 459–472.

Grilleto RR. (1978) Osservazioni sulla Carie Dentaria e sull'Usura dei Denti in una Serie di Crani Egiziani Predinastici. *Dental Cadmos* 46: 66–72.

Grilleto RR. (1979) Comparaison entre les Egyptiens Dynastiques d'Asiut et de Gebelen au Niveau de la Carie et de l'Usure des Dents. In: Reineke WF. (ed.) *First International Congress of Egyptology – Actes, 2–10 October 1976.* Berlin: Akademie Verlag. pp. 249–253.

Grøn O, L Aurdal, F Christensen & A Loska. (2004) Mapping and verifying invisible archaeological sites in agricultural fields by means of multi-spectral satellite images and soil chemistry. In: Wang C. (ed.) *International Conference on Remote Sensing Archaeology.* Beijing: Chinese Center for Remote Sensing Archaeology. pp. 83–90.

Grøn O, S Palmer, F-A Stylegar, K Esbensen, S Kucheryavski & S Aase. (2011) Interpretation of archaeological small-scale features in spectral images. *Archaeometry* 38: 2024–2030.

Halcrow SE & N Tayles. (2011) The bioarchaeological investigation of children and childhood. In: Agarwal SC & BA Glencross. (eds.) *Social Bioarchaeology.* Oxford: Blackwell. pp. 333–360.

Hall A, G Hewitt, V Tuffrey & N de Silva. (2008) A review and meta-analysis of the impact of intestinal worms on child growth and nutrition. *Maternal and Child Nutrition* 4 (Suppl 1): 118–236.

Hallberg L. (1987) Wheat fiber, phytates and iron absorption. *Scandinavian Journal of Gastroenterology* 22: 73–79.

Hammer CU, G Kurat, P Hoppe, W Grum & HB Clausen. (2003) Thera eruption date 1645 BC confirmed by new ice core data? In: Bietak M. (ed.) *The Synchronisation of Civilisations in the Eastern Mediterranean in the Second Millennium BC.* Vienna: Austrian Academy of Sciences Press. pp. 87–94.

Harbeck M, R Schleuder, J Schneider, I Wiechmann, WW Schmahl & G Grupe. (2011) Research potential and limitations of trace analyses of cremated remains. *Forensic Science International* 204: 191–200.

Harris DR. (1986) Plant and animal domestication and the origins of agriculture: the contribution of radiocarbon accelerator dating. In: Gowlett JAJ & REM Hedges. (eds.) *Archaeological Results from Accelerator Dating.* Oxford: Oxford University Committee for Archaeology. pp. 5–12.

Harris EC. (1979) *Principles of Archaeological Stratigraphy.* London: Academic Press.

Harris JE & KR Weeks. (1973) *X-Raying the Pharaohs*. New York: Charles Scribner's Sons.

Harrison FV. (1998) Introduction: Expanding the Discourse on 'Race'. *American Anthropologist* 100: 609–631.

Harter S, M Le Bailly, F Janot & F Bouchet. (2003) First paleoparasitological study of an embalming rejects jar found in Saqqara, Egypt. *Memórias do Instituto Oswaldo Cruz, Rio de Janeiro* 98 (Suppl I): 119–121.

Hassan FA. (1997) The dynamics of a riverine civilization: a geoarchaeological perspective on the Nile Valley, Egypt. *World Archaeology* 29: 51–74.

Hassan FA. (2002) Palaeoclimate, food and cultural change in Africa: an overview. In: Hassan FA. (ed.) *Droughts, Food and Culture: Ecological Change and Food Security in Africa's Later Prehistory*. New York: Kluwer Academic. pp. 11–26.

Hatcher H, MS Tite & JN Walsh. (1995) A comparison of ICP-ES and ICP-AAS analysis on standard reference silicate materials and ceramics. *Archaeometry* 37: 83–94.

Haustein M, C Gillis & E Pernicka. (2010) Tin isotopy – a new method for solving old problems. *Archaeometry* 52: 816–32.

Hawass Z, YZ Gad, S Ismail, R Khairat, D Fathallah, N Hasan, A Ahmed, H Elleithy, M Ball, F Gaballah, S Wasef, M Fateen, H Amer, P Gostner, A Selim, A Zink & CM Pusch. (2010) Ancestry, Pathology and Cause of Death in King Tutankhamun's Family. *Journal of the American Medical Association* 303: 638–647.

Hawkey DE & CF Merbs. (1995) Activity-Induced Musculoskeletal Stress Markers (MSM) and Subsistence Strategy Changes among Ancient Hudson Bay Eskimos. *International Journal of Osteoarchaeology* 5: 324–338.

Hayden B. (1993) *Archaeology: The Science of Once and Future Things*. New York: W.H. Freeman.

Hayes WC. (1953) *The Scepter of Egypt*. New York, MMA.

Henderson J. (1988) Electron microprobe analysis of mixed alkali glasses. *Archaeometry* 30: 77–91.

Henderson J. (1998) Scientific analysis of glass and glaze from Tell Brak and its archaeological implications. In: Oates D, J Oates & H McDonald. (eds.) *Excavations at Tell Brak, Volume 1: The Mitanni and Old Babylonian periods*. Cambridge: McDonald Institute Monographs. pp. 94–100.

Henderson J. (2000) *The science and archaeology of materials*. London: Routledge.

Henderson J, J Evans & K Nikita. (2010) Isotopic evidence for the primary production, provenance and trade of Late Bronze Age glass in the Mediterranean. *Mediterranean Archaeology* 10: 1–24.

Hendrickx S. (1994) *Elkab V. The Naqada III Cemetery*. Brussels: Comité des Fouilles Belges en Égypte et Musées Royaux d'Art et d'Histoire.

Hendrickx S. (2006) Predynastic – Early Dynastic Chronology. In: Hornung E, R Krauss & DA Warburton. (eds.) *Ancient Egyptian Chronology*. Brill: Leiden. pp. 55–93.

Henton E. (2012) The combined use of oxygen isotopes and microwear in sheep teeth to elucidate seasonal management of domestic herds: the case study of Çatalhöyük, central Anatolia. *Journal of Archaeological Science* 39: 3264–3276.

Henton E, W Meier-Augenstein & HF Kemp. (2010) The use of oxygen isotopes in sheep molars to investigate past herding practices at the Neolithic settlement of Çatalhöyük, central Anatolia. *Archaeometry* 52: 429–449.

Hepper FN. (1990) *Pharaoh's Flowers: The Botanical Treasures of Tutankhamun*. London: HMSO.

Herbich T. (2012a) Geophysical methods and landscape archaeology. *Egyptian Archaeology* 41: 11–14.

Herbich T. (2012b) Magnetic survey: Magnetic method in the prospection of sites in the Nile Delta prior to research at Tell el-Farkha. In: Chłodnicki M, KM Ciałowicz & A Mączyńska. (eds.) *Tell el-Farkha 1. Excavations 1998–2011*. Poznań: Poznań Archaeological Museum. pp. 383–392.

Herbich TM. (2014) How deep can we see? Practical observations on the vertical range of fluxgate gradiometers when surveying brick structures in the Nile Valley. In: Jucha MA, J Dębowska-Ludwin & P Kołodziejczyk. (eds.) *Aegyptus Est Imago Caeli*. Kraków: Jagiellonian University in Kraków. pp. 307–317.

Herbich T & I Forstner-Müller (2013) Small harbours in the Nile Delta: The case of Tell el-Daba. *Études et Travaux* 26: 257–272.

Herbich T & C Peeters. (2006) Results of the Magnetic Survey at Deir al-Barsha, Egypt. *Archaeological Prospection* 13: 11–24.

Herbich T & AJ Spencer. (2006) Geophysical survey at Tell el-Balamun. *Egyptian Archaeology* 29: 16–19.

Herz N. (1992) Provenance determination of Neolithic to Classical Mediterranean marbles by stable isotopes. *Archaeometry* 34: 185–194.

Hesse A, P Andrieux, M Atya, C Benech, C Camerlynck, M Dabas, C Fechant, A Jolivet, C Kuntz, P Mechler, C Panissod, L Pastor, A Tabbagh & J Tabbagh. (2002) L'Heptastade d'Alexandrie. In: Empereur J-Y. (ed.) *Alexandrina 2*. Cairo; Institut Français D'Archéologie Orientale.

Higham TFG, L Basell, RM Jacobi, R Wood, C Bronk Ramsey & NJ Conard. (2012), Testing models for the beginnings of the Aurignacian and the advent of figurative art and music: The radiocarbon chronology of Geißenklösterle. *Journal of Human Evolution* 62: 664–676.

Hillier JK, JM Bunbury & A Graham. (2007) Monuments on a migrating Nile. *Journal of Archaeological Science* 34: 1011–1015.

Hillson S. (1992) *Mammal Bones & Teeth: An Introductory Guide to Methods of Identification*. London: UCL Press.

Hillson S. (1996) *Dental Anthropology*. Cambridge: Cambridge University Press.

Hillson S. (2005) *Teeth*. Cambridge: Cambridge University Press.

Hodges FM. (2001) The ideal prepuce in ancient Greece and Rome: Male genital aesthetics and their relation to *lipodermos*, circumcision, foreskin restoration, and the *kynodesmē*. *Bulletin of the History of Medicine* 75: 375–405.

Hoffman MA. (1980) *Egypt Before the Pharaohs*. New York: Dorset Press.

Höflmayer F, A Hassler, W Kutschera & EM Wild. (2013) Radiocarbon Data for Aegean Pottery in Egypt: New Evidence from Saqqara (Lepsius) Tomb 16 and its Importance for LM IB/LH IIA. In: Shortland AJ & C Bronk Ramsey. (eds.) *Radiocarbon and the Chronologies of Ancient Egypt*. Oxford: Oxbow. pp. 110–120.

Holcomb S & L Konigsberg. (1995) Statistical study of sexual dimorphism in the human fetal sciatic notch. *American Journal of Physical Anthropology* 97: 113–125.

Hollimon SE. (2011) Sex and gender in bioarchaeological research. In: Agarwal SC & BA Glencross. (eds.) *Social Bioarchaeology*. Oxford: Blackwell. pp. 149–182.

Hope CA. (1979) Dakhleh Oasis Project – report on the study of pottery and kilns. *Journal of the Society for the Study of Egyptian Antiquities* 9: 187–201.

Hope CA. (1980) Dakhleh Oasis Project – report on the study of pottery and kilns. *Journal of the Society for the Study of Egyptian Antiquities* 10: 283–311.

Hope CA. (1981) Dakhleh Oasis Project – report on the study of pottery and kilns, third season 1980. *Journal of the Society for the Study of Egyptian Antiquities* 11: 233–241.

Hope CA. (1991) Blue-painted and polychrome decorated pottery from Amarna: a preliminary corpus. *Cahier de la Céramique Égyptienne* 2: 17–92.

Hornung E, R Krauss & DA Warburton. (2006a) Royal Annals. In: Hornung E, R Krauss & DA Warburton. (eds.) *Ancient Egyptian Chronology*. Leiden: Brill. pp. 19–25

Hornung E, R Krauss & DA Warburton. (2006b) King Lists & Manetho's Aigyptiaka. In: Hornung E, R Krauss & DA Warburton. (eds.) *Ancient Egyptian Chronology*. Leiden: Brill. pp. 33–36.

Hornung E, R Krauss & DA Warburton. (eds.) (2006c) *Ancient Egyptian Chronology*. Leiden: Brill.

Houlihan PF. (1986) *The Birds of Ancient Egypt*. Warminster: Aris & Phillips.

Howells WW. (1973) *Cranial Variation in Man: A Study by Multivariate Analysis of Patterns of Difference Among Recent Human Populations*. Papers of the Peabody Museum of Archaeology and Ethnology, Harvard University 67. Cambridge, MA: Harvard University Press.

Howells WW. (1989) *Skull Shapes and the Map: Craniometric Analyses in the Dispersion of Modern Homo*. Peabody Museum Papers, Harvard University 79. Cambridge, MA: Harvard University Press.

Howells WW. (1995) *Who's Who in Skulls: Ethnic Identification of Crania from Measurements*. Papers of the Peabody Museum of Archaeology and Ethnology, Harvard University 82. Cambridge, MA: Harvard University Press.

Hughes EE & WB Mann. (1964) The half-life of carbon-14: comments on the mass spectrometric method. *International Journal of Applied Radiation and Isotopes* 15: 97–100.

Humphrey L. (2000) Interpretations of the growth of past populations. In: Derevenski JS. (ed.) *Children and Material Culture*. London: Routledge. pp. 193–205.

Ibrahim EH, MM ElGamili, AGh Hassaneen, MN Soliman & AM Ismael. (2002) Geoelectrical investigation beneath Behbiet ElHigara and ElKom ElAkhder archaeological sites, Samannud Area, Nile Delta, Egypt. *Archaeological Prospection* 9: 105–113.

Ibrahim MA. (1987) *A Study of Dental Attrition and Diet in Some Ancient Egyptian Populations*. Durham: Unpublished PhD thesis, Durham University.

Ikram S. (2000) Meat processing. In: Nicholson PT & I Shaw. (eds.) *Ancient Egyptian Materials and Technology*. Cambridge: Cambridge University Press. pp. 656–672.

Ikram S. (2005) (ed.) *Divine Creatures: Animal mummies in Ancient Egypt*. Cairo: American University in Cairo Press.

Ikram S. (2010) Mummification. In: Dieleman J & W Wendrich. (eds.) *UCLA Encyclopedia of Egyptology*. Los Angeles, CA: http://digital2.library.ucla.edu/viewItem.do?ark=21198/zz001nwz18

Ikram S & A Dodson. (1998) *The mummy in ancient Egypt: Equipment for Eternity*. London: Thames & Hudson.

Imai T, T Sakayama & T Kanemori. (1987) Use of ground-probing radar and resistivity surveys for archaeological investigations. *Geophysics* 52: 137–150.

Irish JD. (1997) Characteristic high- and low-frequency dental traits in sub-Saharan African populations. *American Journal of Physical Anthropology* 102: 455–467.

Irish JD. (1998a) Dental morphological indications of population discontinuity and Egyptian gene flow in post-Paleolithic Nubia. *In:* Lukacs JR. (ed.) *Human Dental Development, Morphology, and Pathology: A Tribute to Albert A. Dahlberg*. University of Oregon Anthropological Papers 54. Eugene: University of Oregon Press. pp 155–172.

Irish JD. (1998b) Diachronic and synchronic dental trait affinities of Late and post-Pleistocene peoples from North Africa. *Homo* 49: 138–155.

Irish JD. (2006) Who were the ancient Egyptians? Dental affinities among Neolithic through postdynastic peoples. *American Journal of Physical Anthropology* 129: 529–543.

Irish JD & R Friedman. (2010) Dental affinities of the C-group inhabitants of Hierakonpolis, Egypt: Nubian, Egyptian, or both? *Homo* 61: 81–101.

İşcan MY, SR Loth & SK Wright. (1984) Metamorphosis at the sternal rib end. A new method to estimate age at death in white males. *American Journal of Physical Anthropology* 65: 147–156.

İşcan MY, SR Loth & RK Wright. (1985) Age estimation from the rib by phase analysis: white females. *Journal of Forensic Sciences* 30: 853–863.

Jackson CM. (2005a) Making colourless glass in the Roman Period. *Archaeometry* 47: 763–780.

Jackson CM. (2005b) Glassmaking in Bronze Age Egypt. *Science* 308: 1750–1752.

Jackson C & P Nicholson. (2013) Glass at el-Amarna. In Janssens KHA. (ed.) *Modern Methods for Analysing Archaeological and Historical Glass*. Chichester: Wiley. eBook.

Jansen-Winkeln K. (2006a) Dynasty 21. In: Hornung E, R Krauss & DA Warburton. (eds.) *Ancient Egyptian Chronology*. Leiden: Brill. pp. 218–233.

Jansen-Winkeln K. (2006b) Third Intermediate Period. In: Hornung E, R Krauss & DA Warburton. (eds.) *Ancient Egyptian Chronology*. Leiden: Brill. pp. 234–264.

Janssen RM & JJ Janssen. (1990) *Growing Up in Ancient Egypt*. London: Rubicon Press.

Janssen RM & JJ Janssen. (2005) *Growing Up and Getting Old in Ancient Egypt*. London: Golden House.

Jasieńska G & PT Ellison. (1998) *Physical work causes suppression of ovarian function in women*. Proceedings of the Royal Society of London B 265: 1847–1851.

Jaskulska E. (2009) Skeletal bilateral asymmetry in a medieval population from Deir an-Naqlun (Nekloni), Egypt. *Bioarchaeology of the Near East* 3: 17–26.

Jeffreys D. (2003) All in the Family? Heirlooms in Ancient Egypt. In: Tait J. (ed.) *'Never had the like occurred': Egypt's view of its past*. London: UCL Press. pp. 197–212.

Johnson A & T Earle. (2000) *The evolution of human societies*. Stanford: Stanford University Press.

Johnson AL & NC Lovell. (1994) Biological differentiation at predynastic Naqada, Egypt: an analysis of dental morphological traits. *American Journal of Physical Anthropology* 93: 427–433.

Johnson M. (1999) *Archaeological theory: an introduction*. Oxford: Blackwell.

Jones A. (2004) Archaeometry and materiality: materials-based analysis in theory and practice. *Archaeometry* 46: 327–338.

Jones B. (2000) Aerial Archaeology around the Mediterranean. In: Pasquinucci M & F Trément. (eds.) *Non-Destructive Techniques Applied to Landscape Archaeology*. The Archaeology of Mediterranean Landscapes Volume 4. Oxford: Oxbow. pp. 49–60.

Jones H, FJ Leigh, I Mackay, MA Bower, LM Smith, MP Charles & W Powell. (2008) Population-based resequencing reveals that the flowering time adaptation of cultivated barley originated east of the Fertile Crescent. *Molecular Biology and Evolution* 25: 2211–2219.

Jones J, TFG Higham, R Oldfield, TP O'Connor & SA Buckley. (2014) Evidence for Prehistoric origins of Egyptian mummification in Late Neolithic burials. *PLoS ONE* 9(8): e103608.

Jurmain R, F Alves Cardoso, C Henderson & S Villotte. (2012) Bioarchaeology's Holy Grail: The Reconstruction of Activity. In: Grauer AL. (ed.) *A Companion to Paleopathology*. Chichester: Wiley-Blackwell. pp. 531–552.

Kaczmarczyk A. (1986) The source of cobalt in ancient Egyptian pigments. In: Olin JS & M Blackman. (eds.) *Proceedings of the 24th International Archaeometry Symposium*. Washington: Smithsonian. pp. 369–376.

Kaczmarczyk A & REM Hedges. (eds.) (1983) *Ancient Egyptian Faience*. Warminster: Aris and Phillips.

Kaiser W. (1956) Stand und probleme der Ägyptischen vorgeschichtsforschung. *Zeitschrift für Ägyptische Sprache und Altertumskunde* 81: 87–109.

Kaiser W. (1957) Zur inneren Chronologie der Naqadakultur. *Archaeologia Geographica* 6: 69–77.

Kamei H, MA Atya, TF Abdallatif, M Mori & P Hemthavy. (2002) Ground-penetrating radar and magnetic survey to the west of Al-Zayyan Temple, Kharga Oasis, Al-Wadi Al-Jadeed (New Valley), Egypt. *Archaeological Prospection* 9: 93–104.

Karafet TM, FL Mendez, MB Meilerman, PA Underhill, SL Zegura & MF Hammer. (2008) New binary polymorphisms reshape and increase resolution of the human Y chromosomal haplogroup tree. *Genome Research* 18: 830–838.

Katzenberg MA. (2012) The Ecological Approach: understanding past diet and the relationship between diet and disease. In: Grauer AL. (ed.) *A Companion to Paleopathology*. Chichester: Wiley-Blackwell. pp. 97–113.

Kearey P, M Brooks & I Hill. (2002) *Introduction to Geophysical Exploration*. Oxford: Blackwell.

Keita SOY. (1992) Further studies of crania from ancient Northern Africa: An analysis of crania from First Dynasty Egyptian Tombs, Using Multiple Discriminant Functions. *American Journal of Physical Anthropology* 87: 245–254.

Keita SOY. (2004) Exploring northeast African metric craniofacial variation at the individual level: a comparative study using principal components analysis. *American Journal of Human Biology* 16: 679–689.

Kemp BJ. (1983) Preliminary report on the el-'Amarna expedition, 1981–2. *Journal of Egyptian Archaeology* 69: 5–24.

Kemp BJ. (1984–85) Report on the Tell el-Amarna expedition 1977–1982. *Annales du Services des Antiquités de l'Égypte* 70: 83–97.

Kemp BJ. (1989) *Ancient Egypt: Anatomy of a Civilization*. London: Routledge.

Kemp BJ. (2000) Soil (including mud-brick architecture). In: Nicholson PT & I Shaw (eds.) *Ancient Egyptian materials and technology*. Cambridge: Cambridge University Press. pp. 78–103.

Kemp B. (2005) Settlement and landscape in the Amarna area in the Late Roman period. In: Faiers J. (ed.) *Late Roman Pottery at Amarna and Related Studies*. London: Egypt Exploration Society. pp. 11–56.

Kemp BJ. (2006) *Ancient Egypt: Anatomy of a Civilization*. Second edition. Abingdon: Routledge.

Kemp B. (2010) Tell el-Amarna, 2010. *Journal of Egyptian Archaeology* 96: 1–29.

Kemp B. (2012a) *City of Akhenaten and Nefertiti: Amarna and its People*. London: Thames and Hudson.

Kemp B. (2012b) Tell el-Amarna from 1914 to today. In: Seyfried F. (ed.) *In the Light of Amarna: 100 Years of the Nefertiti Discovery*. Berlin: Ägyptisches Museum und Papyrussammlung. pp. 50–55.

Kemp BJ & S Garfi. (1993) *A Survey of the Ancient City of el-'Amarna*. Occasional Publications 9. London: Egypt Exploration Society.

Kemp BJ & G Vogelsang-Eastwood. (2001) *The ancient textile industry at Amarna*. Sixty-eighth Excavation Memoir. London: Egypt Exploration Society.

Kemp B, A Stevens, GR Dabbs, M Zabecki & JC Rose. (2013) Life, death and beyond in Akhenaten's Egypt: excavating the South Tombs Cemetery at Amarna. *Antiquity* 87: 64–78.

Kendall T. (1997) *Kerma and the Kingdom of Kush 2500–1500 B.C.* Washington DC: National Museum of African Art, Smithsonian Institution.

Kitchen KA. (1971) Punt and how to get there. *Orientalia* 40: 184–207.

Kitchen KA. (1995) *The Third Intermediate Period in Egypt (1100–650 B.C.)*. 2nd edn. Warminster: Aris & Phillips.

Klemm R & D Klemm. (1993) *Steine and Steinbrüche im Alten Ägypten*. Berlin–Heidelberg: Springer.

Klemm R & D Klemm (2008) *Stone and Stone Quarries in Ancient Egypt*. London: British Museum Press.

Knüsel CJ. (2000) Activity-related skeletal change. In: Fiorato V, A Boylston & C Knüsel. (eds.) *Blood red roses: the archaeology of a mass grave from the Battle of Towton AD 1461*. Oxford: Oxbow. pp 103–118.

Knüsel CJ, CA Roberts & A Boylston. (1996) Brief communication: when Adam delved ... an activity-related lesion in three human skeletal populations. *American Journal of Physical Anthropology* 100: 427–434.

Koller J, U Baumer, Y Kaup, H Etspüler & U Weser. (1998) Embalming was used in Old Kingdom. *Nature* 391: 343–344.

Koritzer RT. (1968) An analysis of the cause of tooth loss in an Ancient Egyptian population. *American Anthropologist* 70: 550–553.

Koutsoudis A, B Vidmar & F Arnaoutoglou. (2013) Performance evaluation of a multi-image 3D reconstruction software on a low-feature artefact. *Journal of Archaeological Science* 40: 4450–4456.

Kozloff AP & BM Bryan. (1992) *Egypt's Dazzling Sun: Amenhotep III and his World*. Cleveland: Cleveland Museum of Art.

Krauss R. (2006) Egyptian Sirius/Sothic Dates and the Question of the Sirius based Lunar Calender. In: Hornung E, R Krauss & DA Warburton. (eds.) *Ancient Egyptian Chronology*. Leiden: Brill. pp. 439–457.

Kristiansen K. (2009) The discipline of archaeology. In: Cunliffe B, C Gosden & RA Joyce. (eds.) *The Oxford Handbook of Archaeology*. Oxford: Oxford University Press. pp. 3–46.

Kromer B. (2009) Radiocarbon and dendrochronology. *Dendrochronologia* 27: 15–19.

Kromer B & M Spurk. (1998) Revision and tentative extension of the tree-ring based 14C calibration, 9200–11,855 cal BP. *Radiocarbon* 40: 1117–1125.

Kruglov AV. (2010) Late Antique Sculpture in Egypt: Originals and Forgeries. *American Journal of Archaeology Online Museum Review* 114.2.

Kuniholm PI. (2001) Dendrochronology and other applications of tree-ring Studies in Archaeology. In: Brothwell DR & AM Pollard. (eds.) *The Handbook of Archaeological Sciences*. London: John Wiley & Sons. pp. 35–46.

Kuniholm PI. (2002) Archaeological dendrochronology. *Dendrochronologia* 20: 63–68.

Kunos CA, SW Simpson, KF Russell & I Hershkovitz. (1999) First rib metamorphosis: its possible utility for human age-at-death estimation. *American Journal of Physical Anthropology* 110: 303–323.

Lahr MM. (1996) *The Evolution of Modern Human Diversity*. Cambridge: Cambridge University Press

Lalremruata A, M Ball, R Bianucci, B Welte, AG Nerlich, JFJ Kun & CM Pusch. (2013) Molecular identification of falciparum malaria and human tuberculosis co-infections in mummies from the Fayum Depression (Lower Egypt). *PLoS One* 8: e60307.

LaMarche VC Jr. & KK Hirschboeck. (1984) Frost rings in trees as records of major volcanic eruptions. *Nature* 307: 121–126.

Lamendin H, E Baccino, JF Humbert, JC Tavernier, RM Nossintchouk & A Zerilli. (1992) A simple technique for age estimation in adult corpses: the two criteria dental method. *Journal of Forensic Sciences* 37: 1373–1379.

Larsen CS. (1995) Biological Changes in Human Populations with Agriculture. *Annual Review of Anthropology* 24: 185–213.

Larsen CS. (1997) *Bioarchaeology*. Cambridge: Cambridge University Press.

Lasaponara R & N Masini. (2006) On the potential of QuickBird data for archaeological Prospection. *International Journal of Remote Sensing* 27: 3607–3614.

Lasaponara R & N Masini. (2007) Detection of archaeological crop marks by using satellite QuickBird multispectral imagery. *Journal of Archaeological Science* 34: 214–221.

Leahy A & I Mathieson. (2002) Late Period Temple Platforms at Saqqara. *Egyptian Archaeology* 21: 14–16.

Leavitt SW & B Bannister. (2009) Dendrochronology and Radiocarbon Dating: The Laboratory of Tree-Ring Research Connection. *Radiocarbon* 51: 373–384.

Lee L & S Quirke. (2000) Painting materials. In: Nicholson P & I Shaw. (eds.) *Ancient Egyptian materials and technology.* Cambridge: Cambridge University Press. pp. 104–120.

Lee LR & D Thicket. (1996) *Selection of Materials for the Storage and Display of Museum Objects.* British Museum Department of Conservation Occasional Paper 111.

Leek FF. (1972a) Bite, Attrition and Associated Oral Conditions as Seen in Ancient Egyptian Skulls. *Journal of Human Evolution* 1: 289–295.

Leek FF. (1972b) Teeth and Bread in Ancient Egypt. *Journal of Egyptian Archaeology* 58: 126–132.

Legge A. (2010) The mammal bones from Grid 12. In: Kemp B & A Stevens. (eds.) *Busy Lives at Amarna: Excavations in the Main City (Grid 12 and the House of Ranefer, N49.18). Volume I: The Excavations, Architecture and Environmental Remains.* London: Egypt Exploration Society. pp. 445–452.

Lepsius CR. (1849–1859) *Denkmæler aus Ægypten und Æthiopien.* Geneva: Éditions de Belles Lettres.

Leroy SAG, K Arpe & U Mikolajewicz. (2011) Vegetation context and climatic limits of the Early Pleistocene hominin dispersal in Europe. *Quaternary Science Reviews* 30: 1448–1463.

Lewis ME. (2007) *The Bioarchaeology of Children: Perspectives from biological and forensic anthropology.* Cambridge: Cambridge University Press.

Lilyquist C & RH Brill. (1993) *Studies in Early Egyptian Glass.* New York: Metropolitan Museum of Art.

Linsalata P, M Eisenbud & EP Franca. (1986) Ingestion estimates of Th and the light rare earth elements based on measurements of human feces. *Health Physics* 50: 163–167.

Linsalata P, R Morse, H Ford, M Eisenbud, EP Franca, MB de Castro, L Nazyo, I Sachett & M Carlos. (1991) Th, U, Ra and rare earth element distribution in farm animal tissues from an elevated natural radiation background environment. *Journal of Environmental Radioactivity* 14: 233–257.

Linseele V, W Van Neer, S Thys, R Phillipps, R Cappers, W Wendrich & S Holdaway. (2014) New archaeozoological data from the Fayum 'Neolithic' with a critical assessment of the evidence for early stock keeping in Egypt. *PLoS ONE* 9(10): e108517.

Livingstone Smith A. (2000) Processing clay for pottery in northern Cameroon: social and technical requirements. *Archaeometry* 42: 21–42.

Loke MH & RD Barker. (1996a) Rapid least-squares inversion of apparent resistivity pseudosections by a quasi-Newton method. *Geophysical Prospection* 44: 131–152.

Loke MH & RD Barker. (1996b) Practical techniques for 3D resistivity surveys and data inversion. *Geophysical Prospecting* 44: 499–523.

Lombard M. (2014) 'In situ' presumptive test for blood residues applied to 62 000-year-old stone tools. *South African Archaeological Bulletin* 69: 80–86.

Loschwitz N. (2012) Cobalt blue pottery painting of the Amarna period. In Seyfried F. (ed.) *In the Light of Amarna: 100 Years of the Nefertiti Discovery.* Berlin: Ägyptisches Museum und Papyrussammlung. pp. 132–135.

Loth S & M Henneberg. (2001) Sexually dimorphic mandibular morphology in the first few years of life. *American Journal of Physical Anthropology* 115: 179–186.

Loth SR & MY İşcan. (1989) Morphological assessment of age in the adult: the thoracic region. In: İşcan MY (ed.) *Age Markers in the Human Skeleton*. Springfield: Charles C Thomas. pp. 105–135.

Lovejoy CO, RS Meindl, TR Pryzbeck & RP Mensforth. (1985) Chronological metamorphosis of the auricular surface of the ilium: a new method for the determination of adult skeletal age at death. *American Journal of Physical Anthropology* 68: 15–28.

Lucas A. (1942) Obsidian. *Annales du Service des Antiquités de l'Egypte, Caire* 41: 271–275.

Lucas A. (1947) Obsidian. *Annales du Service des Antiquités de l'Egypte, Caire* 47: 113–123.

Lucas A & JR Harris. (1962) *Ancient Egyptian Materials and Industries*. London: Dover.

Lucy D, AM Pollard & CA Roberts. (1995) A comparison of three dental techniques for estimating age at death in humans. *Journal of Archaeological Science* 22: 417–428.

Lucy S. (2005) Ethnic and cultural identities. In: Díaz-Andreu M, S Lucy, S Babić & DN Edwards. (eds.) *The Archaeology of Identity*. London: Routledge. pp. 86–109.

Luff R-M. (1984) *Animal Remains in Archaeology*. Princes Risborough: Shire Archaeology.

Luff RM. (1994) Butchery at the Workmen's Village (WV), Tell el-Amarna, Egypt. In: Luff RM & P Rowley-Conwy. (eds.) *Whither environmental archaeology?* Oxbow Monograph 38. Oxford: Oxbow. pp. 158–170.

Lukacs JR. (2012) Oral Health in Past Populations: context, concepts and controversies. In: Grauer AL. (ed.) *A Companion to Paleopathology*. Chichester: Wiley-Blackwell. pp. 553–581.

Lutley K(C) & JM Bunbury. (2008) The Nile on the move. *Egyptian Archaeology* 32: 3–5.

Lythgoe AM. (1965) *The Predynastic Cemetery N 7000, Naga-ed-Dêr*. Berkeley: University of California Press.

MacKenzie WS, CH Donaldson & C Guilford. (1982) *Atlas of the Igneous Rocks and their Textures*. New York: John Wiley.

MacKenzie WS & C Guilford. (1980) *Atlas of the Rock Forming Minerals in Thin Section*. Harlow, Essex: Longman.

Madella M & D Zurro. (eds.) (2007) *Plants, People and Places: Recent Studies in Phytolithic Analysis*. Oxford: Oxbow.

Malainey ME. (2011) *A Consumer's Guide to Archaeological Science*. Heidelberg: Springer.

Malleson C. (in press) Flora. In: Shaw I & E Bloxam. (eds.) *Oxford Handbook of Egyptology*. Oxford: Oxford University Press.

Manetho. (1940) *Aegyptiaca*. (transl. WG Waddell) London: Loeb Classical Library.

Maniatis Y & MS Tite. (1979) Technological examination of Neolithic-Bronze Age pottery from central and south-east Europe and from the Near East. *Journal of Archaeological Science* 8: 59–76.

Manning SW. (2006) Radiocarbon dating and Egyptian chronology. In: Hornung E, R Krauss & DA Warburton. (eds.) *Ancient Egyptian Chronology*. Brill: Leiden. pp. 327–355.

Manning SW, C Bronk Ramsey, C Doumas, T Marketou, G Cadogan & CL Pearson. (2002) New evidence for an early date for the Aegean Late Bronze Age and Thera eruption. *Antiquity* 76: 733–744.

Manning SW, B Kromer, MW Dee, M Friedrich, TFG Higham & C Bronk Ramsey. (2013) Radiocarbon calibration in the Mid to Later 14th Century BC and radiocarbon dating Tell el-Amarna, Egypt. In: Shortland AJ & C Bronk Ramsey. (eds.) *Radiocarbon and the Chronologies of Ancient Egypt*. Oxford: Oxbow. pp. 121–145.

Marcus ES. (2013) Correlating and Combining Egyptian Historical and Southern Levantine Radiocarbon Chronologies at Middle Bronze Age IIA Tel Ifshar, Israel. In: Shortland AJ & C Bronk Ramsey. (eds.) *Radiocarbon and the Chronologies of Ancient Egypt*. Oxford: Oxbow. pp. 182–208.

Marée M. (2010) *The Second Intermediate Period (Thirteenth to Seventeenth Dynasties): Current Research, Future Prospects*. Orientalia Lovaniensia Analecta 192. Leuven: Peeters.

Marlar RA, BL Leonard, BR Billman, PM Lambert & JE Marlar. (2000) Biochemical evidence of cannibalism at a prehistoric Puebloan site in southwestern Colorado. *Nature* 407: 74–78.

Masali M. (1972) Body Size and Proportions as Revealed by Bone Measurements and their Meaning in Environmental Adaptation. *Journal of Human Evolution* 1: 187–197.

Masali M & Chiarelli B. (1966) Risultati preliminari di una indagine sur rapporto sessi sul materiale osteologico delle necropoli di Gebelen e Assiut. *Archivio per l'Anthropologia e l'Etnologia* 98: 111–112.

Mathieson I. (1995) Proton-magnetometer surveys in the main city. In: Kemp BJ. (ed.) *Amarna Reports VI*. London: Egypt Exploration Society. pp. 218–225.

Mathieson I. (2007) Appendix 1: Geophysical Report. In: Nicholson P. (ed.) *Brilliant Things for Akhenaten*. London: Egypt Exploration Society. pp. 161–164.

Mathieson I, E Bettles, S Davies & HS Smith. (1995) A stela of the Persian period from Saqqara. *Journal of Egyptian Archaeology* 81: 23–41.

Mathieson P. (ed.) (2013) *Seeing under the sands of Saqqara: geophysics in the service of Egyptian archaeology. A memoir of the work of Ian J. Mathieson*. Edinburgh: Scottish Egyptian Archaeological Trust.

Mays S. (2010) *The Archaeology of Human Bones*. Second Edition. London: Routledge.

McAleely S. (2013) Garlands from the Deir el-Bahri Cache. In: Bronk Ramsey C & AJ Shortland. (eds.) *Radiocarbon and the Chronologies of Ancient Egypt*. Oxford: Oxbow. pp. 153–166.

McCauley JF, GG Schaber, CS Breed, MJ Grolier, CV Haynes, B Issawi, C Elachi & R Blom. (1982) Subsurface valleys and geoarcheology of the Eastern Sahara revealed by shuttle radar. *Science* 218: 1004–1020.

McDonnell JG. (2005) Pyrotechnology. In: Brothwell DR & AM Pollard. (eds.) *Handbook of archaeological sciences*. Chichester: J. Wiley. pp. 493–505.

McHugh WP, CS Breed, GG Schaber, JF McCauley & BJ Szabo. (1988) Acheulian Sites along the 'Radar Rivers', Southern Egyptian Sahara. *Journal of Field Archaeology* 15: 361–379.

McIntosh J. (1999) *The Practical Archaeologist*. London: Thames & Hudson.

McKinley JI. (2013) Cremation: Excavation, analysis, and interpretation of material from cremation-related contexts. In: Tarlow S & L Nilsson-Stutz. (eds.) *The Oxford Handbook of the Archaeology of Death & Burial*. Oxford: Oxford University Press. pp. 147–171.

McPherron SP, T Gernat & J-J Hublin. (2009) Structured light scanning for high-resolution documentation of in situ archaeological finds. *Journal of Archaeological Science* 36: 19–24.

Meeks ND. (1986) Tin-rich surfaces on bronze – some experimental and archaeological considerations. *Archaeometry* 28: 133–162.

Meeks ND & MS Tite. (1980) The analysis of platinum group element inclusions in gold antiquities. *Journal of Archaeological Science* 7: 267–275.

Megyesi MS, DH Ubelaker & NJ Sauer. (2006) Test of the Lamendin aging method on two historic skeletal samples. *American Journal of Physical Anthropology* 31: 363–367.

Meindl RS & CO Lovejoy. (1985) Ectocranial suture closure: a revised method for the determination of skeletal age at death based on the lateral-anterior sutures. *American Journal of Physical Anthropology* 68: 57–66.

Mercier N, H Valladas, L Froget, J-L Joron, PM Vermeersch, P Van Peer & J Moeyersons. (1999) Thermoluminescence dating of a middle palaeolithic occupation at Sodmein Cave, Red Sea Mountains (Egypt). *Journal of Archaeological Science* 26: 1339–1345.

Meskell LM. (1994) Dying young: the experience of death at Deir el Medina. *Archaeological Review from Cambridge* 13: 35–45.

Meskell L. (2004) *Object Worlds In Ancient Egypt: material biographies past and present*. New York: Berg.

Metwaly M, AG Green, H Horstmeyer, H Maurer, AM Abbas & A-RGh Hassaneen. (2005) Combined seismic tomographic and ultrashallow seismic reflection study of an Early Dynastic mastaba, Saqqara, Egypt. *Archaeological Prospection* 11: 245–246.

Midant-Reynes B. (2000) *The Prehistory of Egypt*. Oxford: Blackwell.

Miles AEW. (2001) The Miles method of assessing age from tooth wear. *Journal of Archaeological Science* 28: 973–982.

Miller ChS, SAG Leroy, G Izon, HAK Lahijani, F Marret, AB Cundy & PL Teasdale. (2013) Palynology: A tool to identify abrupt events? An example from Chabahar Bay, southern Iran. *Marine Geology* 337: 195–201.

Miller J. (2008) *An Appraisal of the Skulls and Dentition of Ancient Egyptians, Highlighting the Pathology and Speculating on the Influence of Diet and Environment*. BAR International Series 1794. Oxford: BAR & Archaeopress.

Millson DCE. (ed.) (2010) *Experimentation and Interpretation: The Use of Experimental Archaeology in the Study of the Past*. Oxford: Oxbow Books.

Milner GR & JL Boldsen. (2012) Skeletal Age Estimation: where we are and where we should go. In: Dirkmaat DC. (ed.) *A Companion to Forensic Anthropology*. Chichester: Wiley-Blackwell. pp. 224–238.

Minnikin DE, GS Besra, O-C Lee, M Spigelman & HD Donoghue. (2011) The interplay of DNA and lipid biomarkers in the detection of tuberculosis and leprosy in mummies and other skeletal remains. In: Gill-Frerking H, W Rosendahl, A Zink & D Piombino-Mascasli. (eds.) *Yearbook of mummy studies*. München: Verlag Dr. Friedrich Pfeil. pp. 109–114.

Mittler DM & DP Van Gerven. (1994) Developmental, diachronic, and demographic analysis of cribra orbitalia in the medieval Christian populations of Kulubnarti. *American Journal of Physical Anthropology* 93: 287–297.

MMA. (ed.) (1999) *Egyptian Art in the Age of the Pyramids*. New York: Metropolitan Museum of Art.

Mohamed W & S Darweesh. (2012) Ancient Egyptian black patinated copper alloys. *Archaeometry* 54: 175–192.

Molleson T. (1994) The eloquent bones of Abu Hureyra. *Scientific American* 271: 70–75.

Molleson T & M Cox. (1993) *The Spitalfields Project, volume 2, The Middling Sort. Research Report No. 86*. York: Council for British Archaeology.

Molleson T, K Cruise & S Mays. (1998) Some sexually dimorphic features of the human juvenile skull and their value in sex determination in immature skeletal remains. *Journal of Archaeological Science* 25: 719–728.

Mommsen H, T Beier, U Diehl & Ch. Podzuweit. (1992) Provenance determination of Mycenaean sherds found in Tell el Amarna by neutron activation analysis. *Journal of Archaeological Science* 19: 295–302.

Montgomery J, JA Evans & MSA Horstwood. (2010) Evidence for long-term averaging of strontium in bovine enamel using TIMS and LA-MC-ICP-MS strontium isotope intra-molar profiles. *Environmental Archaeology* 15: 32–42.

Montserrat D. (1996) *Sex and Society in Graeco-Roman Egypt*. Abingdon: Routledge.

Moorey PRS. (1994) *Ancient Mesopotamian Materials and Industries*. Oxford: Oxford University Press.

Moorrees CFA, EA Fanning & EE Hunt. (1963a) Formation and resorption of three deciduous teeth in children. *American Journal of Physical Anthropology* 21: 205–213.

Moorrees CFA, EA Fanning & EE Hunt. (1963b) Age variation of formation stages for ten permanent teeth. *Journal of Dental Research* 42: 1490–1502.

Moser S. (2006) *Wondrous Curiosities: Ancient Egypt at the British Museum*. Chicago: University of Chicago Press.

Moussa AH, LT Dolphin & G Mokhtar. (1977) *Applications of Modern Sensing Techniques to Egyptology*. Menlo Park: SRI International. http://www.ldolphin. org/egypt/egypt2/

Moussa M. (2001) Gamma-ray spectrometry: a new tool for exploring archaeological sites; a case study from East Sinai, Egypt. *Journal of Applied Geophysics* 48: 137–142.

Mulhern DM & EB Jones. (2005) Test of revised method of age estimation from the auricular surface of the ilium. *American Journal of Physical Anthropology* 126: 61–65.

Mulligan CJ. (2006) Anthropological Applications of Ancient DNA: Problems and Prospects. *American Antiquity* 71: 365–380.

Murnane WJ & CC Van Siclen III. (1993) *The Boundary Stelae of Akhenaten*. Abingdon: Routledge.

Murray M-A. (2009) Questions of continuity: fodder and fuel use in Bronze Age Egypt. In: Fairbairn A & E Weiss. (eds.) *From Foragers to Farmers: Papers in Honour of Gordon C. Hillman*. Oxford: Oxbow, pp. 254–267.

Museum of London Archaeology Service (MoLAS). (1994) *Archaeological site manual*. 3rd edition. London: MoLAS. http://www.

museumoflondonarchaeology.org.uk/NR/rdonlyres/056B4AFD-AB5F-45AF-9097-5A53FFDC1F94/0/MoLASManual94.pdf.

Nenna M-D. (ed.) (2000) *La Route Du Verre: Ateliers primaires et secondaires du second millénaire av. J.-C. au Moyen Âge*. Paris : Maison de l'Orient méditerranéen-Jean Pouilloux.

Nerlich AG, F Parsche, A Von Den Driesch & U Löhrs. (1993) Osteopathological findings in mummified baboons from ancient Egypt. *International Journal of Osteoarchaeology* 3: 189–198.

Nerlich AG, B Schraut, S Dittrich, T Jelinek & AR Zink. (2008) *Plasmodium falciparum* in Ancient Egypt. *Emerging and Infectious Diseases* 14: 1317–1319.

Nesbitt M. (1993) Archaeobotanical evidence for Early Dilmun diet at Saar, Bahrain. *Arabian Archaeology and Epigraphy* 4: 20–47.

Nicholson P. (1985) 8.9 The brick and clay samples. In: Nicholson P & P Rose, Pottery fabrics and ware groups at el-Amarna. *Amarna Reports* II: 133–174.

Nicholson PT. (1993) *Ancient Egyptian Faience and Glass*. London: Shire Egyptology.

Nicholson PT. (1995) Glass-making and glass-working at Amarna: some new work. *Glass Technology* 37: 11–19.

Nicholson PT. (1998) Materials and technology. In: Friedman FD. (ed.) *Gifts of the Nile – Ancient Egyptian faience*. London: Thames & Hudson. pp. 50–64.

Nicholson PT. (2005) The Sacred Animal Necropolis at North Saqqara: The cults and their catacombs. In: Ikram S. (ed.) *Divine Creatures: Animal mummies in Ancient Egypt*. Cairo: American University in Cairo Press. pp. 44–71.

Nicholson PT. (ed.) (2007) *Brilliant Things for Akhenaten*. London: Egypt Exploration Society.

Nicholson PT & J Henderson. (2000) Glass. In: Nicholson PT & I Shaw (eds.) *Ancient Egyptian Materials and Technology*. Cambridge: Cambridge University Press. pp. 195–224.

Nicholson PT & CM Jackson. (1998) 'Kind of blue' – glass of the Amarna period replicated. In: McCray P & WD Kingery. (eds.) *The Prehistory and History of Glassmaking Technology*. Columbus, OH: American Ceramics Society. pp. 251–268.

Nicholson PT & CM Jackson. (2007) The Furnace Experiment. In: Nicholson P. (ed.) *Brilliant Things for Akhenaten*. London: Egypt Exploration Society. pp. 83–99.

Nicholson PT & EJ Peltenburg. (2000) Faience. In: Nicholson PT & I Shaw. (eds.) *Ancient Egyptian Materials and Technology*. Cambridge: Cambridge University Press. pp. 177–194.

Nicholson P & PJ Rose. (1985) Pottery and Ware groups at el-Amarna. In: Kemp BJ. (ed.) *Amarna Reports II*. London: EES. pp. 133–174.

Nicholson PT & I Shaw. (eds.) (2000) *Ancient Egyptian Materials and Technology*. Cambridge: Cambridge University Press.

Nicholson PT, J Harrison, S Ikram, E Early & Y Qin. (2013) Geoarchaeological and environmental work at the Sacred Animal Necropolis, North Saqqara, Egypt. *Studia Quaternaria* 30: 83–89.

Nicholson PT, S Ikram & S Mills. (in press) The Catacombs of Anubis At North Saqqara. *Antiquity*.

Nicholson PT, CM Jackson & KM Trott. (1997) The Ulu Burun glass ingots, cylindrical vessels and Egyptian glass. *The Journal of Egyptian Archaeology* 83: 143–153.

Nockolds SR, RWOB Knox & GA Chinner. (1978) *Petrology For Students*. Cambridge: Cambridge University Press.

Noel M. (1992) Multielectrode Resistivity Tomography for Imaging Archaeology. In: Spoerry P. (Ed.) *Geoprospection in the Archaeological Landscape*. Oxbow Monograph 18. Oxford: Oxbow.

Nolte B. (1968) *Die Glassgefässe Im Alten Ägypten*. Berlin: Verlag Bruno Hessling.

Nordström H-A & J Bourriau. (1993) Ceramic technology: clays and fabrics. In: Arnold D & JD Bourriau. (eds.) *An Introduction to Ancient Egyptian Pottery*. Mainz am Rhein: Philipp von Zabern. pp. 168–182.

Nunn JF. (1996) *Ancient Egyptian Medicine*. London: British Museum Press.

Nystrom KC, A Goff & M Lee Goff. (2005) Mortuary Behaviour Reconstruction through Palaeoentomology: A Case Study from Chachapoya, Perú. *International Journal of Osteoarchaeology* 15: 175–185.

O'Connor D & A Reid. (2003) Introduction – Locating Ancient Egypt in Africa: modern theories, past realities. In: O'Connor D & A Reid (eds.) *Ancient Egypt in Africa*. London: UCL Press. pp. 1–22.

O'Connor T. (2000) *The Archaeology of Animal Bones*. Stroud: Sutton Publishing.

Ogden J. (2000) Metals. In: Nicholson P & I Shaw. (eds.) *Ancient Egyptian Materials and Technology*. Cambridge: Cambridge University Press. pp. 148–176.

O'Neill MC & CB Ruff. (2004) Estimating human long bone cross-sectional geometric properties: a comparison of noninvasive methods. *Journal of Human Evolution* 47: 221–235.

Oppenheim AL, RH Brill, D Barag & A Von Saldern. (1970) *Glass and Glass-Making in Ancient Mesopotamia*. New York: Corning Museum of Glass.

Ortner D. (1998) Male–female immune reactivity and its implications for interpreting evidence in human skeletal paleopathology. In: Grauer AL & P Stuart-Macadam. (eds.) *Sex and Gender in Paleopathological Perspective*. Cambridge: Cambridge University Press. pp. 79–92.

Ortner D & M Ericksen. (1997) Bone changes in the human skull probably resulting from scurvy in infancy and childhood. *International Journal of Osteoarchaeology* 7: 212–220.

Ortner D & W Putschar. (1985) *Identification of Pathological Conditions in Human Skeletal Remains* Washington, DC: Smithsonian University Press.

Ortner D, W Butler, J Cafarella & L Milligan. (2001) Evidence of probable scurvy in sabadults from archaeological sites in North America. *American Journal of Physical Anthropology* 114: 343–351.

Ortner D, Kimmerle E & M Diez. (1999) Probable evidence of scurvy in subadults from archaeological sites in Peru. *American Journal of Physical Anthropology* 108: 321–331.

Orton C. (2000) *Sampling in Archaeology*. Cambridge: Cambridge University Press.

Orton C & M Hughes. (2013) *Pottery in Archaeology*. Second edition. Cambridge: Cambridge University Press.

Osborn DJ & J Osbornová. (1998) *The Mammals of Ancient Egypt*. Warminster: Aris & Phillips.

Paliou E, D Wheatley & G Earl. (2011) Three-dimensional visibility analysis of architectural spaces: iconography and visibility of the wall paintings of Xeste 3 (Late Bronze Age Akrotiri). *Journal of Archaeological Science* 38: 375–386.

Panagiotakopulu E. (1999) An examination of biological materials from coprolites from XVIII Dynasty Amarna, Egypt. *Journal of Archaeological Science* 26: 547–551.

Panagiotakopulu E & P Buckland. (2010) The insect remains. In: Kemp B & A Stevens. (eds.) *Busy Lives at Amarna: Excavations in the Main City (Grid 12 and the House of Ranefer, N49.18). Volume I: The Excavations, Architecture and Environmental Remains.* London: Egypt Exploration Society. pp. 453–465.

Parcak SH. (2009) *Satellite Remote Sensing for Archaeology.* London: Routledge.

Parr RL. (2002) Mitochondrial DNA sequence analysis of skeletal remains from the Kellis 2 cemetery. In: Hope CA & GE Bowen. (eds.) *Dakhleh Oasis Project: Preliminary Reports on the 1994–1995 to 1998–1999 Field Seasons.* Oxford: Oxbow Books. pp. 257–261.

Patrick P. (2007) Overweight and the human skeleton. In: Zakrzewski SR & W White. (eds.) *Proceedings of the Seventh Annual Conference of the British Association for Biological Anthropology and Osteoarchaeology.* Oxford: Archaeopress. BAR International Series 1712. pp. 62–71.

Payne P. (2007) Appendix 2: Report on the Faunal Material from Site O45.1. In: Nicholson P. (ed.) *Brilliant Things for Akhenaten.* London: Egypt Exploration Society. pp. 165–167.

Paynter S & MS Tite. (2001) The evolution of glazing technologies in the Ancient Near East and Egypt. In: Shortland AJ. (ed.) *The Social Context of Technological Change.* Oxford: Oxbow Books.

Peacock D & L Blue. (eds.) (2006) *Myos Hormos – Quseir al-Qadim, Roman and Islamic Ports on the Red Sea, Volume 1: The Survey and Report on the Excavations.* Oxford: Oxbow Books.

Peacock DPS & VA Maxfield. (eds.) (1997) *Mons Claudianus: Survey and Excavation.* Cairo: Institut français d'archéologie orientale du Caire.

Peacock DPS & VA Maxfield. (eds.) (2007) *The Roman Imperial Quarries: Excavations – Survey and Excavation at Mons Porphyrites 1994–1998.* London: Egypt Exploration Society.

Pearson GW & M Stuiver. (1986) High-precision calibration of the radiocarbon time scale, 500–2500 BC. *Radiocarbon* 28(2B): 839–862.

Pearson GW & M Stuiver. (1993) High-precision bidecadal calibration of the radiocarbon time scale, 500–2500 BC. *Radiocarbon* 35(1): 25–33.

Peebles CS & SM Kus. (1977) Some Archaeological Correlates of Ranked Societies. *American Antiquity* 42: 421–448.

Peltenburg EJ. (1987) Early faience: recent studies, origins and relations with glass. In: Bimson M & IC Freestone. (eds.) *Early Vitreous Materials.* London: British Museum Occasional Papers, 56. pp. 5–29.

Pendlebury JDS. (1933a) Preliminary report of the excavations at Tell el-'Amarnah. *Journal of Egyptian Archaeology* 19: 113–118.

Pendlebury JDS. (1933b) A 'monotheistic utopia' of ancient Egypt. *Illustrated London News* 182 No. 4907: 629–633.

Perizonius WRK. (1984) Closing and non-closing sutures in 256 crania of known age and sex from Amsterdam (AD 1883–1909). *Journal of Human Evolution* 13: 201–216.

Petrie WMF. (1891) *Tell el Hesy (Lachish)*. London: Palestine Exploration Fund.

Petrie WMF. (1894) *Tell el-Amarna*. Warminster: Aris & Phillips.

Petrie WMF. (1899) Sequences in prehistoric remains. *Journal of the Anthropological Institute of Great Britain and Ireland* 3: 295–301.

Petrie WMF. (1901) *The Royal Tombs of the Earliest Dynasties*. London: Quaritch & Asher.

Petrie WMF. (1909a) *The Arts and Crafts of Ancient Egypt*. London: T N Foullis.

Petrie WMF. (1909b) *Memphis I*. London: School of Archaeology in Egypt.

Petrie WMF. (1911) The pottery kilns at Memphis. In: Knobel EB, WW Midgeley, JG Milne, MA Murray & WMF Petrie. (eds.) *Historical Studies II*. London: British School of Archaeology in Egypt. pp. 34–37.

Petrie WMF. (1920) *Prehistoric Egypt*. London: Bernard Quaritch.

Petrie WMF. (1921) *Corpus of Prehistoric Pottery and Palettes*. London: Bernard Quaritch.

Petrie WMF. (1953) *Ceremonial slate palettes: Corpus of proto-dynastic pottery*. British School of Egyptian Archaeology 66A&B. London: British School of Archaeology in Egypt.

Phillips DL, JC Rose & WM van Haarlem. (2010) Bioarchaeology of Tell Ibrahim Awad. *Ägypten und Levante / Egypt and the Levant* 19: 157–210.

Phizackerley K. (2010) *DNA shows that KV55 Mummy probably not Akhenaten.* Online article available at: http://www.kv64.info/2010/03/dna-shows-that-kv55-mummy-probably-not.html

Pietrusewsky M. (2008) Metric analysis of skeletal remains: methods and applications. In: Katzenberg MA & SR Saunders. (eds.) *Biological Anthropology of the Human Skeleton* (2nd edition). Hoboken, NJ: Wiley. pp. 487–532.

Pintaudi R. (2008) *Antinoupolis I*. Firenze: Istituto Papirologico 'G. Vitelli'.

Piperno DR. (2006) *Phytoliths: A comprehensive guide for archaeologists and paleoecologists*. Oxford: AltaMira Press.

Podzorski PV. (1990) *Their Bones Shall Not Perish*. New Malden, Surrey: SIA Publishing.

Pollard AM. (2009) What a long strange trip it's has been: lead isotopes and archaeology. In: Shortland AJ, I Freestone & T Rehren (eds.) *From Mine To Microscope: Advances In The Study Of Ancient Technology*. Oxford: Oxbow Books. pp. 181–189.

Pollard AM, & C Heron. (2008) *Archaeological Chemistry*. Second edition. Royal Society of Chemistry: Cambridge.

Powlesland D. (2014) 3D imaging: enhancing the archaeological record. *British Archaeology* 138: 34–39.

Press F & R Siever. (1997) *Understanding Earth*. New York: W.H. Freeman.

Price C. (2009) Geophysical survey east of the Saqqara Step Pyramid. *Egyptian Archaeology* 34: 38–39.

Prowse TL & NC Lovell. (1996) Concordance of cranial and dental morphological traits and evidence for endogamy in ancient Egypt. *American Journal of Physical Anthropology* 101: 237–246.

Pusch E. (1999) Publikationsverzeichnis der Grabung Ramses-Stadt. *Ägypten und Levante / Egypt and the Levant* 9: 193–195.

Pusch E & T Rehren (2007) *Hochtemperatur-Technologie in der Ramses Stadt*. Hildesheim: Gerstenburg.

Querrien A, J Moulin & A Tabbagh. (2009) Confrontation of geophysical survey, soil studies and excavation data to evidence tillage erosion. *ArcheoSciences. Revue d'archéométrie* 33 (suppl.): 195–198.

Quiles A, E Aubourg, B Berthier, E Delque-Količ, G Pierrat-Bonnefois, MW Dee, G Andreu-Lanoë, C Bronk Ramsey & C Moreau. (2013) Bayesian modelling of an absolute chronology for Egypt's 18th Dynasty by astrophysical and radiocarbon methods. *Journal of Archaeological Science* 40: 423–432.

Redford DB. (1986) *Pharaonic King–Lists, Annals and Day–Books.* Mississauga: Benben Publications.

Reeves DM. (1936) Aerial Photography and Archaeology. *American Antiquity* 2: 102–107.

Rehren T & E Pusch. (1997) New Kingdom glass melting crucibles from Qantir-Piramesses, Nile Delta. *Journal of Egyptian Archaeology* 83: 127–142.

Rehren T & E Pusch. (2005) Late Bronze Age glass production at Qantir-Piramesses, Egypt. *Science* 308: 1756–1759.

Reinhard KJ & A Araújo. (2008) *Archaeoparasitology.* University of Nebraska – Lincoln Anthropology Faculty Publications Paper 22. http://digitalcommons. unl.edu/anthropologyfacpub/22

Reiter S. (2008) *Becoming Adult in Ancient Egypt: Social Recognition of Physical or Social Maturity? An Analysis of Predynastic Grave Data from the Hierakonpolis Fort and Naga-ed-Dêr Cemeteries.* Unpublished BA (Hons) dissertation, Archaeology. University of Southampton.

Renfrew C, JR Cann & JE Dixon. (1965) Obsidian in the Aegean. *Annals of the British School of Athens* 60: 225–247.

Renfrew C, JE Dixon & JR Cann. (1966) Obsidian and the early culutral context of the Near East. *Proceedings of the Prehistoric Society* 32: 30–72.

Renz H & RJ Radlanski. (2006) Incremental lines in root cementum of human teeth – a reliable age marker? *Homo* 57: 29–50.

Ridley M. (1993) *Evolution.* Oxford: Blackwell.

Roberts C. (2013) The bioarchaeology of health and well-being: its contribution to understanding the past. In: Tarlow S & L Nilsson-Stutz. (eds.) *The Oxford Handbook of the Archaeology of Death & Burial.* Oxford: Oxford University Press. pp. 79–98.

Roberts C & K Manchester. (1995) *The Archaeology of Disease.* Ithaca, New York: Alan Sutton.

Robins G. (1983) Natural and Canonical Proportions in Ancient Egyptians. *Göttinger Miszellen* 61: 17–25.

Robling A & S Stout. (2000) Histomorphometry of human cortical bone: applications to age estimation. In: Katzenberg M & S Saunders. (eds.) *Biological Anthropology of the Human Skeleton.* New York: Wiley-Liss. pp. 187–213.

Robling A & S Stout. (2003) Histomorphology, geometry, and mechanical loading in past populations. In: Agarwal S.& S Stout. (eds.) *Bone Loss and Osteoporosis: An Anthropological Perspective.* New York: Kluwer Academic/ Plenum Publishers. pp. 189–203.

Rohl D. (1995) *A Test of Time: The Bible—from Myth to History.* London: Century.

Rose JC. (2006) Paleopathology of the commoners at Tell Amarna, Egypt, Akhenaten's capital city. *Memórias do Instituto Oswaldo Cruz, Rio de Janeiro* 101 (Suppl. II): 73–76.

Rose JC, Armelagos GJ & LS Perry. (1993) Dental Anthropology of the Nile Valley. In: Davies WV & R Walker. (eds.) *Biological Anthropology and the Study of Ancient Egypt.* London: British Museum Press. pp. 61–74.

Rose J & M Zabecki. (2010) The commoners of Tell el-Amarna. In: Ikram S & A Dodson. (eds.) *Beyond the horizon: studies in Egyptian art, archaeology and history in honour of Barry J. Kemp, Vol 2.* Cairo: Publications of the Supreme Council of Antiquities. pp. 408–422.

Rose PJ. (2007) *The Eighteenth Dynasty Pottery Corpus from Amarna.* London: Egypt Exploration Society.

Rowland JM. (2009) Building bridges between radiocarbon, relative and historical chronologies: The case of early Egypt. In: Vymazalová H & M Bartá. (eds.) *Chronology and Archaeology in Ancient Egypt (the Third Millennium BC).* Prague: Czech Institute of Egyptology. pp. 37–43.

Rowland JM. (2013) Problems and possibilities for achieving Absolute Dates from Early Dynastic contexts. In: Shortland AJ & C Bronk Ramsey. (eds.) *Radiocarbon and the Chronologies of Ancient Egypt.* Oxford: Oxbow. pp. 235–249.

Rowland JM & M Hamdan. (2012) The Holocene Evolution of the Quesna Turtle Back: Geological Evolution and Archaeological Relation-ships Within the Nile Delta. In: Kabaciński J, M Chłodnicki & M Kobusiewicz. *Prehistory of Northeastern Africa: New Ideas and Discoveries.* Poznan: Poznan Archaeological Museum. pp. 8–20.

Rowland JM & S Ikram. (2013) The Falcon necropolis at Quesna. *Egyptian Archaeology* 42: 5–7.

Rowland JM & KD Strutt. (2011) Geophysical survey and sub-surface investigations at Quesna and Kom el-Ahmar (Minuf), Governorate of Minufiyeh: an integrated strategy for mapping and understanding the sub-surface remains of mortuary, sacred and domestic contexts. In: Belova G. (ed.) *Achievements and Problems in Modern Egyptology: Proceedings of the International Conference held in Moscow on September 29–October 2, 2009.* Moscow: Russian Academy of Sciences, Center for Egyptological Studies. pp. 332–349.

Rowland JM, S Ikram, GJ Tassie & L Yeomans. (2013) The Sacred Falcon Necropolis of Djedhor(?) at Quesna: Recent investigations from 2006–2012. *Journal of Egyptian Archaeology* 99: 53–84.

Rowland J, S Inskip & S Zakrzewski. (2010) The Ptolemaic–Roman Cemetery at the Quesna Archaeological Area. *Journal of Egyptian Archaeology* 96: 31–48.

Rösing FW. (1990) *Qubbet el Hawa und Elephantine.* Stuttgart: Gustav Fischer Verlag.

Ruff CB. (1999) Skeletal structure and behavioral patterns of prehistoric Great Basin populations. In: Hemphill BE & CS Larsen. (eds.) *Understanding Prehistoric Lifeways in the Great Basin Wetlands. Bioarchaeological Reconstruction and Interpretation.* Salt Lake City, UT: University of Utah Press. pp. 290–320.

Ruff CB. (2000a) Biomechanical analyses of archaeological human skeletons. In: Katzenberg A & SR Saunders. (eds.) *Biological Anthropology of the Human Skeleton.* New York: Alan R. Liss. pp. 71–102.

Ruff CB. (2000b) Body size, body shape, and long bone strength in modern humans. *Journal of Human Evolution* 38: 269–290.

Ruff CB. (2005) Growth tracking of femoral and humeral strength from infancy through late adolescence. *Acta Paediatrica* 94: 1030–1037.

Ruff CB. (2007) Body size prediction from juvenile skeletal remains. *American Journal of Physical Anthropology* 133: 698–716.

Ruffer A. (1920) Study of Abnormalities and Pathology of Ancient Egyptian Teeth. *American Journal of Physical Anthropology* 3: 335–382.

Russell KF, SW Simpson, J Genovese, MD Kinkel, RS Meindl & CO Lovejoy. (1993) Independent test of the fourth rib aging technique. *American Journal of Physical Anthropology* 92: 53–62.

Ryholt K. (2006) The Turin King-List or so-called Turin Canon (TC) as a source for chronology. In: Hornung E, R Krauss & D Warburton. (eds.) *Ancient Egyptian Chronology. Handbook of Oriental Studies, Section 1.* Leiden: EJ. Brill. pp. 26–32.

Said R. (1981) *The geological evolution of the River Nile.* New York: Springer Verlag.

Said R. (ed.) (1990) *Geology of Egypt.* Rotterdam: A.A.Balkema.

Said R. (1993) *The River Nile: Geology, Hydrology and Utilization.* Oxford: Pergamon Press.

Samuel D. (1993) Ancient Egyptian Bread and Beer: An Interdisciplinary Approach. In: Davies WV & R Walker. (eds.) *Biological Anthropology and the Study of Ancient Egypt.* London: British Museum. pp. 156–164.

Samuel D. (1994) Cereal food processing in ancient Egypt, a case study of integration. In: Luff RM & P Rowley-Conwy. (eds.) *Whither environmental archaeology?* Oxbow Monographs 38. Oxford: Oxbow, 153–158.

Samuel D. (1999) Bread making and social interactions at the Amarna Workmen's Village, Egypt. *World Archaeology* 31: 121–144.

Samuel D. (2000) Brewing and baking. In: Nicholson PT & I Shaw (eds.) *Ancient Egyptian Materials and Technology.* Cambridge: Cambridge University Press. pp. 537–576.

Sandford KS & WJ Arkell. (1929) *Paleolithic man and the Nile–Faiyum divide: a study of the region during Pliocene and Pleistocene times.* Chicago: University of Chicago Press.

Saunders SR & DL Rainey. (2008) Nonmetric Trait Variation in the Skeleton: Abnormalities, Anomalies, and Atavisms. In: Katzenberg MA & SR Saunders. (eds.) *Biological Anthropology of the Human Skeleton* (2nd edition). Hoboken, NJ: Wiley. pp. 533–560.

Saunders SR, AHW Chan, B Kahlon, HF Kluge & CM FitzGerald. (2007) Sexual dimorphism of the dental tissues in human permanent mandibular canines and third premolars. *American Journal of Physical Anthropology* 133: 735–740.

Säve-Söderbergh T & IU Olson. (1970) 14C dating and Egyptian chronology. In: Olsson IU. (ed.) *Radiocarbon variations and absolute chronology.* Stockholm: Almgvist & Wiksell. pp. 35–55.

Saxe AA. (1971) Social dimensions of mortuary practices in a Mesolithic population from Wadi Halfa, Sudan. In: Brown JA. (ed.) *Approaches to the Social Dimensions of Mortuary Practices.* Washington DC: Memoir of the Society for American Archaeology. pp. 39–57.

Sayer D. (2010) *Ethics and Burial Archaeology.* London: Duckworth.

Scheuer L & S Black. (2000) *Developmental Juvenile Osteology.* San Diego: Academic Press.

Scheuer L & S Black. (2004) *The Juvenile Skeleton*. London: Elsevier.

Scheuer L, JH Musgrave & SP Evans. (1980) The estimation of late fetal and perinatal age from limb bone lengths by linear and logarithmic regression. *Annals of Human Biology* 7: 257–265.

Schiestl R & T Herbich. (2013) Kom el-Gir in the western Delta. *Egyptian Archaeology* 42: 28–29.

Schild R & F Wendorf. (2010) Late Palaeolithic hunter-gatherers in the Nile Valley of Nubia and Upper Egypt. In: Garcea E. (ed.) *South-eastern Mediterranean peoples between 130,000 and 10,000 years ago*. Oxford: Oxbow. pp. 89–125.

Schmidt A. (2004) Remote Sensing and Geophysical Prospection. *Internet Archaeology* 15. http://intarch.ac.uk/journal/issue15/9/toc.html

Schmidt A. (2013) *Earth Resistance for Archaeologists*. Lanham: Altamira Press.

Schmidt CM, MS Walton & K Trentelman. (2009) Characterization of Lapis Lazuli Pigments Using a Multitechnique Analytical Approach: Implications for Identification and Geological Provenancing. *Analytical Chemistry* 81: 8513–8518.

Schutkowski H. (1993) Sex determination of infant and juvenile skeletons. I. Morphological features. *American Journal of Physical Anthropology* 90: 199–205.

Scollar I, A Tabbagh, A Hesse & I Herzog. (1990) *Archaeological Prospecting and Remote Sensing*. Cambridge: Cambridge University Press.

Scott DA, S Warmlander, J Mazurek & S Quirke. (2009) Examination of some pigments, grounds and media from Egyptian cartonnage fragments in the Petrie Museum, University College London. *Journal of Archaeological Science* 36: 923–932.

Scott GR. (2008) Dental anthropology. In: Katzenberg MA & SR Saunders. (eds.) *Biological Anthropology of the Human Skeleton* (2nd edition). Hoboken, NJ: Wiley. pp. 265–298.

Sekkina MA, MA El Fiki, SA Nossair & NR Khalil. (2003) Thermoluminescence Archaeological Dating of Pottery in the Egyptian Pyramids Zone. *Ceramics – Silikáty* 47: 94–99.

Sengupta A, RP Shellis & DK Whittaker. (1998) Measuring Root Dentine Translucency in Human Teeth of Varying Antiquity. *Journal of Archaeological Science* 25: 1221–1229.

Serpico M. (1996) *Mediterranean Resin Trade in New Kingdom Egypt*, Unpublished PhD thesis, University College London: London.

Serpico M & R White. (2000) The botanical identity and transport of incense during the Egyptian New Kingdom. *Antiquity* 74: 884–897.

Service ER. (1962) *Primitive social organization: An evolutionary perspective*. New York: Random House.

Seyfried F. (2012a) The workshop complex of Thutmosis. In Seyfried F. (ed.) *In the Light of Amarna: 100 Years of the Nefertiti Discovery*. Berlin: Ägyptisches Museum und Papyrussammlung. pp. 170–187.

Seyfried F. (ed.) (2012b) *In the Light of Amarna: 100 Years of the Nefertiti Discovery*. Berlin: Ägyptisches Museum und Papyrussammlung.

Shaaban F & A Shaaban. (2001) Use of two-dimensional electric resistivity and ground penetrating radar for archaeological prospecting at the ancient capital of Egypt. *Journal of African Earth Sciences* 33: 661–671.

Shanklin E. (1998) The profession of the color blind: sociocultural anthropology and racism in the 21st Century. *American Anthropologist* 100: 669–679.

Shaw IME. (1986) A survey at Hatnub. In: Kemp BJ. (ed.) *Amarna reports* **III**. London: Egypt Exploration Society. pp. 189–212.

Shaw I & E Bloxam. (1999) Survey and excavation at the ancient gneiss quarrying site of Gebel el-Asr. Lower Nubia. *Sudan and Nubia* 3: 13–20.

Shaw I & P Nicholson. (1995) *The Dictionary of Ancient Egypt.* London: British Museum Press.

Shaw I, R Jameson & J Bunbury. (1999) Emerald mining in Roman and Byzantine Egypt. *Journal of Roman Archaeology* 12: 203–215.

Shaw R & A Corns. (2011) High resolution LIDAR specifically for archaeology: are we fully exploiting this valuable resource? In: Cowley D. (ed.) *Remote Sensing for Archaeological Heritage Management.* Budapest: Europae Archaeologiae Consilium. pp. 77–86.

Shennan S. (1997) *Quantifying Archaeology.* Edinburgh: Edinburgh University Press.

Sheridan SG, DM Mittler, DP Van Gerven & HH Covert. (1991) Biomechanical association of dental and temporomandibular pathology in a medieval Nubian population. *American Journal of Physical Anthropology* 85: 201–205.

Sherratt A & S Sherratt. (1991) From luxuries to commodities: the nature of Mediterranean Bronze Age trading systems. In: Gale NH. (ed.) *Bronze Age trade in the Mediterranean.* Jonserad: Paul Astroms Forlag. pp. 351–386.

Shoeib ASA, M Roznerska & K Boryk-Jòzefowicz. (1990) Weathering effects on an ancient Egyptian limestone, which has been affected by salt. In: Zezza F. (ed.) *The conservation of monuments in the Mediterranean Basin: the influence of coastal environment and salt spray on limestone and marble.* Brescia, Italy: Grafo Edizioni. pp. 203–208.

Shortland AJ. (2000a) The number, extent and distribution of the vitreous materials workshops at Amarna. *Oxford Journal of Archaeology* 19: 115–134.

Shortland AJ. (2000b) *Vitreous Materials at Amarna: the production of glass and faience in 18th Dynasty Egypt.* BAR International Series S827. Oxford: BAR & Archaeopress.

Shortland AJ. (2002) The use and origin of antimonate colorants in early Egyptian glass. *Archaeometry* 44: 517–531.

Shortland AJ. (2004) Evaporites of the Wadi Natrun: Seasonal and annual variation and its implication for ancient exploitation. *Archaeometry* 46: 497–516.

Shortland AJ. (2005) Shishak, King of Egypt: the challenges of Egyptian calendrical chronology. In: Levy TE & T Higham. (eds.) *The Bible and Radiocarbon Dating: Archaeology, Text and Science.* London: Equinox. pp. 43–54.

Shortland AJ. (2008) The raw materials of early glass and their implications for the trade in glass between Egypt, Mesopotamia and the Aegean. In: Gillis C, C Risberg & B Sjöberg. (eds.) *Trade and production 8: Crossing borders, Proceedings of the 8th International Workshop, Athens 1998.* Gothenberg: SIMA pocket book. pp. 241–257.

Shortland AJ. (2013) An Introduction to Egyptian Historical Chronology. In: Shortland AJ & C Bronk Ramsey. (eds.) *Radiocarbon and the Chronologies of Ancient Egypt.* Oxford: Oxbow. pp. 19–28.

Shortland AJ & C Bronk Ramsey. (eds.) (2013) *Radiocarbon and the Chronologies of Ancient Egypt.* Oxford: Oxbow.

Shortland AJ & K Eremin. (2006) The analysis of second millennium glass from Egypt and Mesopotamia, Part 1: new WDS analyses. *Archaeometry* 48: 581–605.

Shortland AJ & MS Tite. (2000) Raw materials of glass from Amarna and implications for the origins of Egyptian glass. *Archaeometry* 42: 141–153.

Shortland AJ, P Degryse, M Walton, M Geer, V Lauwers & L Salou. (2010) The evaporitic deposits of Lake Fazda (Wadi Natrun, Egypt) and their use in Roman glass production. *Archaeometry* 52: 380–388.

Shortland AJ, PT Nicholson & CM Jackson. (2000) Lead isotopic analysis of 18th dynasty Egyptian eyepaints and lead antimonate colourants. *Archaeometry* 42: 153–159.

Shortland AJ, N Rogers & K Eremin. (2007) Trace element discriminants between Egyptian and Mesopotamian Late Bronze Age glasses. *Journal of Archaeological Science* 34: 781–789.

Shortland AJ, L Schachner, IC Freestone & MS Tite. (2005) Natron as a flux in the early vitreous materials industry: sources, beginnings and reasons for decline. *Journal of Archaeological Science* 33: 1–10.

Shugar AN & JL Mass. (2012) *Handheld XRF for Art and Archaeology* (Studies in Archaeological Sciences). Leuven: Leuven University Press.

Sillar B. (2000) Dung by preference: the choice of fuel as an example of how Andean pottery production is embedded within wider technical, social, and economic practices. *Archaeometry* 42: 43–60.

Sillar B & MS Tite. (2000) The challenge of technological choices for materials science approaches in archaeology. *Archaeometry* 42: 2–20.

Skibo JM. (2013) *Understanding Pottery Function.* New York: Springer.

Smeaton JE & G Burns. (1988) Physicochemistry of the Tomb of Nefertari, Egypt. *MRS Proceedings* 123: 299–304. doi: 10.1557/PROC-123-299.

Smedley A. (1998) 'Race' and the Construction of Human Identity. *American Anthropologist* 100: 690–702.

Smith AL. (2000) Processing clay for Pottery in Northern Cameroon: Social and technical requirements. *Archaeometry* 42: 21–42.

Smith BH. (1991) Standards of tooth formation and dental age assessment. In: Kelley MA & CS Larsen. (eds.) *Advances in Dental Anthropology.* New York: Wiley-Liss. pp. 143–168.

Smith W. (2003) *Archaeobotanical investigations of agriculture at Late Antique Kom el-Nana (Tell el-Amarna).* London: Egypt Exploration Society.

Sofaer J. (2005) The materiality of age: osteoarchaeology, objects, and the contingency of human development. *Ethnographisch-Archäologischen Zeitschrift* 45: 165–180.

Sofaer J. (2006a) *The Body as Material Culture. A Theoretical Osteoarchaeology.* Cambridge: Cambridge University Press.

Sofaer J. (2006b) Gender, bioarchaeology and human ontogeny. In: Gowland R & C Knüsel. (eds.) *Social Archaeology of Funerary Remains.* Oxford: Oxbow. pp. 155–167.

Sofaer J. (2011) Towards a social bioarchaeology of age. In: Agarwal SC & BA Glencross. (eds.) *Social Bioarchaeology.* Oxford: Blackwell. pp. 285–311.

Solheim T. (1989) Dental root translucency as an indicator of age. *Scandinavian Journal of Dental Research* 97: 189–197.

Spence KE. (2000) Ancient Egyptian chronology and the astronomical orientation of pyramids. *Nature* 408: 320–324.

Spence K. (2010) Settlement structure and social interaction at el-Amarna. In: Bietak M, E Czerny & I Forstner-Müller. (eds.) *Cities and Urbanism in Ancient Egypt.* Vienna: Österreichische Akademie der Wissenschaften. pp. 289–298.

Stanley DJ & AG Warne. (1993a) Nile Delta: recent geological evolution and human impact. *Science* 260: 628–634.

Stanley DJ & AG Warne. (1993b) Sea level and initiation of Predynastic culture in the Nile delta. *Nature* 363: 435–438.

Starkie A, W Birch, R Ferllini & TJU Thompson. (2011) Investigation into the Merits of Infrared Imaging in the Investigation of Tattoos Postmortem. *Journal of Forensic Sciences* 56: 1569–1573.

Steele DG & C Bramblett. (1988) *The Anatomy and Biology of the Human Skeleton.* College Station, TX: Texas A&M University Press.

Stern B, C Heron, L Corr, M Serpico & J Bourriau. (2003) Compositional variations in aged and heated pistacia resin found in Late Bronze Age Canaanite amphorae and bowls from Amarna, Egypt. *Archaeometry* 45: 457–469.

Stevens A. (ed.) (2012a) *Akhenaten's Workers. The Amarna Stone Village Survey, 2005–2009. Volume I: The Survey, Excavations and Architecture.* London: Egypt Exploration Society.

Stevens A. (ed.) (2012b) *Akhenaten's Workers. The Amarna Stone Village Survey, 2005–2009. Volume II: The Faunal and Botanical Remains, and Objects.* London: Egypt Exploration Society.

Stevens C & A Clapham. (2010) The botanical samples. In: Kemp B & A Stevens. (eds.) *Busy Lives at Amarna: Excavations in the Main City (Grid 12 and the House of Ranefer, N49.18). Volume I: The Excavations, Architecture and Environmental Remains.* London: Egypt Exploration Society. pp. 427–443.

Stevenson A. (2006) *Gerzeh, a cemetery shortly before History.* London: Golden House.

Stevenson A. (2009) Social Relationships in Predynastic Burials. *Journal of Egyptian Archaeology* 95: 175–192.

Stock JT. (2002) A test of two methods of radiographically deriving long bone cross-sectional properties compared to direct sectioning of the diaphysis. *International Journal of Osteoarchaeology* 12: 335–342.

Stock JT, M O'Neill, CB Ruff, M Zabecki, L Shackelford & J Rose. (2011) Body size, skeletal biomechanics, mobility and habitual activity from the Late Palaeolithic to mid-Dynastic Nile Valley. In: Pinhasi R & JT Stock. (eds.) *Human Bioarchaeology of the Transition to Agriculture.* Chichester: Wiley-Blackwell. pp. 347–370.

Stocks D. (1993) Making stone vessels in ancient Mesopotamia and Egypt. *Antiquity* 67: 596–603.

Stocks D. (2003) *Experiments in Egyptian archaeology: stoneworking technology in ancient Egypt.* London: Routledge.

Stodder AL, AJ Osterholtz, K Mowrer & JP Chuipka. (2010) Processed human remains from the Sacred Ridge Site: Context, taphonomy, interpretation. In: Perry EM, ALW Stodder & C Bollong. (eds.) *Animas–La Plata Project: Bioarchaeology.* SWCA Anthropological Research Papers No 10, vol. XV. Phoenix AZ: SWCA Environmental Consultants. pp. 279–415.

Stokes MA & TL Smiley. (1968) *An introduction to tree-ring dating.* Chicago: University of Chicago Press.

Stone AC, GR Milner, S Paabo & M Stoneking. (1996) Sex determination in ancient human skeletons using DNA. *American Journal of Physical Anthropology* 99: 231–238.

Stos ZA. (2009) Across wine dark seas... sailor tinkers and royal cargoes in the Late Bronze Age Eastern Mediterranean. In: Shortland AJ, I Freestone & T Rehren. (eds.) *From Mine To Microscope: Advances In The Study Of Ancient Technology.* Oxford: Oxbow Books. pp. 163–180.

Stos-Gale Z, N Gale & J Houghton. (1995) The Origin of Egyptian Copper: Lead Isotope Analysis of Metals from El-Amarna. In: Davies WV & L Schofield. (eds.) *Egypt, the Aegean and the Levant.* London: British Museum. pp. 127–135.

Strouhal E. (1992) *Life in Ancient Egypt.* London: Opus.

Strouhal E & J Jungwirth. (1979) Paleogenetics of the late Roman-early Byzantine cemeteries at Sayala, Egyptian Nubia. *Journal of Human Evolution* 8: 699–703.

Strouhal E & J Jungwirth. (1980) Paleopathology of the late Roman-early Byzantine cemeteries at Sayala, Egyptian Nubia. *Journal of Human Evolution* 9: 61–70.

Strutt K, J Heidel & A Graham. (2012) New geophysical survey of the city and necropolis at Antinoupolis, Middle Egypt. *The Newsletter of the International Society for Archaeological Prospection* 32: 15–17.

Strutt KD, J Heidel & A Graham. (2014) The 2012 geophysical survey at Antinoupolis. In: Pitaudi R. (ed.) *Antinoupolis II.* Firenze: Edizioni dell'Istituto Papirologico 'G. Vitelli'. pp. 99–120.

Stuart-Macadam P. (1992a) Anemia in past human populations. In: Stuart-Macadam P & S Kent. (eds.) *Diet, Demography, and Disease: Changing Perspectives on Anemia.* New York: Aldine de Gruyter. pp. 151–170.

Stuart-Macadam P. (1992b) Porotic hyperostosis: a new perspective. *American Journal of Physical Anthropology* 87: 39–47.

Stuiver M & GW Pearson. (1986) High-precision calibration of the radiocarbon time scale, AD 1950–500 BC. *Radiocarbon* 28(2B): 805–838.

Stuiver M & GW Pearson. (1993) High-precision bidecadal calibration of the radiocarbon time scale, AD 1950–500 BC and 2500–6000 BC. *Radiocarbon* 35: 1–23.

Suess HE. (1978) La Jolla measurements of radiocarbon in tree-ring dated wood. *Radiocarbon* 20: 1–18.

Sultan SA. (2004) Geoelectrical mapping and tomography for archaeological prospection at Al Ghouri mausoleum, Islamic Cairo, Egypt. *International Journal of Earth Observation and Geoinformation* 6: 143–156.

Swedlund AC & GJ Armelagos. (1969) Une Recherche en Paleodemographie: la Nubia Soudanaise. *Annales: Economies, Societies, Civilizations* 24: 1287–1298.

Swedlund AC & GJ Armelagos. (1976) *Demographic Anthropology. Elements of Anthropology Series.* Dubuque, IA: W.C. Brown.

Swetnam TW. (1993) Fire history and climate change in giant sequoia groves. *Science* 262: 885–889.

Swetnam TW & AM Lynch. (1993) Multicentury, regional-scale patterns of western spruce budworm outbreaks. *Ecological Monographs* 11: 399–424.

Switsur VR. (1984) Radiocarbon date calibration using historically dated specimens from Egypt and new radiocarbon determinations from el-Amarna. *Amarna Reports* I: 178–188.

Szpakowska K. (2008) *Daily Life in Ancient Egypt.* Oxford: Blackwell Publishing.

Taconis WK & GJR Maat. (2005) Radiological findings in the human mummies and human heads. In: Raven MJ & WK Taconis. (eds.) *Egyptian mummies: radiological atlas of the collections in the National Museum of Antiquities at Leiden.* Turnhout: Brepols Publishers. pp. 53–80.

Tahar AG. (1999) Inland saline lakes of Wadi El-Natrun depression, Egypt. *International Journal of Salt Lake Research* 8: 149–169.

Tainter JA. (1978) Mortuary practices and the study of prehistoric social systems. In: Schiffer MB. (ed.) *Advances in Archaeological Method and Theory* 1: 105–137. New York: Academic Press.

Tallet P & G Marouard. (2012) An early pharaonic harbour on the Red Sea Coast. *Egyptian Archaeology* 40: 40–43.

Tanner JM. (1963) The Regulation of Human Growth. *Child Development* 34: 817–847.

Tanner JM. (1989) *Foetus into Man: Physical Growth from Conception to Maturity.* Ware: Castlemead.

Tapp E. (1984) Disease and the Manchester mummies – the pathologist's role. In: David AR & E Tapp. (eds.) *Evidence embalmed.* Manchester: Manchester University Press. pp. 78–95.

Tassie GJ. (2003) Identifying the Practice of Tattooing in Ancient Egypt and Nubia. *Papers from the Institute of Archaeology* 14: 85–101.

Tassie GJ & LS Owens. (2010) *Standards of Archaeological Excavation.* London: Golden House Publications.

Taylor GM, S Blau, S Mays, M Monot, OY-C Lee, DE Minnikin, GS Besra, ST Cole & P Rutland. (2009) Mycobacterium leprae genotype amplified from an archaeological case of lepromatous leprosy in Central Asia. *Journal of Archaeological Science* 36: 2408–2414.

Taylor JH. (2013) New Radiocarbon Dates for the 21st Dynasty. In: Shortland AJ & C Bronk Ramsey. (eds.) *Radiocarbon and the Chronologies of Ancient Egypt.* Oxford: Oxbow. pp. 167–173.

Taylor RE. (1987) *Radiocarbon Dating: An archaeological perspective.* London: Academic Press.

Täckholm V. (1974) *Students' Flora of Egypt.* Cairo: Cairo University.

Täckholm V & M Drar. (1950) *Flora of Egypt. Vol. II.* Cairo: Fouad I University Press.

Täckholm V & M Drar. (1954) *Flora of Egypt Vol III.* Cairo: Fouad I University Press.

Täckholm V & M Drar. (1969) *Flora of Egypt Vol IV.* Cairo: Fouad I University Press.

Täckholm V & G Täckholm. (1941) *Flora of Egypt Vol I.* Cairo: Fouad I University Press.

Telford WM, LP Geldart & RE Sheriff. (1990) *Applied Geophysics.* Second Edition. Cambridge: Cambridge University Press.

Templeton AR. (1998) Human Races: A Genetic and Evolutionary Perspective. *American Anthropologist* 100: 632–650.

Thanheiser U. (1995) Electrostatic extraction of archaeological plant remains from soil: A new method. *Acta Palaeobotanica* 35: 117–119.

Thompson AH, MP Richards, A Shortland & SR Zakrzewski. (2005) Isotopic palaeodiet studies of Ancient Egyptian fauna and humans. *Journal of Archaeological Science* 32: 451–463.

Thulin M. (1993) *Flora of Somalia: Angiospermae (Tiliaceae-Apiaceae)*. Kew: Royal Botanic Gardens.

Thulin M. (1995) *Flora of Somalia: Angiospermae (Hydrocharitaceae-Pandanaceae)*. Kew: Royal Botanic Gardens.

Tite MS. (1987) Characterisation of early vitreous materials. *Archaeometry* 29: 21–34.

Tite MS & M Bimson. (1986) Faience: an investigation of microstructures associated with the different methods of glazing. *Archaeometry* 28: 69–78.

Tite MS & Y Maniatis. (1975) Examination of ancient pottery using the scanning electron microscope. *Nature* 257: 122–123.

Tite MS, IC Freestone & M Bimson. (1983) Egyptian faience: an investigation of the methods of production. *Archaeometry* 25: 17–27.

Tite MS, IC Freestone, ND Meeks & M Bimson. (1982a) The use of scanning electron microscopy in the technological examination of ancient ceramics. In: Olin JS & AD Franklin. (eds.) *Archaeological ceramics*. Washington DC: Smithsonian Institution Press. pp. 109–120.

Tite MS, Y Maniatis, ND Meeks, M Bimson, MJ Hughes & SC Leppard. (1982b) Technological studies of ancient ceramics from the Near East, Aegean and Southeast Europe. In: Wertime TA & SF Wertime. (eds.) *Early pyrotechnology*. Washington DC: Smithsonian Institution Press. pp. 61–71.

Tite MS, AJ Shortland & Y Maniatis. (2006) The composition of soda-rich and mixed alkali plant ashes used in the production of glass. *Journal of Archaeological Science* 33: 1284–1292.

Tocheri MW, TL Dupras, P Sheldrick & JE Molto. (2005) Roman Period fetal skeletons from the east cemetery (Kellis 2) of Kellis, Egypt. *International Journal of Osteoarchaeology* 15: 326–341.

Todd TW. (1920) Age Changes in the Pubic Bone, I: The Male White Pubis. *American Journal of Physical Anthropology* 3: 285–334.

Torres JC, G Arroyo, C Romo & J De Haro. (2012) 3D Digitization using Structure from Motion. In: Navazo I & G Patow. (eds.) *Proceedings of the Spanish Computer Graphics Conference*. CEIG, Eurographics.

Trigger BG. (1989) *A History of Archaeological Thought*. Cambridge: Cambridge University Press.

Trueman CN. (1999) Rare earth element geochemistry and taphonomy of terrestrial vertebrate assemblages. *PALAIOS* 14: 555–568.

Trueman CN & N Tuross. (2002) Trace elements in recent and fossil bone apatite. *Reviews in Mineralogy and Geochemistry* 48: 489–521.

Trueman CN, AK Behrensmeyer, R Potts & N Tuross. (2006) High-resolution records of location and stratigraphic provenance from the rare earth element composition of fossil bones. *Geochimica et Cosmochimica Acta* 70: 4343–4355.

Turner WES. (1956a) Studies in ancient glasses and glass making processes. Part III. The chronology of glass making constituents. *Journal of the Society of Glass Technology* 40: 39–52.

Turner WES. (1956b) Studies in ancient glasses and glass making processes. Part IV. The chemical composition of ancient glasses. *Journal of the Society of Glass Technology* 40: 162–184.

Tütken T, TW Vennemann & H-U Pfretzschner. (2008) Early diagenesis of bone and tooth apatite in fluvial and marine settings: Constraints from combined oxygen isotope, nitrogen and REE analysis. *Palaeogeography, Palaeoclimatology, Palaeoecology* 266: 254–268.

Ubelaker DH. (1978) *Human skeletal remains: excavation, analysis, interpretation.* Chicago: Aldine.

Ubelaker DH. (2009) The forensic evaluation of burned skeletal remains: A synthesis. *Forensic Science International* 183: 1–5.

Ungar PS, RS Scott, JR Scott & M Teaford. (2008) Dental microwear analysis: historical perspectives and new approaches. In: Irish JD & GC Nelson. (eds.) *Technique and Application in Dental Anthropology.* Cambridge: Cambridge University Press. pp. 389–425.

van de Mieroop M. (2010) *The Eastern Mediterranean in the Age of Ramesses II.* Oxford: Blackwell.

van der Plicht J. (2004) Calibration at all ages? In: Higham T., Bronk Ramsey C., & C Owen. (eds.). *Proceedings of the Fourth Symposium, 14C and Archaeology, Oxford 2002.* Oxford: Oxford University School of Archaeology. pp. 1–8.

van der Veen M. (1985) Carbonised seeds, sample size and on-site sampling. Fieller NRJ, DD Gilbertson & NGA Ralph. (eds.) *Palaeoenvironmental investigations: Research design, methods and data analysis.* BAR International Series 258. Oxford: BAR. pp. 165–174.

van der Veen M. (2001) The botanical evidence. In: Maxfield VA & DPS Peacock. (eds.) *Survey and Excavations at Mons Claudianus 1987–1993, The Excavations: Part 1.* Cairo: Institut Français d'Archéologie Orientale du Caire. pp. 174–247.

van der Veen M & N Fieller. (1982) Sampling seeds. *Journal of Archaeological Science* 9: 287–298.

Vandiver P. (1982) Technological change in Egyptian faience. In: Olin JS & AD Franklin. (eds.) *Archaeological Ceramics.* Washington: Smithsonian Institution. pp. 167–179.

Vandiver P. (1983) Appendix A: Egyptian faience technology. In : Kaczmarczyk A & REM Hedges. (eds.) *Ancient Egyptian Faience.* Warminster: Aris and Phillips. pp. A1–A137.

Vandiver P. (1998) A review and the proposal of new criteria for the production technologies of Egyptian faience. In: Colinart S & M Menu. (eds.) *La couleur dans la peinture et l'émaillage de l'Égypte ancienne.* Ravello: Centro Universitario Europeo. pp. 121–139.

Van Driel-Murray C. (2000) Leatherwork and skin products. In: Nicholson PT & I Shaw. (eds.) *Ancient Egyptian Materials and Technology.* Cambridge: Cambridge University Press. pp. 299–319.

Van Gerven DP, GJ Armelagos & A Rohr. (1977) Continuity and Change in Cranial Morphology of Three Nubian Archaeological Populations. *Man* 12: 270–277.

Van Gerven DP, DS Carlson & GJ Armelagos (1973) Racial History and Bio-Cultural Adaptation of Nubian Archaeological Populations. *Journal of African History* 14: 555–564.

Van Peer P, P Vermeersch & E Paulissen. (2010) *Chert Quarrying, Lithic Technology, and a Modern Human Burial at the Palaeolithic Site of Taramsa 1, Upper Egypt.* Egyptian Prehistory Monographs 5. Leuven: Leuven University Press.

Velde B & IC Druc. (1999) *Archaeological Ceramic Materials: Origin and Utilization.* Berlin: Springer.

Vercoutter J. (ed.) (1970) *Mirgissa I.* Paris: CNRS.

Verhoeven G. (2011) Taking computer vision aloft – archaeological three-dimensional reconstructions from aerial photographs with photoscan. *Archaeological Prospection* 18: 67–73.

Vermeersch PM. (1992) The Upper and Late Palaeolithic of Northern and Eastern Africa. In: Klees F & R Kuper. (eds.) *New Light on the Northeast African Past: Current Prehistoric Research.* Africa Praehistorica 5. Köln: Hienrich Barth Institute. pp. 99–154.

Vermeersch PM. (2006) The Nile valley and the Eastern Sahara: a prehistoric population expanding and retracting according to the changing climate. *Comptes Rendus Palevol* 5: 255–262.

Vermeersch PM, E Paulissen, P Van Peer, S Stokes, C Charlier, C Stringer & W Lindsay. (1998) A Middle Palaeolithic burial of a modern human at Taramsa Hill, Egypt. *Antiquity* 72: 475–484.

Viner S, J Evans, U Albarella & M Parker Pearson (2010) Cattle mobility in prehistoric Britain: strontium isotope analysis of cattle teeth from Durrington Walls (Wiltshire, Britain). *Journal of Archaeological Science* 37: 2812–2820.

Vogelsang-Eastwood G. (2000) Textiles. In: Nicholson PT & I Shaw. (eds.) *Ancient Egyptian Materials and Technology.* Cambridge: Cambridge University Press. pp. 268–298.

von den Driesch A. (1976) *A Guide to the Measurement of Animal Bones from Archaeological Sites.* Peabody Museum Bulletin 1. Cambridge: Harvard University.

Walker PL & D Collins Cook. (1998) Gender and sex: Vive la difference. *American Journal of Physical Anthropology* 106: 255–259.

Walker PL, KWP Miller & R Richman. (2008) Time, temperature, and oxygen availability: an experimental study of the effects of environmental conditions on the color and organic content of cremated bone In: Schmidt CW & SA Symes. (eds.) *The Analysis of Burned Human Remains.* London: Academic Press. pp. 129–135.

Walker R. (1985) *A Guide to Post-cranial Bones of East African Animals.* Norwich: Hylochoerus Press.

Walton M, AJ Shortland, S Kirk & P Degryse. (2009) Evidence for the trade of Mesopotamian and Egyptian glass to Mycenean Greece. *Journal of Archaeological Science* 36: 1496–1503.

Waterbolk HT. (1971) Working with Radiocarbon Dates. *Proceedings of the Prehistoric Society* 37: 15–33.

Weatherhead FJ. (2007) *Amarna Palace Paintings.* London: Egypt Exploration Society.

Weatherhead FJ & A Buckley. (1989) Artists' pigments from Amarna. *Amarna Reports* V: 202–240.

Weber GW & FL Bookstein. (2011) *Virtual Anthropology: A guide to a new interdisciplinary field*. Wien: Springer-Verlag.

Weeks KR. (1980) Ancient Egyptian Dentistry. In: Harris JE & EF Wente. (eds.) *An X-Ray Atlas of the Royal Mummies*. Chicago: University of Chicago Press. pp. 99–121.

Weiss E. (2015) *Paleopathology in Perspective*. London: Rowman & Littlefield.

Wendorf F & R Schild. (1976) *Prehistory of the Nile Valley*. New York: Academic Press.

Wendorf F & R Schild. (1980) *Prehistory of the Eastern Sahara*. New York: Academic Press.

Wendorf F & R Schild. (1984) The Emergence of Food Production in the Egyptian Sahara. In: Clark JD & SA Brandt. (eds.) *From Hunters to Farmers: The Causes and Consequences of Food Production in Africa*. London: University of California Press. pp. 93–101.

Wendorf F, R Schild & AE Close. (1984) *Cattle-Keepers of the Eastern Sahara, the Neolithic of Bir Kiseiba*, Southern Methodist University, Dallas.

Wendorf F, R Schild, N El Hadidi, AE Close, M Kobusiewicz, H Wieckowska, B Issawi & H Haas. (1979) Use of barley in the Egyptian late paleolithic. *Science* 205: 1341–1347.

Wendrich W, RE Taylor & J Southon. (2010) Dating stratified settlement sites at Kom K and Kom W: Fifth millennium BCE radiocarbon ages for the Fayum Neolithic. *Nuclear Instruments and Methods in Physics Research Section B: Beam Interactions with Materials and Atoms* 268: 999–1002.

Wendrich WZ. (2000) Basketry. In: Nicholson PT & I Shaw. (eds.) *Ancient Egyptian Materials and Technology*. Cambridge: Cambridge University Press. pp. 254–267.

Weyl WA. (1951) *Coloured glasses*. Sheffield: Society of Glass Production.

Wheeler SM, L Williams, P Beauchesne & TL Dupras. (2013) Shattered lives and broken childhoods: Evidence of physical child abuse in ancient Egypt. *International Journal of Paleopathology* 3: 71–82.

Whitbread IK. (2005) Ceramic petrology, clay geochemistry and ceramic production. In: Brothwell DR & AM Pollard. (Eds.) *Handbook of Archaeological Sciences*. Chichester: J. Wiley. pp. 449–459.

White CD. (1993) Isotopic determination of seasonality in diet and death from Nubian mummy hair. *Journal of Archaeological Science* 20: 657–666.

White CD & GJ Armelagos. (1997) Osteopenia and stable isotope ratios in bone collagen of Nubian female mummies. *American Journal of Physical Anthropology* 103: 185–199.

Whitehouse H. (1989) Egyptology and Forgery in the Seventeeth Century: The case of the Bodleian shabti. *Journal of the History of Collections* 1: 187–195.

Whittle EH. (1975) Thermoluminiscent dating of Egyptian Predynastic pottery from Hemamie and Qurna-Tarif. *Archaeometry* 17: 119–122.

Wilkinson JG. (1837) [1854] *Manners and Customs of the Ancient Egyptians*. Volume 1. London: John Murray.

Wilkinson RH. (1994) *Symbol and Magic in Egyptian Art*. Thames & Hudson: London.

Wilkinson TH. (2000) *The Royal Annals of Ancient Egypt*. London: Routledge.

Willcocks W. (1904) *The Nile in 1904*. London: E & FN Spon.

Williams LJ. (2008) *Investigating Seasonality of Death at Kellis 2 Cemetery Using Solar Alignment and Isotopic Analysis of Mummified Tissues*. Unpublished Doctoral dissertation, University of Western Ontario: London, Ontario, Canada.

Williams LJ, SM Wheeler, TL Dupras & P Sheldrick. (2012) *Unusual Modes of Reproduction: Seasonal Birthing Cycle from Kellis 2 Cemetery, Dakhleh Oasis, Egypt*. Poster presented at 7th International Conference of the Dakhleh Oasis Project. Leiden: Netherlands.

Wils THG, UGW Sass-Klaassen, Z Eshetu, A Bräuning, A Gebrekirstos, C Couralet, I Robertson, R Touchan, M Koprowski, D Conway, KR Briffa & H Beeckman. (2010) Dendrochronology in the dry tropics: the Ethiopian case. *Trees* 25: 345–354.

Wilson AS & MTP Gilbert. (2007) Hair and nail. In: Thompson T & S Black. (eds.) *Forensic Human Identification: An Introduction*. Boca Raton, FL: CRC Press. pp. 147–174.

Wilson LA, N MacLeod & LT Humphrey. (2008) Morphometric criteria for sexing juvenile human skeletons using the ilium. *Journal of Forensic Sciences* 53: 269–278.

Wilson L & AM Pollard. (2005) The provenance hypothesis. In: Brothwell DR & AM Pollard. (eds.) *Handbook of Archaeological Sciences*. Chichester: J. Wiley. pp. 507–517.

Wilson MA, MA Carter, C Hall, WD Hoff, C Ince, SD Savage, B McKay & IM Betts. (2009) Dating fired clay ceramics using long-term power law rehydroxylation kinetics. *Proceedings of the Royal Society* A 465: 2407–2415.

Wilson P. (ed.) (2006) *The Survey of Sais (Sa el-Hagar), 1997–2002*. London: Egypt Exploration Society.

Wilson P & G Gilbert. (2003) The Prehistoric period at Saïs (Sa el-Hagar). *Archéo-Nil* 13: 65–72.

Wilson P & D Grigoropoulos. (eds.) (2009) *The West Delta Regional Survey, Beheira and Kafr el-Sheikh Provinces*. Excavation Memoir 86. London: Egypt Exploration Society.

Wittwer-Backofen U & H Buba. (2002) Age estimation by tooth cementum annulation: perspectives of a new validation study. In: Hoppa RD & JW Vaupel. (eds.) *Paleodemography*. Cambridge: Cambridge University Press. pp. 107–128.

Wittwer-Backofen U, J Gampe & JW Vaupel. (2004) Tooth cementum annulation for age estimation: results from a large known-age validation study. *American Journal of Physical Anthropology* 123: 119–129.

Wodzińska A. (2009a) *A Manual of Egyptian Pottery. Volume 1: Fayum A–Lower Egyptian Culture*. AERA Field Manual Series 1. Boston: Ancient Egypt Research Associates. http://www.aeraweb.org/publications/

Wodzińska A. (2009b) *A Manual of Egyptian Pottery. Volume 2: Naqada III–Middle Kingdom*. AERA Field Manual Series 1. Boston: Ancient Egypt Research Associates. http://www.aeraweb.org/publications/

Wodzińska A. (2010a) *A Manual of Egyptian Pottery. Volume 3: Second Intermediate Period–Late Period*. AERA Field Manual Series 1. Boston: Ancient Egypt Research Associates. http://www.aeraweb.org/publications/

Wodzińska A. (2010b) *A Manual of Egyptian Pottery. Volume 4: Ptolemaic Period–Modern*. AERA Field Manual Series 1. Boston: Ancient Egypt Research Associates. http://www.aeraweb.org/publications/

Wolff J. (1892) (transl 1896) *The Law of Bone Remodeling*. Berlin: Springer.

Wood JW, GR Milner, HC Harpending & KM Weiss. (1992) The Osteological Paradox. *Current Anthropology* 33: 343–370.

Workshop of European Anthropologists. (1980) Recommendations for age and sex diagnoses of skeletons. *Journal of Human Evolution* 9: 517–549.

Writing Committee of the WHO Consultation. (2010) Clinical Aspects of Pandemic (H1N1) 2009 Influenza. *New England Journal of Medicine* 362: 1708–1719.

Wunderlich J. (1993) The natural conditions for pre-and early Dynastic settlement in the western Nile delta around Tell el-Fara'in, Buto. In: Krzyżaniak L, M Kobusiewicz & JA Alexander. (eds.) *Environmental change and human culture in the Nile basin and northern Africa until the second millennium BC*. Poznan: Poznan Archaeological Museum. pp. 259–266.

Wüst RAJ & C Schlüchter. (2000) The Origin of Soluble Salts in Rocks of the Thebes Mountains, Egypt: The damage potential to ancient Egyptian wall art. *Journal of Archaeological Science* 27: 1161–1172.

Wyart J, P Bariand & J Filippi. (1981) Lapis Lazuli from Sar-e-Sang, Badakhshan, Afghanistan. *Gems and Gemmology* 17: 184–190.

Zabecki M. (2008) 3.7 Human bones from the South Tombs Cemetery. In: Doling W. (ed.) *Report on the South Tombs Cemetery*. http://www.amarnaproject.com/pages/recent_projects/excavation/south_tombs_cemetery/2008.shtml#3_7

Zabecki M. (2009) *Late Predynastic Egyptian Workloads, Musculoskeletal Stress Markers at Hierakonpolis*. Fayetteville, AR: Unpublished PhD Dissertation, University of Arkansas, Fayetteville.

Zakrzewski SR. (2003) Variation in Ancient Egyptian Stature and Body Proportions. *American Journal of Physical Anthropology* 121: 219–229.

Zakrzewski SR. (2007) Population continuity or population change: Formation of the ancient Egyptian state. *American Journal of Physical Anthropology* 132: 501–509.

Zakrzewski SR. (2012) Dental morphology, dental health and its social implications. In: Kabaciński J, M Chłodnicki & M Kobusiewicz (eds.) *Prehistory of Northeastern Africa: New Ideas and Discoveries*. Poznan: Poznan Archaeological Museum. pp. 125–140.

Zivie A & R Lichtenberg. (2005) The cats of the goddess Bastet. In: Ikram S. (ed.) *Divine Creatures: Animal mummies in Ancient Egypt*. Cairo: American University in Cairo Press. pp. 106–119.

Zohary M. (1973) *Geobotanical foundations of the Middle East*. Stuttgart: Taylor & Francis.

Zohary D & M Hopf. (1988) *Domestication of Plants in the Old World*. Oxford: Oxford University Press.

Zohary D & M Hopf. (1994) *Domestication of Plants in the Old World. The Origin and Spread of Cultivated Plants in West Asia, Europe and the Nile Valley*. Oxford: Clarendon Press.

Zohary D & M Hopf. (2000) *Domestication of Plants in the Old World. The Origin and Spread of Cultivated Plants in West Asia, Europe and the Nile Valley*. Oxford: Oxford University Press.

Zohary D, M Hopf & E Weiss. (2012) *Domestication of Plants in the Old World. The Origin and Spread of Domesticated Plants in Southwest Asia, Europe and the Mediterranean Basin*. Oxford: Oxford University Press.

Index

For Product Safety Concerns and Information please contact our EU
representative GPSR@taylorandfrancis.com
Taylor & Francis Verlag GmbH, Kaufingerstraße 24, 80331 München, Germany

www.ingramcontent.com/pod-product-compliance
Ingram Content Group UK Ltd.
Pitfield, Milton Keynes, MK11 3LW, UK
UKHW020936180425
457613UK00019B/429